Scaffolding

Trainee Guide
Second Edition

PEARSON

Boston Columbus Indianapolis New York San Francisco Upper Saddle River
Amsterdam CapeTown Dubai London Madrid Milan Munich Paris Montreal Toronto
Delhi Mexico City São Paulo Sydney Hong Kong Seoul Singapore Taipei Tokyo

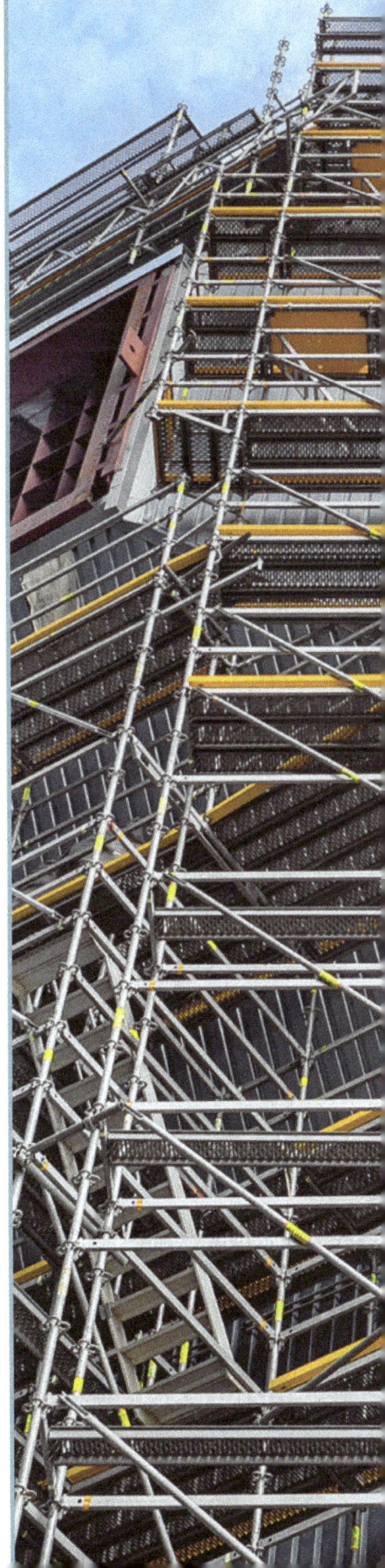

NCCER

President: Don Whyte
Director of Product Development: Daniele Dixon
Scaffolding Project Manager: Jamie Carroll
Senior Manager of Production: Tim Davis
Quality Assurance Coordinator: Debie Hicks

Desktop Publishing Coordinator: James McKay
Permissions Specialist: Adrienne Payne
Production Specialist: Adrienne Payne
Editor: Tanner Yea

Writing and development services provided by S4Carlisle Publishing Services, Dubuque, IA

Project Managers: Barb Tucker, Michael B. Kopf
Writers: Brian E. Light, Paul Lagasse
Art Development: S4Carlisle Publishing Services

Permissions Specialists: Kim Schmidt, Lauren McLean, Jodie Bernard
Copy Editor: Michael H. Toporek

Pearson Education, Inc.

Director, Global Employability Solutions: Jonell Sanchez
Head of Associations: Andrew Taylor
Editorial Assistant: Douglas Greive
Program Manager: Alexandrina B. Wolf
Project Manager: Janet Portisch
Operations Supervisor: Deidra M. Skahill
Art Director: Diane Ernsberger
Directors of Marketing: David Gesell, Margaret Waples
Field Marketers: Brian Hoehl, Stacey Martinez

Composition: NCCER

Text Fonts: Palatino and Univers

Credits and acknowledgments for content borrowed from other sources and reproduced, with permission, in this textbook appear at the end of each module.

ScoutAutomatedPrintCode

PEARSON

Perfect bound   ISBN-13: 978-0-13-383081-1
                    ISBN-10:   0-13-383081-0

Preface

To the Trainee

Scaffolding is a specialized and unique field that is critical to the success of a variety of construction projects. Scaffold builders have many opportunities to work in residential, business, and industrial construction, and are charged with the vital tasks of inspecting, properly storing, lifting, assembling, and maneuvering scaffolding. The nature of the job requires that much of the work be performed at heights; therefore, safety is a critical component of the trade. There is a need for continuous training to upgrade skills, and to keep abreast of the changing codes, regulations, and industry developments.

This second edition of *Scaffolding* provides the fundamentals of the scaffolding trade. It has been revised by industry subject matter experts from across the nation to update the curriculum with modern techniques. *Scaffolding* presents an apprentice approach to the trade, and will help you to be knowledgeable, safe, and effective on the job.

We wish you the best for an exciting and promising career as you begin your training. The revised *Scaffolding* curriculum will enable you to enter the workforce with the knowledge and skills needed to perform productively as a scaffold builder.

New with *Scaffolding*

NCCER is proud to release this edition of *Scaffolding* in full color, with updates to the curriculum that will engage you and provide the best training possible. In this edition, the layout has changed to better align with the learning objectives. There are also new end-of-section review questions to compliment the module review. The text, graphics, and special features have been enhanced to reflect advancements in scaffolding technology and techniques.

Safety is a critical component of the scaffolding industry due to the increased risks associated with working from heights. To address and teach best safety practices, this edition of *Scaffolding* features expanded safety content, including Warning boxes throughout the text. *Trade Safety* (31102-15) now identifies OSHA's "Fatal Four" occurrences as they relate to scaffolding. Another change to this edition is reflected in the title change of *Stationary Scaffolds* (31105) to *Supported Scaffolds* (31105-15). This new edition has five additional instructional hours, and provides the requisite information for those wishing to pursue an exciting career in the scaffolding industry.

We invite you to visit the NCCER website at **www.nccer.org** for information on the latest product releases and training, as well as online versions of the *Cornerstone* magazine and Pearson's NCCER product catalog.

Your feedback is welcome. You may email your comments to **curriculum@nccer.org** or send general comments and inquiries to **info@nccer.org**.

NCCER Standardized Curricula

NCCER is a not-for-profit 501(c)(3) education foundation established in 1996 by the world's largest and most progressive construction companies and national construction associations. It was founded to address the severe workforce shortage facing the industry and to develop a standardized training process and curricula. Today, NCCER is supported by hundreds of leading construction and maintenance companies, manufacturers, and national associations. The NCCER Standardized Curricula was developed by NCCER in partnership with Pearson, the world's largest educational publisher.

Some features of the NCCER Standardized Curricula are as follows:

- An industry-proven record of success
- Curricula developed by the industry for the industry
- National standardization providing portability of learned job skills and educational credits
- Compliance with the Office of Apprenticeship requirements for related classroom training *(CFR 29:29)*
- Well-illustrated, up-to-date, and practical information

NCCER also maintains a Registry that provides transcripts, certificates, and wallet cards to individuals who have successfully completed a level of training within a craft in NCCER's Curricula. *Training programs must be delivered by an NCCER Accredited Training Sponsor in order to receive these credentials.*

Special Features

In an effort to provide a comprehensive, user-friendly training resource, this curriculum incorporates many different features for your use. Whether you are a visual or hands-on learner, this book will provide you with the information to you will need to develop skills and techniques required of scaffold builders.

Introduction

This page is found at the beginning of each module and lists the Objectives, Performance Tasks, and Trade Terms for that module. The Objectives list the skills and knowledge you will need in order to complete the module successfully. The Performance Tasks give you an opportunity to apply your knowledge to real-world tasks you will undertake as a scaffold builder. The list of Trade Terms identifies important terms you will need to know by the end of the module.

Special Features

Features provide a head start for those learning scaffold building by presenting technical tips and professional practices. These features often include real-life scenarios similar to those you might encounter on the job site.

Wind Loads

Wind loads are unique. A wind load is simply another form of distributed load acting on the exposed area of the scaffold. The force of wind applied to the scaffold can change rapidly. The forces can act to push a scaffold toward the structure, away from the structure, or even lift planking or platforms off the scaffold. Wind loads and the steps taken to satisfy those loads must be determined by a professional engineer.

Color Illustrations and Photographs

Full-color illustrations and photographs are used throughout each module to provide vivid detail. These figures highlight important concepts from the text and provide clarity for complex instructions. Each figure reference is denoted in the text in *italic type* for easy reference.

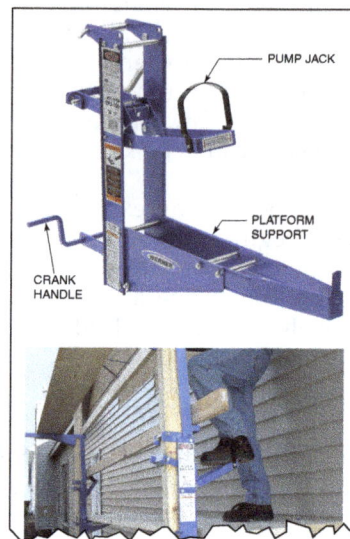

Notes, Cautions, and Warnings

Safety features are set off from the main text in highlighted boxes and are organized into three categories based on the potential danger of the issue being addressed. Notes simply provide additional information on the topic area. Cautions alert you of a danger that does not present potential injury but may cause damage to equipment. Warnings stress a potentially dangerous situation that may cause injury to you or a co-worker.

> **NOTE**
> It is not uncommon for trade terms to vary from one region of the country to another. For example, what might be called a putlog in the Midwest is known as a truss elsewhere.

> **CAUTION**
> Do not use the claw end of a hammer to break steel straps or banding. Using any tool in a manner for which it is not designed poses a safety hazard.

> **WARNING!**
> To avoid the possibility of electrical shock while erecting or working from a scaffold, maintain a safe distance from electrical lines. Electrical shock could cause serious injury or death.

Step-by-Step Instructions

Step-by-step instructions are used throughout to guide you through technical procedures and tasks from start to finish. These steps show you not only how to perform a task but also how to do it safely and efficiently.

Perform the following steps to erect this system area scaffold:

Step 1 Gather and inspect all scaffold equipment for the scaffold arrangement.

Step 2 Place appropriate mudsills in their approximate locations.

Step 3 Attach the screw jacks to the mudsills.

Step 4 Adjust the screw jacks to near their lowest position.

Step 5 Determine the location of the highest base.

Trade Terms

Each module presents a list of Trade Terms that are discussed within the text and defined in the Glossary at the end of the module. These terms are denoted in the text with bold, blue type upon their first occurrence. To make searches for key information easier, a comprehensive Glossary of Trade Terms from all modules is located at the back of this book.

- When platforms are being moved from one level to the next level, the existing platform shall be left undisturbed until the new bearers have been set in place and braced prior to receiving the new platforms.
- Cross bracing, also called transverse bracing, forming an "X" across the width of the scaffold, must be installed at the scaffold ends and at least at every third set of legs horizontally (measured from only one end) and every fourth runner vertically. Bracing shall extend diagonally from the inner to the outer legs or runners upward to the next outer or inner legs or runners. *Figure 1* shows diagonal bracing and cross bracing. Cross bracing increases the sta-

Section Review

The *Section Review* features helpful additional resources and review questions related to the objectives in each section of the module.

Additional Resources

"Scaffolding." OSHA. www.osha.gov/SLTC/scaffolding/index.html

Fall Protection and Scaffolding Safety: An Illustrated Guide. 2000. Grace Drennan Ganget. Government Institutes.

2.0.0 Section Review

1. To minimize the possibility of elbow damage, use a hammer that weighs no more than _____.
   a. 23 ounces
   b. 28 ounces
   c. 32 ounces
   d. 36 ounces

2. A circular-saw blade should be in contact with the work before the cut is started.
   a. True
   b. False

Review Questions

Review Questions are provided to reinforce the knowledge you have gained. This makes them a useful tool for measuring what you have learned.

Review Questions

1. One step to ensure that scaffolds are safe is _____.
   a. annual inspections
   b. regular cleaning
   c. proper storage
   d. posting warning signs

2. When storing scaffold equipment, remove items from the job site that have a _____.
   a. red tag
   b. yellow tag
   c. green tag
   d. multicolored tag

3. Ultimate responsibility for avoiding the use of damaged equipment rests with the _____.
   a. manufacturer
   b. OSHA inspector
   c. supervisor
   d. scaffold erector and user

4. Verifying that rivets are not loose or rusty is a step involved when inspecting _____.
   a. tubing frame legs
   b. bracket latches
   c. braces and guardrails
   d. tube-and-clamp scaffolds

5. Any alteration or modification of scaffold equipment must be approved by a _____.
   a. licensed engineer
   b. qualified manufacturer's representative
   c. supervisor
   d. site safety officer

6. The most common length for a pocket tape is _____.
   a. 10 feet
   b. 15 feet
   c. 20 feet
   d. 25 feet

7. When setting bracing, a helpful tool is a level with a bubble tube that can be used to check an angle of _____.
   a. 33 degrees
   b. 45 degrees
   c. 66 degrees
   d. degrees

8. String lines used to establish baselines when laying out scaffolds are usually made of _____.
   a. nylon
   b. cotton
   c. polyester
   d. braided fine wire

9. For cutting toeboards, the most convenient tool to use is the _____.
   a. circular saw
   b. saber saw
   c. handsaw
   d. reciprocating saw

10. Hammers with the heavier head weights are not recommended for scaffold work because they can _____.
    a. cause elbow damage
    b. be hard to control
    c. rapidly fatigue the user
    d. damage equipment

11. Another name for a crowbar is a _____.
    a. wrecking bar
    b. power bar
    c. long bar
    d. Wonder Bar™

12. A flat, 12"–18" bar used primarily for pulling nails is referred to as a _____.
    a. cat's paw
    b. Wonder Bar™
    c. short bar
    d. nail grabber

13. For twisting and cutting wire, the tool that is usually used is the _____.
    a. side cutter
    b. needle-nose pliers
    c. lineman's pliers
    d. end nipper

14. The box-end wrench _____.
    a. doesn't grip bolt heads as well as an open-end wrench
    b. has a ratcheting action
    c. ed d shoulders of a
    d. size

NCCER Standardized Curricula

NCCER's training programs comprise more than 80 construction, maintenance, pipeline, and utility areas and include skills assessments, safety training, and management education.

Boilermaking
Cabinetmaking
Carpentry
Concrete Finishing
Construction Craft Laborer
Construction Technology
Core Curriculum:
    Introductory Craft Skills
Drywall
Electrical
Electronic Systems Technician
Heating, Ventilating, and
    Air Conditioning
Heavy Equipment Operations
Highway/Heavy Construction
Hydroblasting
Industrial Coating and Lining
    Application Specialist
Industrial Maintenance Electrical
    and Instrumentation Technician
Industrial Maintenance
    Mechanic
Instrumentation
Insulating
Ironworking
Masonry
Millwright
Mobile Crane Operations
Painting
Painting, Industrial
Pipefitting
Pipelayer
Plumbing
Reinforcing Ironwork
Rigging
Scaffolding
Sheet Metal
Signal Person
Site Layout
Sprinkler Fitting
Tower Crane Operator
Welding

Maritime

Maritime Industry Fundamentals
Maritime Pipefitting
Maritime Structural Fitter

Green/Sustainable Construction

Building Auditor
Fundamentals of Weatherization
Introduction to Weatherization
Sustainable Construction
    Supervisor
Weatherization Crew Chief
Weatherization Technician
Your Role in the Green
    Environment

Energy

Alternative Energy
Introduction to the Power Industry
Introduction to Solar Photovoltaics
Introduction to Wind Energy
Power Industry Fundamentals
Power Generation Maintenance
    Electrician
Power Generation I&C
    Maintenance Technician
Power Generation Maintenance
    Mechanic
Power Line Worker
Power Line Worker: Distribution
Power Line Worker: Substation
Power Line Worker: Transmission
Solar Photovoltaic Systems
    Installer
Wind Turbine Maintenance
    Technician

Pipeline

Control Center Operations, Liquid
Corrosion Control
Electrical and Instrumentation
Field Operations, Liquid
Field Operations, Gas
Maintenance
Mechanical

Safety

Field Safety
Safety Orientation
Safety Technology

Supplemental Titles

Applied Construction Math
Tools for Success

Management

Fundamentals of Crew Leadership
Project Management
Project Supervision

Spanish Titles

Acabado de concreto: nivel uno,
    nivel dos
Aislamiento: nivel uno, nivel dos
Albañilería: nivel uno
Andamios
Aparejamiento básico
Aparajamiento intermedio
Aparajamiento avanzado
Carpintería:
    Introducción a la carpintería,
    nivel uno; Formas para
    carpintería, nivel tres
Currículo básico: habilidades
    introductorias del oficio
Electricidad: nivel uno, nivel dos,
    nivel tres, nivel cuatro
Encargado de señales
Especialista en aplicación de
    revestimientos industriales:
    nivel uno, nivel dos
Herrería: nivel uno, nivel dos,
    nivel tres
Herrería) de refuerzo: nivel uno
Instalación de rociadores: nivel
    uno
Instalación de tuberías: nivel uno,
    nivel dos, nivel tres, nivel cuatro
Instrumentación: nivel uno, nivel
    dos, nivel tres, nivel cuatro
Orientación de seguridad
Mecánico industrial: nivel uno,
    nivel dos, nivel tres, nivel
    cuatro, nivel cinco
Paneles de yeso: nivel uno
Seguridad de campo
Soldadura: nivel uno, nivel dos,
    nivel tres

Portuguese Titles

Currículo essencial: Habilidades
    básicas para o trabalho
Instalação de encanamento
    industrial: nível um, nível dois,
    nível três, nível quatro

Acknowledgments

This curriculum was revised as a result of the farsightedness and leadership of the following sponsors:

ABC Merit Shop Training Program Inc.
The Brock Group
Ellis Fall Safety Solutions LLC
Greater Baton Rouge Industrial Mgrs. Assn.
JV Industrial Companies, Ltd.

Miken Specialties, Ltd.
Safety Council of Texas City
Scaffold Resource
Starcon
Turner Industries Group, LLC

This curriculum would not exist were it not for the dedication and unselfish energy of those volunteers who served on the Authoring Team. A sincere thanks is extended to the following:

Gustavo Castillo
David Coleman
Rene DeLatte
Bill Drexel
Dr. Nigel Ellis

Steve Freimuller
Carlos Gomez
Keith Green
Enrique Mendez Jr.
Brett Richardson

Mark Rich
Trey Rivet
Ronald Sokol
Craig Thornton-Wright
Keith Vernon

NCCER Partners

American Fire Sprinkler Association
Associated Builders and Contractors, Inc.
Associated General Contractors of America
Association for Career and Technical Education
Association for Skilled and Technical Sciences
Carolinas AGC, Inc.
Carolinas Electrical Contractors Association
Center for the Improvement of Construction Management and Processes
Construction Industry Institute
Construction Users Roundtable
Construction Workforce Development Center
Design Build Institute of America
GSSC – Gulf States Shipbuilders Consortium
Manufacturing Institute
Mason Contractors Association of America
Merit Contractors Association of Canada
NACE International
National Association of Minority Contractors
National Association of Women in Construction
National Insulation Association
National Ready Mixed Concrete Association
National Technical Honor Society
National Utility Contractors Association

NAWIC Education Foundation
North American Technician Excellence
Painting & Decorating Contractors of America
Portland Cement Association
SkillsUSA®
Steel Erectors Association of America
U.S. Army Corps of Engineers
University of Florida, M. E. Rinker School of Building Construction
Women Construction Owners & Executives, USA

NCCER Business Partners

ACT ProV JUDGMENT INDEX

PEARSON

NACB ISN CONSTRUCT NET INTERNATIONAL

Contents

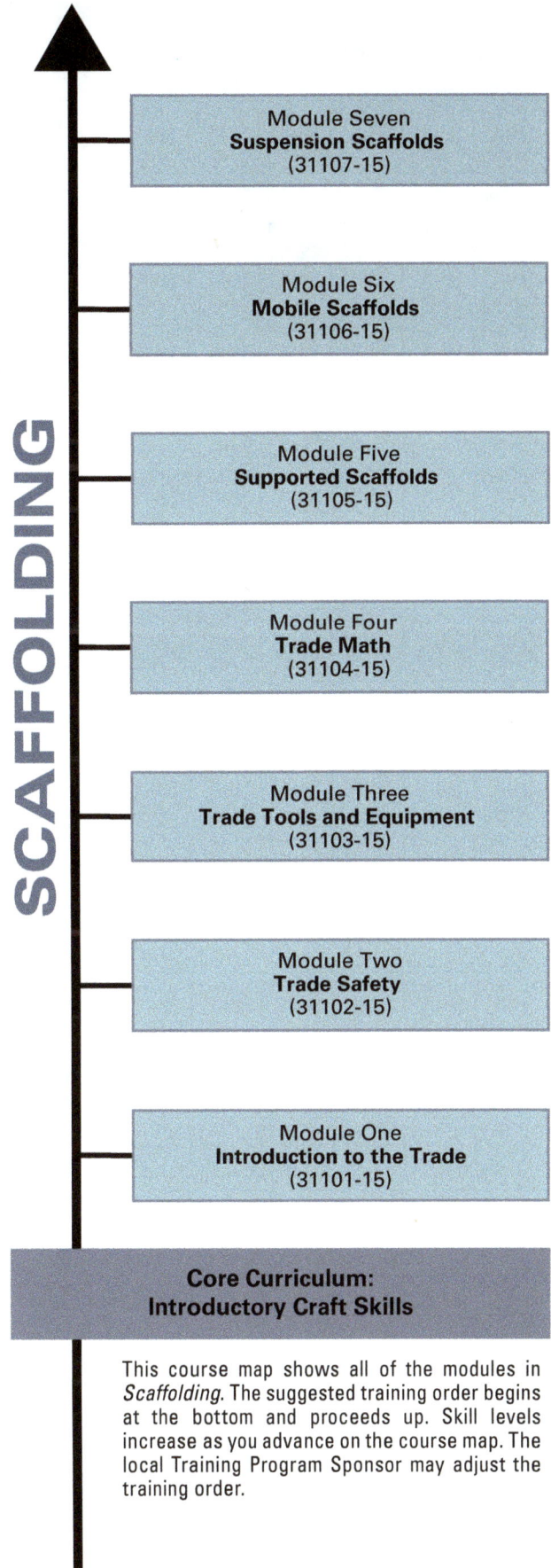

Module Seven

Suspension Scaffolds

Identifies the types of equipment used with suspension scaffolds. Describes the rigging of suspension scaffolds. (Module ID 31107-15; 7.5 hours)

Glossary

Index

SCAFFOLDING

Module Seven
Suspension Scaffolds
(31107-15)

Module Six
Mobile Scaffolds
(31106-15)

Module Five
Supported Scaffolds
(31105-15)

Module Four
Trade Math
(31104-15)

Module Three
Trade Tools and Equipment
(31103-15)

Module Two
Trade Safety
(31102-15)

Module One
Introduction to the Trade
(31101-15)

**Core Curriculum:
Introductory Craft Skills**

This course map shows all of the modules in *Scaffolding*. The suggested training order begins at the bottom and proceeds up. Skill levels increase as you advance on the course map. The local Training Program Sponsor may adjust the training order.

31101-15

Introduction to the Trade

OVERVIEW

The scaffolding trade offers numerous career opportunities, from a dedicated scaffold builder to a painter, a mason, or a carpenter. Scaffold builders have opportunities to work in residential, commercial, and industrial construction, as well as in and around some of the most unique and historical architectural structures. During their work, scaffold builders are required to work safely to ensure their own personal safety and the safety of others on the job site.

Module One

Trainees with successful module completions may be eligible for credentialing through the NCCER Registry. To learn more, go to **www.nccer.org** or contact us at **1.888.622.3720**. Our website has information on the latest product releases and training, as well as online versions of our *Cornerstone* magazine and Pearson's product catalog.

   Your feedback is welcome. You may email your comments to **curriculum@nccer.org**, send general comments and inquiries to **info@nccer.org**, or fill in the User Update form at the back of this module.

   This information is general in nature and intended for training purposes only. Actual performance of activities described in this manual requires compliance with all applicable operating, service, maintenance, and safety procedures under the direction of qualified personnel. References in this manual to patented or proprietary devices do not constitute a recommendation of their use.

Objectives

When you have completed this module, you will be able to do the following:

1. Explain the scaffolding trade as well as the trade math, regulations, and standards associated with the trade.
   a. Describe the scaffolding trade.
   b. Summarize the math applications used in the scaffolding trade.
   c. Identify the regulatory agencies in the scaffolding trade and their basic standards.
2. Identify commonly used scaffold systems and the safety guidelines associated with each type of system.
   a. Identify the safety guidelines, characteristics, and applications of supported-scaffold systems.
   b. Identify the safety guidelines, characteristics, and applications of mobile-scaffold systems.
   c. Identify the safety guidelines, characteristics, and applications of suspension-scaffold systems.
3. Identify personal qualities that contribute to job success.
   a. Describe the responsibilities of a scaffold builder.
   b. Describe the attributes of a good scaffold builder.
4. Explain the apprenticeship training process.
   a. Describe the types of formal craft training available in the scaffolding industry.
   b. Describe the standards associated with an apprenticeship program.
   c. Identify the functions of the Bureau of Apprenticeship and Training (BAT).
   d. Identify the advantages and benefits of today's apprenticeship training programs.

Performance Tasks

This is a knowledge-based module; there are no performance tasks.

Trade Terms

Apprentice
Competent person
Leveling jack

Qualified person
Related instruction

Industry-Recognized Credentials

If you're training through an NCCER-accredited sponsor, you may be eligible for credentials from NCCER's Registry. The ID number for this module is 31101-14. Note that this module may have been used in other NCCER curricula and may apply to other level completions. Contact NCCER's Registry at 888.622.3720 or go to **www.nccer.org** for more information.

Code Note

Codes vary among jurisdictions. Because of the variations in code, consult the applicable code whenever regulations are in question. Referring to an incorrect set of codes can cause as much trouble as failing to reference codes altogether. Obtain, review, and familiarize yourself with your local adopted code.

Contents

Topics to be presented in this module include:

Figures and Tables

SECTION ONE

1.0.0 SCAFFOLDING TRADE

Objective

Explain the scaffolding trade as well as the trade math, regulations, and standards associated with the trade.
  a. Describe the scaffolding trade.
  b. Summarize the math applications used in the scaffolding trade.
  c. Identify the regulatory agencies in the scaffolding trade and their basic standards.

Trade Terms

Competent person: Defined by OSHA as one who is capable of identifying existing and predictable hazards in the surroundings or working conditions that are unsanitary, hazardous, or dangerous to employees, and who has authorization to take prompt corrective measures to eliminate them.

Qualified person: Defined by OSHA as one who, by possession of a recognized degree, certificate, or professional standing, or who by extensive knowledge, training, and experience, has successfully demonstrated the ability to solve or resolve problems related to the subject matter, the work, or the project.

To be well-rounded in your chosen craft, you should have a basic understanding of its development. An understanding of past developments will help you to understand why things are done as they are, and gives some indication of developments that may be in store for the future.

WARNING!

Because life safety is involved, it is imperative that prior to erecting or dismantling any scaffold, the manufacturers' recommendations/instructions are understood, and will be implemented.

1.1.0 Overview of the Trade

The erection, use, and disassembly of scaffold may be performed by a dedicated scaffold builder or as an additional duty performed by a painter, mason, carpenter, etc. This section of the module describes the history of scaffolds, the scaffolding trade, trade math, and the regulations and standards associated with scaffolds.

1.1.1 Development of the Trade

Scaffolds are temporary platforms, supported above the ground or existing level, that support people and/or materials. Scaffolds have been in existence since structures were first built more than head high. Scaffolds have been built to work from, to elevate a person to address a crowd, or to support works of art.

For thousands of years, scaffolds were constructed of wood. It was easy to obtain, easy to shape, and had enough strength to support the loads placed on it. Where wood was in short supply, such as in Asia, bamboo was tied, braced, and placed in close spans.

When the first high-rise buildings were built in the early twentieth century, there was a demand for lighter, stronger, and more durable scaffolds. Engineers began designing systems using steel and aluminum to meet these needs. The first of these scaffolds was imported from England in the early 1900s; it was of the tube-and-clamp design, but such scaffolds were expensive and time consuming to erect.

The first scaffold frame was designed in 1935 by Rheinholt Uecker and David Beatty. The frame was made from angle iron with a short piece of tubing welded to the top and bottom of the frame, allowing the frames to be stacked vertically. Threaded bolts were welded to each frame to hold braces for stability and strength. Seamless tubing, manufactured by Globe Steel, eventually replaced angle iron and became the basic building component of the frame scaffold.

During the Depression, little new construction took place in America. The need to maintain existing buildings required scaffolds to become more versatile. This resulted in the design and testing of a number of scaffold accessories such as side brackets, screw jacks, access ladders, and putlogs.

During World War II, scaffolds were used during the construction of ships and aircraft. With the restrictions on steel use during the war, Republic Steel made mechanical tubing for scaffolds from railroad rails that were melted down. Unfortunately, the carbon content in the steel varied greatly and was so high at times that the steel became brittle.

System scaffolds were introduced into America in the late 1970s. System scaffolds consist of uniformly spaced posts with uniformly spaced connection points to accept runners, bearers, and diagonal bracing. System scaffolds are the most common scaffolds used today. They are well suited for applications where there is limited access, obstructions, uneven surfaces, and non-rectangular-shaped construction.

Scaffolding technology will likely continue to advance, with scaffolds becoming more versatile and economical to assemble and disassemble. It is the responsibility of each employer and scaffold builder to obtain the training to use these products safely as they are developed.

1.1.2 Basic Job Description

Scaffolding is a special construction-related activity that requires skills that are not used in other construction tasks. Each scaffold builder must understand that there is a degree of risk; to reduce these risks, they must understand the potential hazards and be able to maintain constant safety awareness.

The nature of the scaffold-builder tasks requires that much of the work be performed at heights. A scaffold builder must be physically fit to perform the climbing and lifting required in the assembly of a scaffold. There is also a need for continuous training to upgrade skills, and to keep abreast of the changing codes, regulations, and industry developments.

Each site has unique challenges when constructing scaffolds. Scaffold builders must be prepared to deal with varying conditions in a responsible way. They must be able to conduct a hazard assessment and to interpret safety and hazard warnings. Based on these hazards and any special permits or local requirements, they must be able to develop a safe working plan.

Regulations apply to overall safety considerations when performing scaffold erection, as well as local variations to the codes and regulations that must be met. The competent person must always be aware of the full range of regulations, as well as company policy and procedure.

Scaffold builders must also be aware of the limitations of the many types of scaffolds to provide each customer with a safe scaffold that will meet their needs. The requirements of each application must be clearly understood before a scaffold type is selected.

The scaffold equipment and tools must be inspected on a continual basis. Any sections, braces, and other components found to be defective must be immediately removed from inventory. The responsibility of every scaffold builder is to ensure that the entire scaffold is safe.

1.1.3 Types of Scaffolds

There are many kinds of scaffolds, each designed for a specific use and job-site condition. No matter who the manufacturer is, scaffolds are divided into three basic types:

- Supported scaffolds
- Mobile scaffolds
- Suspension scaffolds

Each of these types of scaffolds will be discussed in later sections in this module. Each type is also covered in detail in a separate module that will provide detailed descriptions, assembly and disassembly guidelines, specific applications, and safety considerations.

The qualified person must consider a number of factors when selecting the type of scaffold for an individual task. Some of these factors include availability, cost, site conditions, loads expected, and the experience of the competent person available to assemble the structure.

Table 1 lists a number of activities that use scaffolds and the type of scaffold that might be used for each activity.

Table 1 Activities for Scaffold Use

Activity	Tubular Welded-Frame	Tube-And-Clamp	Suspension	Other
Barricade	X	X		Horse Scaffold
Bleachers	X	X		Caster-Built
Boiler Inspection	X	X	X	
Brick and Stone	X	X		Multipoint
Bricklaying	X	X		Bricklayer's Mast Climbing
Camera Tower	X			
Ceiling Work	X	X		Horse Scaffold / Interior Hung
Crane Inspection, Manufacturing Plant	X	X		Rolling Tower

Activity	Tubular Welded-Frame	Tube-And-Clamp	Suspension	Other
Drywall/Stucco	X	X		Rolling Tower / Single-Pole Scaffold
Electrical Work	X	X	X	
Exterior-Wall Work (New)	X			Pump-Jack
Grandstand	X		X	Caster-Built
Gutter Work	X			Ladder Jack /Pump Jack
Hvac Repair	X	X		Rolling Tower
Insulation Installation	X	X		
Joist/Roof Repair	X	X		Catenary
Lighting	X			Rolling Tower
Lighting (Theater)	X			Ladder Scaffold
Material-Hoist Tower	X	X		
Metal Siding			X	Multilevel
Painting	X	X	X	Trestle Ladder
Piping	X	X		Rolling Scaffold Tower
Roof Perimeter Guardrail	X			
Shipbuilding / Maintenance		X	X	Hoists
Shoring	X			Hoists
Siding Installation	X		X	Pump-Jack
Sign / Banner Support	X	X		Wood Pole
Skylight Installation	X			Rolling Scaffold Tower
Stair Tower	X	X		
Steel Beam / Column Assembly	X	X		Float Scaffold
Stock Work	X	X		Hoists
Structural Work		X	X	Needle Beam
Stud-Wall Construction	X		X	Rolling Scaffold Tower
Temporary Shelters	X	X		Hoists
Theatrical Stage	X	X		Caster-Built
Trash-Chute Tower	X	X		Wood Pole
Tuckpointing	X			Outrigger
Valve Replacement Refinery	X	X		
Window Installation	X	X	X	
Window Repair			X	Boatswain's Chair
Window Replacement	X		X	Ladder
Window Work	X		X	Window Jack

1.2.0 Math Applications

Each scaffold builder is responsible for building a scaffold system that is safe for use as intended. To ensure that the scaffold is safe, the scaffold builder must be able to determine the loading on the scaffold as a whole and on an individual span.

1.2.1 Load Calculations

To calculate the weight of a load on a scaffold, combine the total weight of material that will be placed on the platforms with the estimated weight of the workers. The average worker with standard tools is considered to weigh 250 pounds. Weight calculations are also important when considering the footings for the scaffold legs. The entire weight of the scaffold assembly as well as the loads on each platform must be supported on these footings.

1.2.2 Load Limits

Depending on the design, scaffolds are classified for light-duty, medium-duty, or heavy-duty use. This means that they can safely support loads up to 25, 50, or 75 pounds per square foot, respectively. To determine if the scaffold will be safe for the intended use, the weight of all personnel, equipment, and materials must be considered.

The most common scaffold platform size is 5 feet wide and 7 feet long ($5 \times 7 = 35$ square feet) on a light-duty scaffold frame. This can safely support up to 875 pounds (35 square feet $\times$ 25 psf [pounds per square foot] = 875 pounds). On a heavy-duty frame this limit rises to 2,625 pounds.

1.3.0 Regulations and Standards

The Occupational Safety and Health Administration (OSHA) is the primary regulatory agency covering scaffolds. In addition there are a number of codes, such as the *International Building Code*, that affect the design and use of scaffolds. It is important to be aware of the regulations, codes, and standards relating to scaffolds so that they can be used safely without presenting a hazard to other workers or the public. Each scaffold builder is expected to work around the scaffold equipment in a safe and responsible manner.

1.3.1 National Agencies

The codes and OSHA regulations attempt to cover all safety aspects of scaffold use and misuse. OSHA regulations are divided into sections, with each industry covered in a separate section, as listed below:

- General Industry Standard
- Maritime Standard (Shipyard Use)
- Construction Standard
- Corps of Engineers

Each regulation is designed to have jurisdiction over an industry using scaffolds.

1.3.2 Additional Regulatory Organizations

In addition to federal, state, and local regulations, organizations working with scaffolds have developed standards that are useful to the scaffold builder. The main sources of these standards are:

- American National Standards Institute (ANSI)
- Army Corps of Engineers (ACE)
- Associated Builders and Contractors (ABC)
- Association of General Contractors (AGC)
- Scaffold and Access Industry Association (SAIA)
- Scaffold, Shoring, and Forming Institute (SSFI)
- Various scaffold manufacturers' company manuals

1.0.0 Section Review

1. Steel scaffold was introduced in the United States in the _____.

    a. mid-nineteenth century
    b. late nineteenth century
    c. early twentieth century
    d. mid-twentieth century

2. The primary regulatory agency covering scaffolds is the American National Standards Institute (ANSI).

    a. True
    b. False

3. A medium-duty scaffold can safely support a per-square-foot load of _____.

    a. 25 pounds
    b. 50 pounds
    c. 75 pounds
    d. 100 pounds

SECTION TWO

2.0.0 SCAFFOLD SYSTEMS AND SAFETY GUIDELINES

Objective

Identify commonly used scaffold systems and the safety guidelines associated with each type of system.

  a. Identify the safety guidelines, characteristics, and applications of supported-scaffold systems.
  b. Identify the safety guidelines, characteristics, and applications of mobile-scaffold systems.
  c. Identify the safety guidelines, characteristics, and applications of suspension-scaffold systems.

Trade Term

Leveling jack: A threaded, adjustable screw jack inserted between scaffold legs and their base plates or caster wheels, used to bring scaffolds into a level position when being erected on uneven surfaces.

Employment is an obligation, like a promise or a contract. In exchange for the benefits of your employment and your own well-being, you agree to work safely. In other words, you are obligated to work safely. You are also obligated to make sure anyone you happen to supervise or work with is working safely. Your employer is also obligated to maintain a safe workplace for all employees. Safety is everyone's responsibility.

Some employers will have safety committees. If you work for such an employer, you are then obligated to that committee to maintain a safe working environment. This means two things:

- Follow the safety committee's rules for proper working procedures and practices.
- Report any unsafe equipment and conditions directly to the committee or your supervisor.

> **NOTE**
>
> Here is a basic rule to follow every working day: If you see something that is not safe, report it! Do not ignore it; it will not correct itself. You have an obligation to report it. Your employer knows that the short time lost in making conditions safe again is nothing compared with shutting down the whole job because of a major disaster. Occupational Safety and Health Administration (OSHA) regulations require you to report hazardous conditions.

Safety practices apply to every trade in the construction industry. Whether you work for a large contractor or a small subcontractor, you are obligated to report unsafe conditions. The easiest way to do this is to tell your supervisor. If that person ignores the unsafe condition, report it to the next-highest supervisor. If it is the owner who is being unsafe, let that person know your concerns. If nothing is done about it, report it to OSHA. If you are worried about your job being on the line, think about it in terms of your life, or someone else's, being on the line.

The US Congress passed the Occupational Safety and Health Act in 1970. This act also created OSHA, which is part of the US Department of Labor. The job of OSHA is to set occupational safety and health standards for all places of employment, enforce these standards, ensure that employers provide and maintain a safe workplace for all employees, and provide research and educational programs to support safe working practices.

OSHA requires each employer to provide a safe and hazard-free working environment. OSHA also requires that employees comply with OSHA rules and regulations that relate to their conduct on the job.

According to OSHA standards, you are entitled to on-the-job safety training. As a new employee, you must be:

- Shown how to do your job safely
- Provided with the required personal protective equipment (PPE)
- Warned about specific hazards
- Supervised for safety while performing the work

The Occupational Safety and Health Act was adopted with the stated purpose "to assure as far as possible every working man and woman in the nation safe and healthful working conditions and to preserve our human resources."

The enforcement of this act of Congress is provided by the federal and state safety inspectors who have the legal authority to make employers pay fines for safety violations. The law allows states to have their own safety regulations and agencies to enforce them, but they must first be approved by the US Secretary of Labor. For states that do not develop such regulations and agencies, federal OSHA standards must be obeyed.

These standards are listed in *OSHA Safety and Health Standards for the Construction Industry*, sometimes called the *OSHA Standards*. The most important general requirements that OSHA places on employers in the construction industry are:

- The employer must have a competent person perform frequent and regular job-site inspections of equipment.
- The employer must instruct all employees to recognize and avoid unsafe conditions, and to know the regulations that pertain to the job so they may control or eliminate any hazards.
- No one may use any tools, equipment, machines, or materials that do not comply with OSHA regulations.
- The employer must ensure that only qualified individuals operate tools, equipment, and machines.

Scaffold systems vary widely in components, erection guidelines, and safety in the way they are designed to be employed. Strict compliance to the following is needed for the life safety of all on the construction site.

2.1.0 Supported-Scaffold Systems

Supported scaffolds are assembled from the ground up or suspended from the top down. Scaffolds higher than 125 feet must be designed by a professional engineer. The height limit of 125 feet for supported scaffolds may be limited by the number of levels to be used. Most supported scaffolds are manufactured scaffolds made from tubular steel that can be fitted together to form complete sections. For light-duty use, scaffolds are sometimes made from aluminum or wood. The main types of supported scaffolds manufactured from steel are tubular welded-frame scaffolds, tube-and-clamp scaffolds, and system (modular) scaffolds. *Table 2* shows the maximum number of planked levels for each scaffold rating.

Table 2 Maximum Number of Planked Levels for Each Scaffold Rating

Number of Working Levels	Maximum Number Of Additional Planked Levels			Maximum Height of Scaffold (In Feet)
	Light-Duty	Medium-Duty	Heavy-Duty	
1	16	11	6	125
2	11	1	0	125
3	6	0	0	125
4	1	0	0	125

To ensure your personal safety and the safety of your co-workers, always comply with the following guidelines when erecting and using tubular scaffolds:

- Inspect all scaffold parts before assembly. Never use parts that are broken, damaged, or deteriorated. Be cautious of excessively rusted materials.
- Follow the manufacturer's recommendations for the proper way to erect and use scaffolds. Do not force braces or other parts to fit.
- Adjust the level of the scaffold until the connections can be made easily.
- Provide adequate sills or underpinnings for all scaffolds built on filled or soft ground. Compensate for uneven ground by using adjusting screws or leveling jacks. Do not use boxes, concrete blocks, bricks, or other similar objects to support a scaffold. Be sure scaffolds are plumb and level at all times.
- Follow the proper spacing and positioning requirements for the parts of the scaffold. Anchor or tie-in scaffolds to the building at proper intervals.
- Use ladders rather than cross braces to climb the scaffold. Position ladders with caution to prevent the scaffold tower from tipping.
- Keep scaffolds free of clutter and any slippery material.
- Do not work on a scaffold that is more than 10 feet high without guardrails and toeboards on open sides and ends unless fall protection is provided.
- Lock the casters of a mobile scaffold when it is positioned for use. Do not ride a mobile scaffold.
- Avoid building scaffolds near power lines.

2.1.1 Tubular Welded-Frame Systems

Tubular welded-frame (fabricated frame) scaffolds are used in accessible places with fairly level ground conditions. They use one or more manufactured platforms supported by welded-end frame sections, horizontal bearers, and intermediate members. They are made in various heights and widths that are joined with horizontal and diagonal cross braces and secured by pins. The braces, which have a fixed length, automatically square and vertically align vertical members so the erected scaffold is always plumb, square, and rigid. The scaffold is extended by adding braces and frames until the desired length is reached. Stacking end frames on top of each other increases scaffold height. The bottoms of the legs of the upper end frames slide into the tops of the legs of the lower end frames, and are joined together with drop locks and coupling pins. *Figure 1* shows a tubular welded-frame scaffold system.

2.1.2 Tube-and-Clamp Systems

Tube-and-clamp scaffolds can be either supported or suspension. For this type of scaffold, tubing is erected using couplings.

These scaffolds are typically used in difficult or inaccessible areas. The tubes that form the uprights, braces, bearers, and runners are attached together with right-angle or swivel clamps to build up a scaffold on uneven ground and around unusual shapes, such as tanks. End fittings are used to connect the end of one tube with another. *Figure 2* shows an assembled tube-and-clamp scaffold system and identifies some of the basic components.

Metal base plates attached to the tubular legs or posts of both tube-and-clamp and welded frame scaffolds serve as a foundation for supported scaffolds. On dirt and similar surfaces, planks called

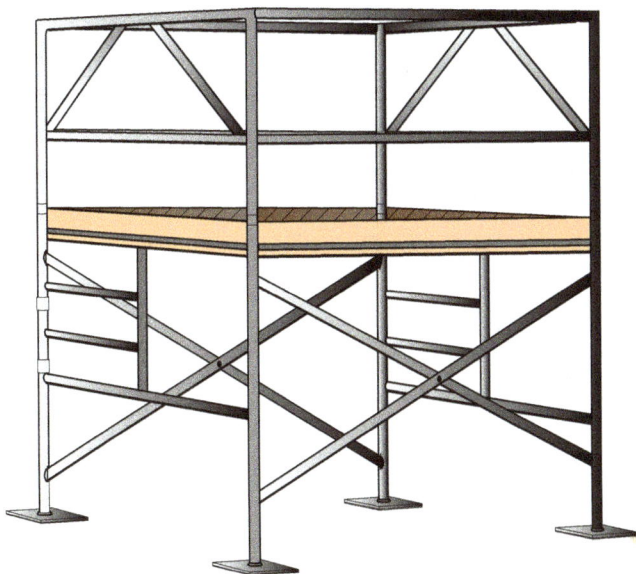

31101-14_F01.EPS

Figure 1 Tubular welded-frame scaffold system.

sills, or mudsills, are placed under the base plates to provide a secure and level footing for the scaffold. Leveling jacks are often used with the base plates to help level the scaffold.

To prevent movement, supported scaffolds are secured to the structure at vertical intervals not to exceed 20 feet for scaffolds with a base 3 feet wide or less and, 26 feet for scaffolds with a base over 3 feet wide. The horizontal interval is 30 feet regardless of the width of the base.

2.1.3 Outrigger Systems

An outrigger scaffold consists of a platform of scaffold boards supported by beams that extend out a window. The supporting beams must be rigidly secured to the structure inside the building. On the outside, the platform is treated just like any other scaffold platform. The area must be fully planked, with guardrails and toeboards installed. Never exceed the standard scaffold board length. *Figure 3* shows a cut-away of an outrigger scaffold.

2.1.4 Pump-Jack Systems

A pump-jack scaffold is supported on a square wooden pole and braced against a structure. The platform is built on a bracket that slips around the supporting pole. Guardrails and end guardrails are attached to this specialized supporting bracket. Toeboards must be attached to the platform. The scaffold is raised manually by pressing down on a lever with your foot. This action raises the scaffold similar to the way a bumper jack raises an automobile. Both ends of the scaffold must be raised at the same time to keep the scaffold level and to prevent binding. *Figure 4* shows a pump-jack scaffold.

2.2.0 Mobile-Scaffold Systems

Mobile scaffolds are portable, can be powered or manual, and are mounted on casters or wheels.

2.2.1 Mobile Scaffolds

Tubular welded-frame scaffolds are often equipped with casters on the bottom of the frame legs to form a freestanding mobile scaffold tower. Mobile scaffolds should only be used on level, smooth surfaces that are free of obstructions and openings. Regulations require that mobile scaffold casters support four times the maximum intended load. They must also have a positive locking device to hold the scaffold in place.

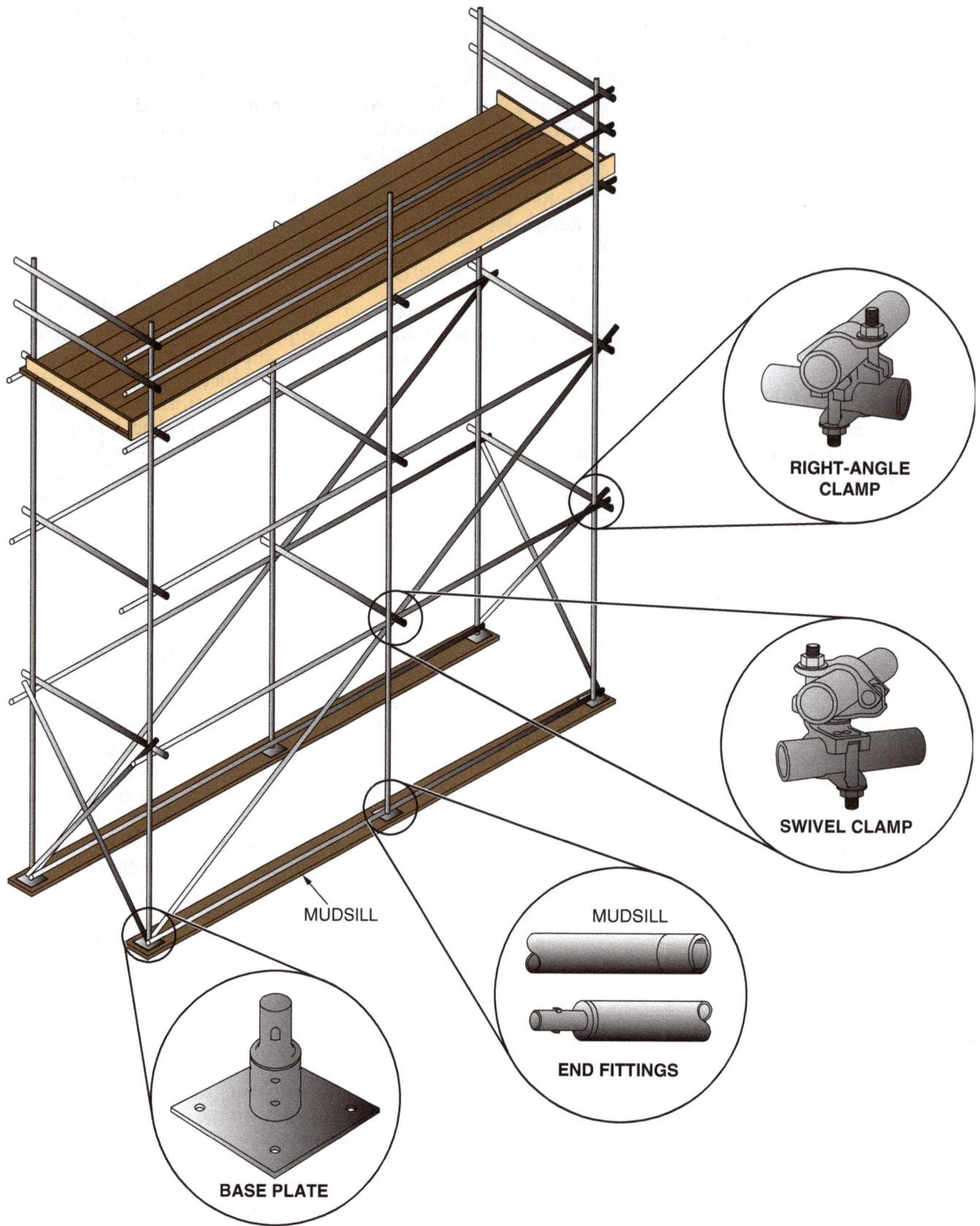

RIGHT-ANGLE
CLAMP

SWIVEL CLAMP

MUDSILL

MUDSILL

END FITTINGS

BASE PLATE

Figure 2 Tube-and-clamp scaffold system.

31101-14_F02.EPS

NCCER – *Scaffolding* 31101-15

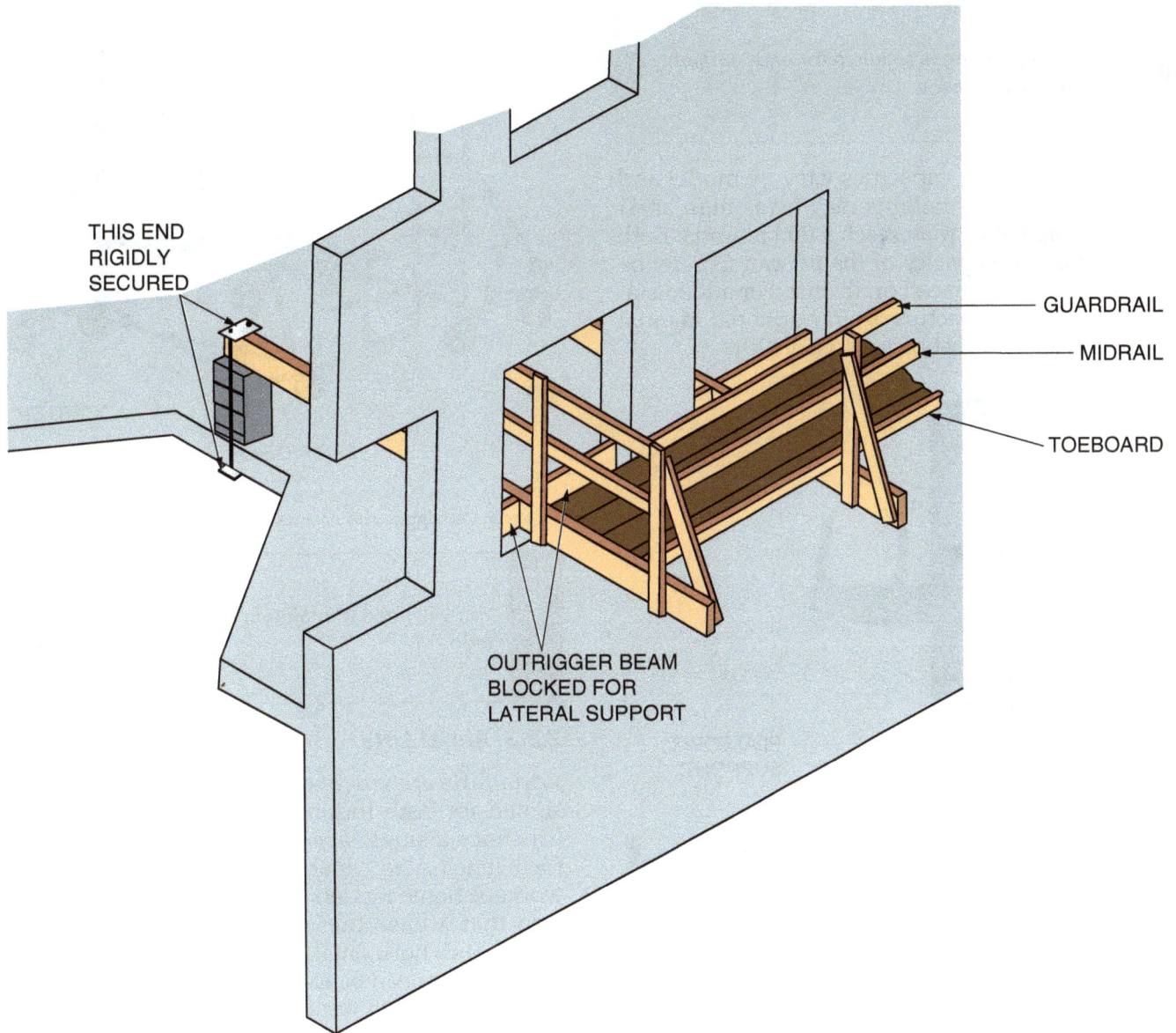

THIS END
RIGIDLY
SECURED

GUARDRAIL

MIDRAIL

TOEBOARD

OUTRIGGER BEAM
BLOCKED FOR
LATERAL SUPPORT

31101-14_F03.EPS

Figure 3 Cut-away of an outrigger scaffold.

The height of mobile towers cannot exceed four times the smallest base dimension. For example, if the scaffold's base is 3 feet by 6 feet, its height should be no more than 12 feet (3 feet × 4). Outriggers, which are supports that must attach to both sides of the scaffold base, as shown in *Figure 5*, can be used to increase the size of the base dimension. When moving a mobile scaffold, force should be applied as close to the base as possible to avoid tipping it over.

Figure 5 shows a scaffold with an outrigger and casters.

2.2.2 Scissors Lifts

Scissors lifts are powered personnel lifts. They use hydraulic rams to raise and lower the platform. Power may be supplied to the hydraulic system by a battery, a gasoline engine, or a liquefied petroleum gas (LPG)-powered engine. The controls for the lift are located at an operator's station on the platform itself. Some models also include an operator's station at the ground and platform level. In addition to controlling the height of the platform, the operator is also able to move the unit to the front or rear and steer it as it moves.

Specific training is required for each different type of equipment.

Lift heights and capacities vary by model and manufacturer. Lift heights may be as high as 41 feet and capacities may reach 2,000 pounds. Both the height and capacity of the lift can usually be found on a decal placed on the lift. For added stability, some manufacturers offer optional manual outriggers. *Figure 6* shows a scissors lift.

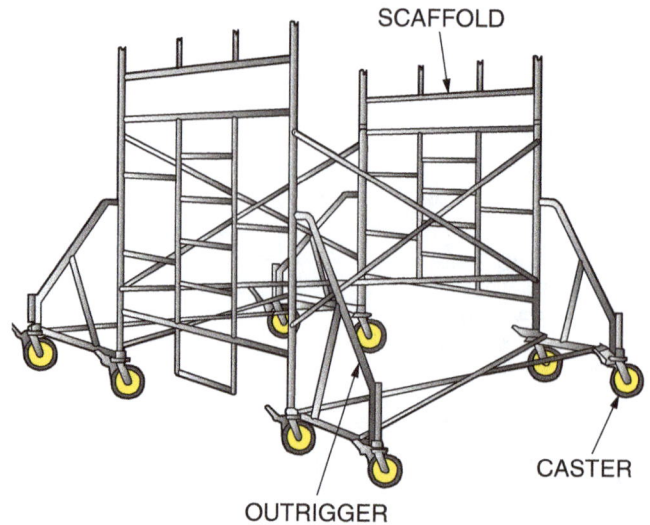

Figure 5 Outrigger and casters used on mobile scaffold.

31101-14_F05.EPS

You will need to maintain a 2:1 height-to-base ratio.

2.2.3 Aerial Lifts

Aerial lifts are truck- or trailer-mounted lifts designed for both indoor and outdoor use. Aerial lifts have a single arm that extends a work platform/enclosure capable of holding one or two workers. Some models have a jointed (articulated) arm that allows the work platform to be positioned both horizontally and vertically. The aerial and platform can be extended beyond the base of the unit. Some of the units have special oscillating axles that allow them to be maneuvered over rough terrain. These lifts can be driven from the platform. *Figure 7* shows a selection of aerial lifts.

2.2.4 Mast-Climbing Platforms

Mast-climbing platforms can be either mobile or fixed. The platforms are supported by one or two masts or towers. Once in position, the platforms can be raised into working position using controls on the platform. The platforms are complete with decking and guardrails. These provide a long platform next to the work site where several people can work at the same time on a wall being built or some other similar task. The maximum height of these platforms may be greater than 300 feet when they are properly braced to an existing structure. These platforms are not intended to be moved while personnel are working on the platform. *Figure 8* shows a typical use of a mast-climbing platform.

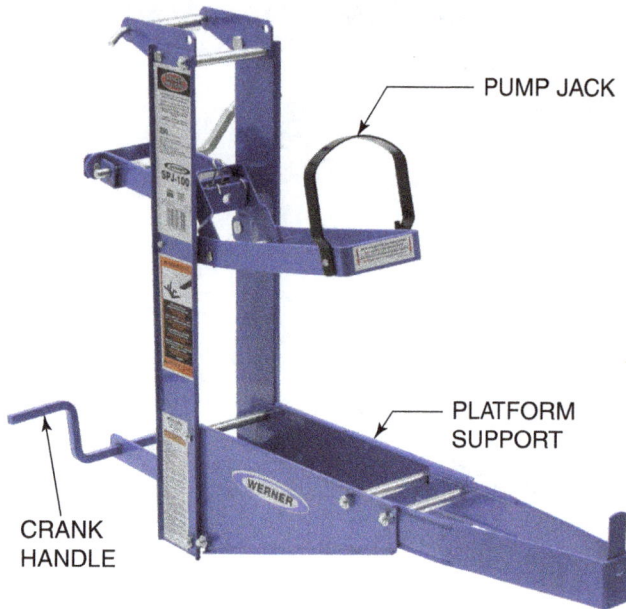

Figure 4 Pump-jack scaffold.

31101-14_F04.EPS

Figure 6 Scissors lift.

2.3.0 Suspension-Scaffold Systems

Suspension scaffolds do not rest on the ground. They are attached to an overhead support. They can be hung by ropes or cables so that the platform can be raised or lowered, or they may be attached by clamps or rigging devices to I-beams.

2.3.1 Swing Scaffolds

A swing scaffold, sometimes called a two-point suspension scaffold, is suspended from a building or structure by two ropes or cables that allow it to be raised or lowered as needed. It consists of a platform or stage, hangers, hoisting mechanism, ropes or cables, and rigging devices. The platform or stage provides a work area 24 to 36 inches wide that is securely attached to the hangers.

The platform may be metal or wood. Suspension-scaffold platforms must have the required guardrails, midrails, and toeboards on all open sides and ends.

Steel hangers, often called stirrups or triangles, support each end of the platform. These hangers are attached to wire cables or ropes, made of wire, fiber, or synthetic material, that are suspended from rigging devices that are secured to the building or structure. The suspension system and rigging devices (including their supports on the structure) must be able to hold four times the rated load of the hoist. OSHA requires that the ropes or cables support at least six times the rated load. Safety lines must be used when working on the swing scaffold. These lines should always be tied off to different anchor points from those used to support the scaffold.

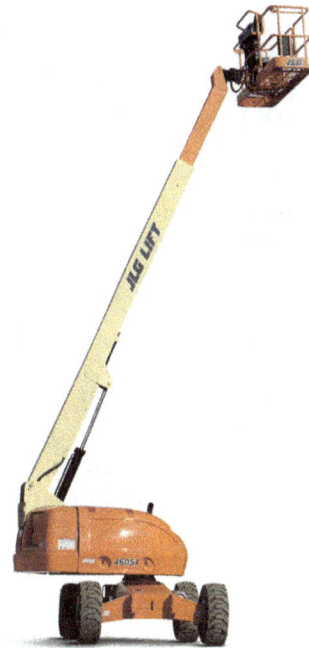

TELESCOPING AERIAL LIFT

ARTICULATING AERIAL LIFT

31101-14_F07.EPS

Figure 7 Aerial lifts.

Cornice hooks, parapet clamps, or outrigger beams are rigging devices commonly used to support swing scaffolds. A cornice hook is a steel device that hooks to a roof, parapet, or structural support. A parapet clamp fits over and clamps onto a parapet that runs along the perimeter of a roof. An outrigger beam is an arm that extends beyond the edge of the roof. Some outrigger devices have wheels so that they can roll along the roof as the work progresses. Rigging devices are anchored and secured to the building or structure with tieback lines installed at a right angle (perpendicular) to the face of the structure. Tieback lines must be as strong as the suspension ropes or cables.

Cornice hooks or parapet clamps are typically used when working on older buildings. For this reason, it is extremely important to make sure that the cornice or parapet where they attach is structurally sound and capable of supporting the combined weight of the scaffold and workers. If there is any doubt, always have an engineer or other competent person check the structural integrity of the cornice or parapet before using it to support the scaffold. *Figure 9* shows typical rigging devices used with swing scaffolds.

The scaffold's hoisting mechanism enables the worker to raise or lower the scaffold. For manually operated scaffolds, the mechanism consists of ropes or cables rigged through tackle blocks and pulleys. Many scaffolds have hoist mechanisms powered by electricity or compressed air. American National Standards Institute (ANSI) standards limit these powered hoists to a maximum speed of 35 feet per minute. They must also have speed reducers and both primary and secondary brakes. Safety devices are also built into powered hoist systems to prevent operation of the scaffold if it is overloaded and to lower the unit without power if necessary. *Figure 10* shows an electric-powered hoist.

31101-14_F08.EPS

Figure 8 Typical use of a mast-climbing platforms.

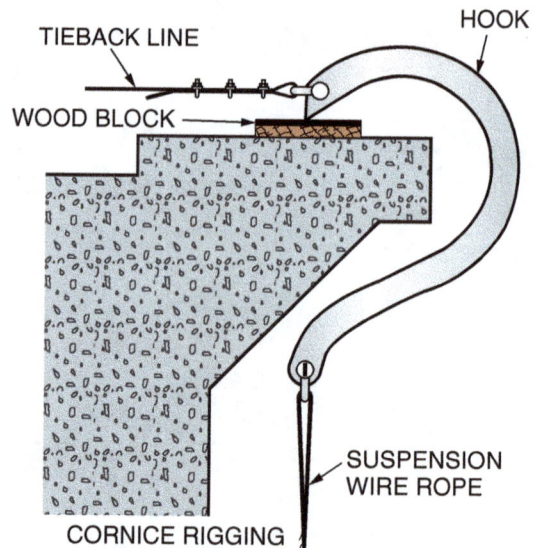

CORNICE RIGGING

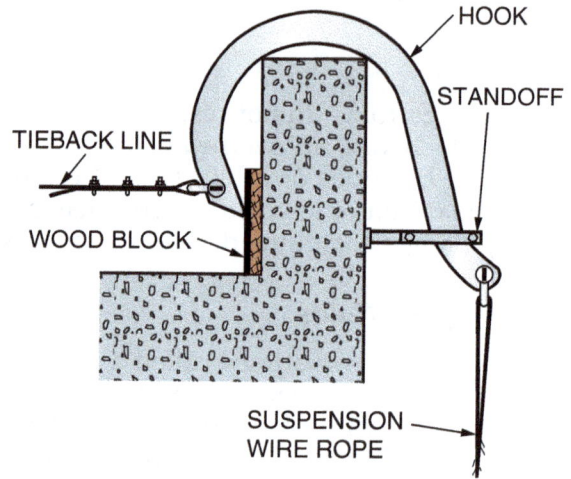

PARAPET RIGGING

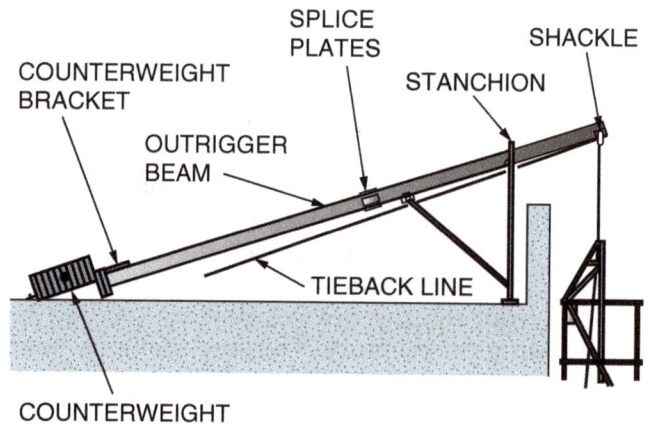

OUTRIGGER

31101-14_F09.EPS

Figure 9 Typical rigging devices used with swing scaffolds.

Figure 10 Electric-powered hoist.

31101-14_F10.EPS

A boatswain's chair is basically a seat attached to a suspension cable or rope. The chair is typically made of nylon rigging with a plywood seat. The worker is restricted to a sitting position with little or no room for tools or equipment on the chair. If necessary, these items can be tied to the chair.

A work cage is a single-point adjustable suspension scaffold. The platform is enclosed with guardrails and toeboards. It is large enough to allow the operator to work standing up. Two work-cage units are sometimes rigged to form a two-point suspension (swing) scaffold. *Figure 11* shows a boatswain's chair and a work-cage scaffold.

2.3.3 Beam-Suspended Scaffolds

Beam-suspended scaffolds are typically used when working on bridges or other industrial steel structures. They are suspended with clamp or roller-type rigging devices that hook onto the lower flanges of two parallel I-beams, each having a bar or bracket to which the scaffold is attached. The roller-type device has wheels that allow the scaffold to be moved along the I-beam flanges as work progresses. This allows the job to continue without having to stop to lower the scaffold and reposition the rigging. *Figure 12* shows a beam-suspended scaffold.

Follow these guidelines when installing or using beam-suspended scaffolds:

- Make sure the I-beam is level so the rollers will not move. If the I-beam is not level, secure the rigging devices to prevent accidental movement.
- Never use this type of equipment on open-ended beams that could allow the scaffold to roll off the supports.
- Install stops if the flange is interrupted anywhere along the beam.
- Make sure that the width of the flange is uniform along the entire length of the beam. A roller set to a particular width could derail if the width of the flange narrows.

2.3.4 Multiple-Point Suspension Scaffolds

Multiple-point suspension scaffolds are suspension scaffolds that are supported by two or more ropes from overhead. These platforms are also equipped with a means of raising or lowering the platform to the desired work levels. The platform must be constructed with full decking, guardrails, midrails, and toeboards. Additional cables or lines must be rigged for fall protection for all personnel working on the platform. A typical application for a multiple-point suspension scaffold is a chimney hoist. *Figure 13* shows a multiple-point suspension scaffold.

Regulations mandate that no more than two workers are permitted to work at one time on a swing or other suspension scaffold designed for a working load of 500 pounds (225 kilograms). With a working load of 750 pounds (337.5 kilograms), no more than three workers are permitted to work at one time. All workers must also wear the proper fall protection equipment when working on suspension scaffolds.

2.3.2 Boatswain's Chairs and Work Cages

Boatswain's (pronounced *bosun's*) chairs and work cages are single-point suspension scaffolds. This means they are suspended from an overhead support by a single cable or rope. Normally, they are used by one worker at a time. Both types provide good maneuverability and accessibility for light-duty work in areas not accessible by other types of scaffolds. Depending on the design, the operator can raise or lower the platform manually with a block and tackle or use an electric or compressed air-powered hoist.

(A) BOATSWAIN'S CHAIR

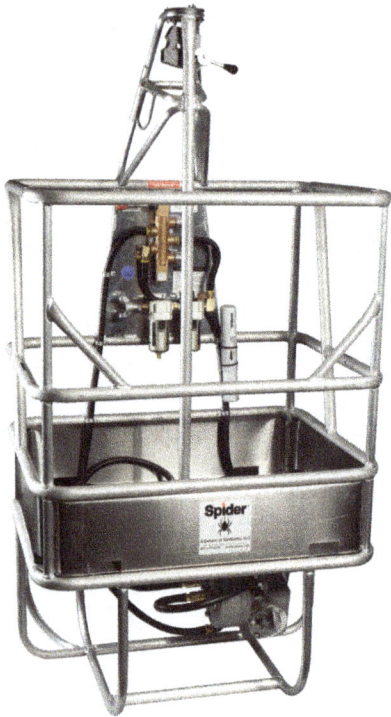

(B) WORK-CAGE SCAFFOLD

31101-14_F11.EPS

Figure 11 Boatswain's chair and work-cage scaffold.

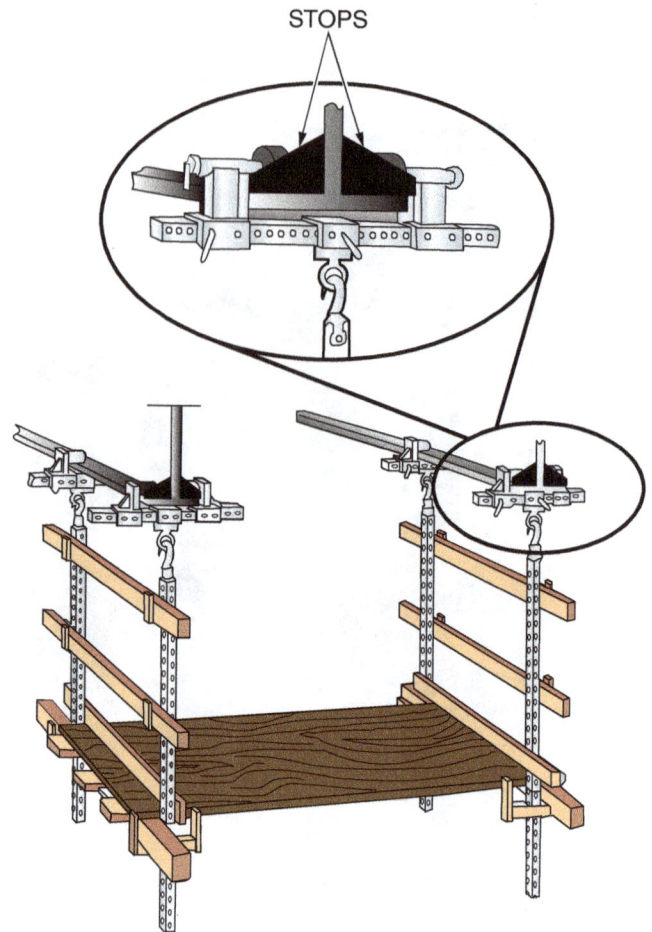

STOPS

31101-14_F12.EPS

Figure 12 Beam-suspended scaffold.

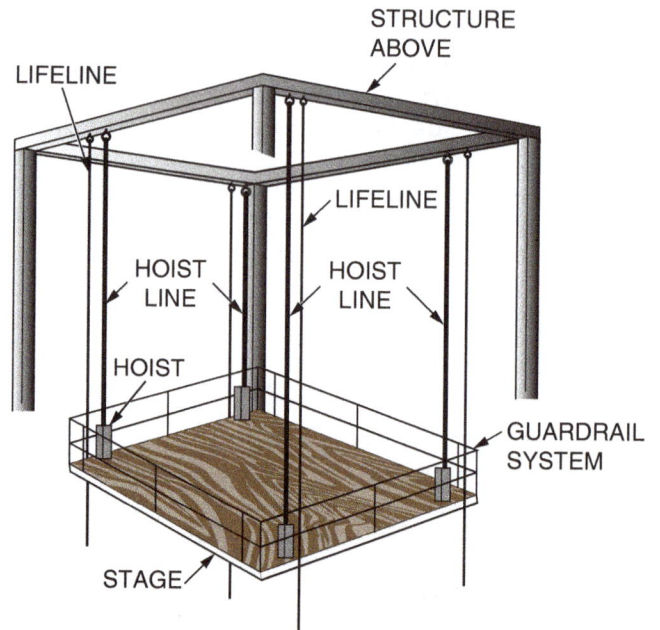

STRUCTURE ABOVE

LIFELINE

LIFELINE

HOIST LINE

HOIST LINE

HOIST

GUARDRAIL SYSTEM

STAGE

31101-14_F13.EPS

Figure 13 Multiple-point suspension scaffold.

2.3.5 Catenary Scaffolds

Catenary scaffolds consist of a platform that is suspended by two horizontal and parallel ropes attached to structural members. Additional support is sometimes provided by vertical pick-ups from the horizontal ropes to the structure above. Hook stops are also installed to prevent the platform from sliding on the support ropes. Fall protection must be provided separately if the platform is more than 10 feet above the next lowest level. *Figure 14* shows an example of a catenary scaffold.

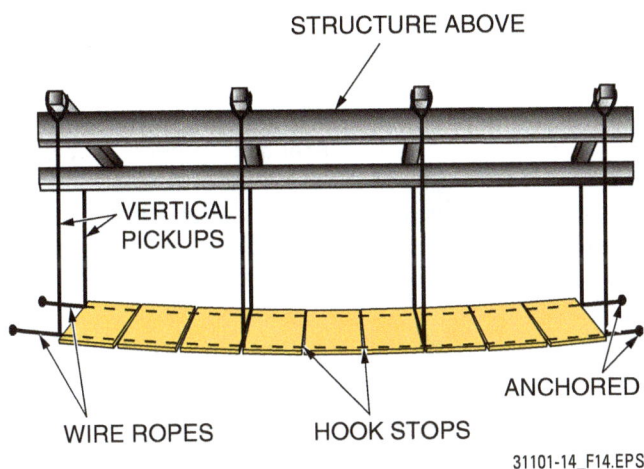

Figure 14 Catenary scaffold.

2.3.6 Float Scaffolds

A float or ship scaffold is a braced platform supported on two parallel bearers. It is similar to the catenary in that it is suspended from overhead by ropes that are a fixed length. Depending on the size of the platform, additional deck bracing may be needed under the platform. *Figure 15* shows a float scaffold.

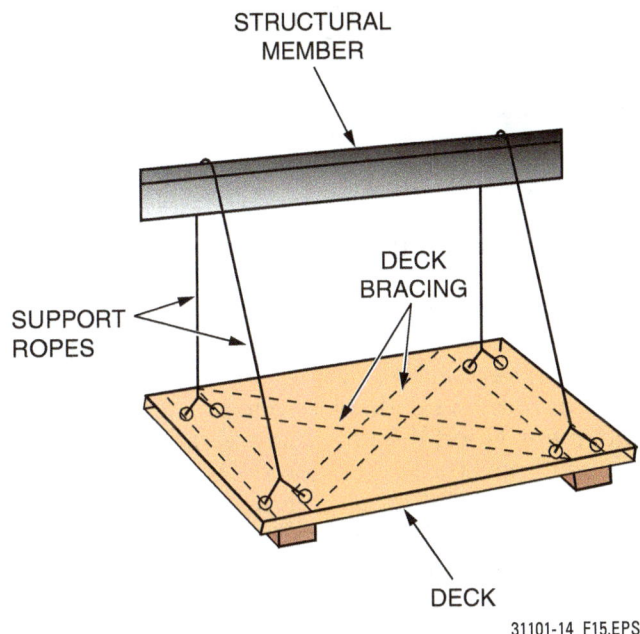

Figure 15 Float scaffold.

2.0.0 Section Review

1. Regulations that require reporting hazardous conditions have been established by the _____.

   a. National Safety Council
   b. US Department of Health and Human Services
   c. American Red Cross
   d. Occupational Safety and Health Administration

2. To prevent movement, supported scaffolds are secured to the structure at intervals of _____.

   a. 10 feet
   b. 20 feet
   c. 30 feet
   d. 40 feet

3. A suspension scaffold designed for a working load of 500 pounds can be used at any one time by no more than _____.

   a. one person
   b. two persons
   c. three persons
   d. four persons

3.0.0 PERSONAL QUALITIES

Objective

Identify personal qualities that contribute to job success.

    a. Describe the responsibilities of a scaffold builder.

    b. Describe the attributes of a good scaffold builder.

Research indicates that most people either quit or lose their jobs because of poor work attitudes, inability to get along with co-workers and supervisors, lack of loyalty to their employers, or a disregard of company policies. All of these reasons, which can be summarized as poor human relations, lead to unsuccessful employment. Good human relations skills on the job are as important as good job skills.

It is important to keep in mind that your employer has provided you with a place to work, tools and equipment, training that will prepare you for a job that can lead to economic security, a salary, and other benefits. OSHA requires, as applicable, that you receive training from a competent person in erecting, disassembling, moving, operating, repairing, maintaining, and inspecting a scaffold.

Your employer has a right to expect you to satisfy fitness-for-duty requirements, such as passing drug tests, etc., and your cooperation in making the business operate efficiently and profitably.

OSHA's General-Duty Clause

The Williams-Steiger Occupational Safety and Health Act of 1970 has become known as the "general-duty clause." It is a catchall for citations if OSHA identifies unsafe conditions to which a regulation does not exist. OSHA 5B of this clause states, "Each employee shall comply with occupational safety and health standards and all rules, regulations, and orders issued pursuant to this Act which are applicable to his own actions and conduct."

It is not possible to list all of the personal qualities that would guarantee job success for every employee. There are too many differences in jobs, surroundings, and people. However, studies of persons in many different work settings have indicated some important qualities and attitudes that are related to job stability and advancement.

3.1.0 Scaffold-Builder Responsibilities

A scaffold builder must be a responsible person with a high degree of concern for the safety of all workers on the job site, and the quality of the work.

3.1.1 Willingness to Take Responsibility

Every scaffold builder should take responsibility for working safely. After being properly trained and instructed, most contractors expect their employees to be aware of what needs to be done. Once responsibility has been delegated, valued employees will take the initiative and continue to perform their duties without further direction.

3.1.2 Rules and Regulations

People can work together well only if there is some understanding about what work is to be done, how it will accomplished, when it will be done, and who will do it. Adequate communication of rules and regulations are a necessity in any work situation.

3.1.3 Tardiness and Absenteeism

Tardiness is being late for work, and absenteeism is not showing up at work when expected to do so. Repeated tardiness and unapproved absences are an indication of unprofessional conduct, a lack of concern for your fellow workers, and insufficient commitment to your contractor.

Work life is governed by the clock. All members of a scaffold crew are required to be at work at a specific time. Failure to arrive on time contributes to a disorganized start to the workday, and resentment on the part of those who do arrive on time. In addition, frequent tardiness or absenteeism will no doubt lead to penalties, including dismissal. When accepting a job with a contractor, you agree to the terms of work, which include arriving in plenty of time to begin the day's work at the scheduled time. Supervisors cannot keep track of people if their employees come in any time they please. It is not fair to others for your employer to ignore tardiness. Your failure to be on time can hold up the work of your fellow workers and other crafts. Better planning of your morning routine will often keep you from being delayed and thus prevent a stressful, late arrival. Arriving a little early will indicate your interest and enthusiasm for your work, and will be appreciated by contractors. The habit of being late is one thing that will stand in the way of promotion.

It is sometimes necessary to take time off from work. No one should be expected to work when sick or when there is a serious issue at home. However, it is possible to get into the habit of letting unimportant and unnecessary matters keep you from the job. This results in lost production and hardship on those who try to carry on the work with less help. The contractor that hires you has a right to expect you to be on the job unless there is good reason for being absent. Do not let a trivial reason keep you home. Staying up late at night can contribute to tardiness, and will affect your productivity during the workday. If you are ill, use your time at home to recover quickly. This, is no more than what you would expect of a person you hired, and on whom you depend to do a certain job.

The most frequent causes of absenteeism are illness or death in the family, accidents, personal business, and dissatisfaction with the job. Some of the causes are legitimate and unavoidable, while others can be controlled. For most situations, you can carry on most personal business affairs after working hours. Frequent absences will reflect unfavorably on a worker when being considered for promotion.

Unforeseen issues will arise from time to time. If this should happen, phone the office as early in the day as possible so your supervisor can find someone to replace you for the day, if needed. Some workers remain at home without contacting the contractor, which is a very unprofessional way to handle the matter. If you fail to inform your employer about an upcoming absence, it will affect your fellow employees in a negative way. They have no way of knowing whether you have merely been held up and will be in later, or whether immediate steps should be taken to assign your work to someone else. Courtesy demands that you let the supervisor know if you cannot come to work.

Contractors sometimes resort to docking pay, demotion, and even dismissal in an effort to control tardiness and absenteeism. No contractor likes to impose penalties of this kind. However, in fairness to those workers who do come on time and show up for work, a contractor is sometimes forced to discipline those who will not follow the rules.

3.2.0 Scaffold-Builder Characteristics

In addition to taking on the responsibilities assigned to a scaffold builder by a supervisor, each member of the crew should also have strong personal characteristics to be successful on the job. These characteristics include:

- Professionalism
- Honesty
- Loyalty
- Willingness to learn
- Cooperation
- Positive attitude

The Customer/Owner

When you are on a job site, you are working for your contractor as well as the facility owner. When you are honest and maintain a professional attitude when interacting with customers, everyone will benefit. Your contractor will be pleased with your performance, and the customer will be happy with the work that is being done. A good, professional attitude goes a long way toward building trust in your abilities to handle the job and ensuring repeat business.

Late for Work

Showing up on time is mandatory. Your contractor rightly expects your arrival at a set time. Legitimate emergencies may arise that can cause you to be late or miss work. However, consistent tardiness is always a poor choice. You could face serious consequences as a result of repeated tardiness and absenteeism.

Ethical Principles for Members of the Construction Trades

Honesty – Be honest and truthful in all your dealings. The result of one dishonest act is that it will take a long time for you to regain the trust of everyone involved. Therefore, conduct business according to the highest professional standards. Faithfully fulfill all contracts and commitments, and never deliberately mislead or deceive others.

Integrity – Integrity calls for adherence to a code of ethical behavior. Demonstrate personal integrity and do not allow peer pressure to cause you to compromise what is right. Never sacrifice your principles for expediency, or act in an unscrupulous manner.

Loyalty – Be worthy of trust. Demonstrate your commitment in fulfilling what is expected of you by your company, your crew, as well as other contractors, trades, and organizations.

Fairness – Be fair and just in all dealings. Treat others as you want them to treat you. Do not take advantage of the mistakes or difficulties of others. Fairness calls for commitment to the courteous and equal treatment of all individuals, and acceptance of diversity in the sex, race, and personality traits of others.

Law abiding – Comply with all laws, rules, and regulations relating to both personal and business activities.

Commitment to excellence – Pursue excellence in performing your duties, be well-informed and -prepared, and constantly endeavor to increase your proficiency by gaining new skills and knowledge.

Leadership – By your own conduct, seek to be a positive role model for others.

3.2.1 Professionalism

The word *professionalism* is a broad term that describes the desired overall behavior and attitude expected in the workplace. Professionalism is too often absent from the construction site and the various trades. Most people would argue that professionalism must start at the top in order to be successful. It is true that management's support of professionalism is important to success in the workplace, but it is even more important to recognize the importance of professionalism. You will have more interaction with the client, and other trades on the job site, than most members of your company's management team.

Professionalism can be demonstrated in a variety of ways every minute an employee is on the job site. It can be noted in the way you dress, how you carry yourself, and how you relate to others. Most important, do not encourage the unprofessional behavior of co-workers. Do your utmost to demonstrate the characteristics of professional behavior, setting the proper example for others.

Professionalism is a personal responsibility that benefits both the contractor and the worker. The cumulative image of our industry is affected by each individual. If you maintain a professional attitude, the industry image will be enhanced.

3.2.2 Honesty

Honesty and personal integrity are very important characteristics and will contribute to your success over time. Professionals pride themselves in superior job performance, punctuality, and dependability. Honesty requires that each job is completed in a professional way, never by cutting corners or substituting inferior materials. Professionalism requires that you are not concerned with only yourself, but with protecting property, such as tools and materials belonging to contractors, customers, and other trades, from damage or theft.

Honesty is not simply a choice between right and wrong, but it is a choice between success and failure. The unauthorized taking of materials, tools, or equipment from the job site or lying about the scope of your work will always become known in time. When such improper conduct is discovered, it usually results in termination. A reputation as a dishonest person will certainly hamper future employment opportunities.

If your desire is to be successful and enjoy continuous employment, consistent earnings, and to be sought after as a desired employee, then make a personal commitment to honesty in the workplace and you will reap the benefits. Honesty means more than giving a fair day's work for a fair day's pay; it means following through on what you have agreed to do; it means that your words are not half-truths, but convey what has actually taken place. Contractors place a high value on an employee who is honest.

3.2.3 Loyalty

You will rightfully expect your employer to look out for your best interests, to provide you with steady employment, and to promote you as better job openings occur. Contractors feel that they, too, have the right to expect you to be loyal to them, to keep their best interests in mind; to speak positively of them to others; to keep minor issues strictly within the company; and to keep absolutely confidential all matters that pertain to the business. Both contractors and workers should keep in mind that loyalty is not something that can be demanded; rather, it is something to be earned.

3.2.4 Willingness to Learn

Every contractor has their own way of doing things. Each will expect their employees to be willing to learn the company way, and you must be willing to learn company-specific methods and procedures as quickly as possible. The use of new tools or equipment could make it necessary for even experienced scaffold builders to adapt. It is often the case that skilled workers resent having to accept new procedures because of the retraining that is involved. However, contractors have a right to expect employees to put forth the necessary effort to employ current methods so as to keep pace with competitors and show a profit. Profit enables the contractor to continue in business and provide you with steady employment.

3.2.5 Cooperation

To cooperate means to work together, and has been described as an "I can do that" attitude. Your goal should be to work as a member of a team with your contractor, supervisor, and fellow workers in a common effort to get the work done efficiently, safely, and on time. Many people underestimate the importance of cooperating with others.

The term *human relations* is often associated with the willingness to cooperate. It involves being friendly, pleasant, courteous, cooperative, adaptable, and sociable. Good human relations demands more than getting people to like you; it is also knowing how to handle difficult situations as they arise.

Human relations involves knowing how to work with supervisors who can be demanding and may sometimes seem unfair. It means understanding the personality traits of others — as well as your own. You may have to restore good working relationships that have deteriorated for one reason or another, perhaps through no fault of your own. You will need to learn how to handle frustrations without alienating others. Human relations is building and maintaining relationships with all kinds of people, regardless of whether they are easy to get along with or not.

Effective human relations contribute directly to job productivity. If there is dissension on any crew, productivity is diminished. Productivity is the key to business success. There are work schedules to be met and jobs that must be completed. Whether you are newly hired or have years of experience, you will be evaluated not only by the work you generate, but by how the crew as a whole is operating. Always perform your tasks in a way that does not detract from the unity of the

Working with Other Trades

Cooperation among the various trades at the job site and respect for their desire to work effectively is essential in achieving a smooth-running project that operates on schedule. Time is money.

On many well-run jobs, a sense of respect and cooperation soon develops, and the trades can work together with efficiency. Many times, there is even a trade-off of activities that allows the project to progress at a uniform pace. For example, the scaffold builders may alter the location of where they intended to begin work, on a given day, for the benefit of the electricians who may need to reach an elevated area quickly. Such practices are fine as long as they can be implemented without additional cost. This type of cooperation will be repaid to your crew by other trades as the job progresses.

crew. There are very few construction jobs that are performed alone. You will have to learn to work smoothly with others, or you will soon find yourself unemployed.

3.2.6 Positive Attitude

A positive attitude is essential to a successful career. Being positive means being energetic, highly motivated, attentive to detail, and alert. A positive attitude is essential to safety on the job. A worker with a good attitude has a positive effect on others. Whatever your demeanor, it will be transmitted to others on the job. A persistent negative attitude can spoil the positive attitudes of others. It is very difficult to maintain a high level of productivity while working next to a person with a negative attitude.

People favor a person who is positive. Being positive makes a person's job more interesting and exciting. The kind of attitude transmitted to management has a great deal to do with a worker's future success in the company. Supervisors will be paying close attention to your attitude, demonstrated to them by your approach to the job, reaction to directives, and the way you handle problems.

This does not mean you will be expected to walk around in a perpetual state of glee with a smile on your face. As a matter of fact, some people transmit a positive attitude even though they seldom smile. They do this by the way they treat others, the way they look at their responsibilities, and the approach they take when faced with problems.

Here are a few suggestions that will help you to maintain a positive attitude:

- Remember that your attitude follows you wherever you go. If you make a greater effort to be a more positive person in your social and personal lives, it will automatically help you on the job. The reverse is also true: one effort will complement the other.
- Negative comments are seldom welcomed by fellow workers on the job; neither are they welcome socially. The solution is to talk about positive things and be complimentary. Constant complainers do not build healthy and fulfilling relationships.

Teamwork

Many of us like to follow all sorts of different teams: racing teams, baseball teams, football teams, and soccer teams. Just as in sports, a job site is made up of a team. As a part of that team, you have a responsibility to your teammates. What does teamwork really mean on the job? As employees we all must sincerely do everything we can to build strong, professional working relationships with fellow workers, supervisors, and customers.

- Look for the good things in people on the job, especially in your supervisor. Nobody is perfect, but almost everyone has a few worthwhile qualities. If you dwell on people's good features, it will be easier to work with them.
- Look for the good things where you work. Are there factors that make it a good place to work? Is it the hours, the physical environment, the people, the actual work being done, or is it the atmosphere? Keep in mind that you are not expected to like everything. No work assignment is perfect, but if you concentrate on the good things, the negative factors will seem less important and bothersome.
- Look for the good things in the contractor. Just as there are no perfect assignments, there are no perfect contractors. Nevertheless, almost all organizations have good features. Is the contractor progressive? What about promotional opportunities? Are there chances for self-improvement? What about the wage and benefit package? Is there a good training program? Don't expect to have everything you would like, but there should be enough to keep you positive. In fact, if you decide to stay with a contractor for a long period of time, it is wise to look at the good features and focus on them. If you think positively, you will act positively.
- You may not be able to change the negative attitude of another worker, but you can prevent your own attitude from becoming negative.

Trade Coordination

Scaffold builders are generally on the job before most other trades. Therefore, an important part of trade coordination is ensuring that the work done by the scaffold builders will accommodate the other trades. For example, plans for location of heating and air conditioning duct runs, plumbing fixtures and piping, and electrical components and wiring runs must be continually reviewed by the superintendent or supervisor to ensure that scaffold builders are allocating space, in the correct locations, for these facilities.

3.0.0 Section Review

1. Repeated tardiness is an indication of lack of concern for your fellow employees.

   a. True
   b. False

2. All the following are characteristics leading to success on the job, *except* _____.

   a. willingness to learn
   b. honesty
   c. positive attitude
   d. allowing unsafe conditions

SECTION FOUR

4.0.0 APPRENTICESHIP

Objective

Explain the apprenticeship training process.

a. Describe the types of formal craft training available in the scaffolding industry.
b. Describe the standards associated with an apprenticeship program.
c. Identify the functions of the Bureau of Apprenticeship and Training (BAT).
d. Identify the advantages and benefits of today's apprenticeship training programs.

Trade Terms

Apprentice: A person that agrees to work for another for a specific amount of time in return for instruction in a trade, art, or business.

Related instruction: Classroom instruction that contributes to training received on the job site.

Transferring practical knowledge from one generation to another goes far back in human history. Provisions for the teaching and training of new craftsmen are found in the Babylonian Code of Hammurabi from 4,000 years ago. In ancient Rome, the high level of craftsmanship that produced aqueducts, great public buildings like the Coliseum, and a large network of roads was the result of craftsmen and artisans whose skills were handed down from generation to generation.

During the Middle Ages (approximately 500 AD to 1500 AD), there occurred the development of two large classes: merchants and skilled artisans. Both organized themselves into guilds in order to control their respective fields, stem competition, and provide training for apprentices. The guilds established the indenture system. In this system, sons of journeymen would be indentured to a master who, in turn, would train the apprentice in the craft, as well as provide food, clothing, and shelter. The length of apprenticeships varied from 2 to 10 years, depending on the trade.

4.1.0 Formal Craft Training

Over the past 30 years, the rate of formal training within the construction industry has been in decline. Until the establishment of NCCER (National Center for Construction Education and Research), the only opportunity for formal construction training was through the US Department of Labor, Bureau of Apprenticeship and Training (BAT). The federal government established registered apprenticeship training through the *Code of Federal Regulations* (*CFR*) Title 29, Part 29, which dictates specific requirements for apprenticeship, and *CFR* Title 29, Part 30, which dictates specific guidelines for recruitment, outreach, and registration into BAT-approved apprenticeship programs.

Compared to the overall employment in the construction industry, the percentage of enrollment in BAT-style programs has been below 5 percent for the past decade. BAT programs rely on mandatory classroom instruction and on-the-job training (OJT). The classroom instruction required is 144 hours per year while the OJT requirement is 2,000 hours per year. A typical four-year BAT-approved program requires 8,000 hours of OJT and 576 hours of related classroom training in order to receive a BAT journeyman certificate.

Craft training through the BAT has not changed for 30 years, which is believed to be one reason for the lack of use of this program in the construction industry today. Education and training throughout the country are undergoing significant changes. As educational, political, financial, and student factions argue over the direction and future of education, educators and researchers are learning and applying new techniques to adjust to how today's students learn and apply their education.

NCCER is an independent, not-for-profit, educational foundation founded and funded by the construction industry to solve the training problem plaguing the industry today. The basic idea of NCCER is to replace governmental control and credentialing of the construction workforce with industry-driven training and education programs. NCCER enhances traditional classroom learning through its competency-based training approach. Competency-based training means that instead of requiring specific hours of classroom training, you simply have to prove that you know what is required and demonstrate that you can perform the specific skill. All completion information for every trainee is sent to NCCER and kept within the Registry. The Registry can then confirm training and skills for workers as they move from company to company, state to state, or even within their own company.

The dramatic shortage of skills within the construction workforce, combined with the shortage of new workers coming into the industry, is forcing the industry to design and implement new training initiatives to combat the problem. Whether you enroll in a BAT program, an NCCER program, or both, it is critical that you work for an employer that supports a national, standardized training program that includes credentials to confirm your skill development.

4.2.0 Apprenticeship Program and Standards

Apprentice training goes back thousands of years, but its basic principles have not changed in that time. First, it is a means for individuals entering the craft to learn from those who have mastered the craft. Second, it focuses on learning by doing: real skills versus theory. Although some theory is presented in the classroom, it is always presented in a way that helps the trainee understand the purpose behind the skill that is to be learned.

All apprenticeship standards prescribe certain work-related or on-the-job training. This on-the-job training is broken down into specific tasks in which the apprentice receives hands-on training during the period of the apprenticeship. In addition, a specified number of hours is required in each task. In a competency-based program, it may be possible to shorten this time by testing out of specific tasks through a series of performance exams.

In a traditional program, the required on-the-job training is usually measured in increments of hours per year. The apprentice must log all work time and turn it in to the Apprenticeship Committee so that accurate time control can be maintained.

The classroom instruction and work-related training will not always run concurrently due to such reasons as layoffs, type of work needed to be done in the field, etc. Apprentices with special job experience or course work may obtain credit toward their classroom requirements. This reduces the total time required in the classroom while maintaining the total on-the-job training hours

requirement. These special cases will depend on the type of program, the regulations, and standards under which it operates.

Informal on-the-job training provided by employers is usually less thorough than that provided through a formal apprenticeship program. The degree of training and supervision in this type of program often depends on the size of the employing firm. A small contractor who specializes in home building may provide training in only one area, such as rough framing. In contrast, a large general contractor may be able to provide training in several areas.

For those entering an apprenticeship program, a high school or technical school education is desirable, as are courses in shop, mechanical drawing, and general mathematics. Manual dexterity, good physical condition, a good sense of balance, and a lack of fear of working in high places are important. The ability to solve arithmetic problems quickly and accurately and to work closely with others is essential.

The prospective apprentice must submit certain information to the apprenticeship committee. This may include the following:

- Aptitude test (General Aptitude Test Battery or GATB Form Test) results (usually administered by the local job-assistance agency)
- Proof of educational background (candidate should have school[s] send transcripts to the committee)
- Letters of reference from past employers and friends
- Results of a physical examination
- Proof of age
- If the candidate is a veteran, a copy of Form DD214
- A record of technical training received that relates to the construction industry and/or a record of any pre-apprenticeship training
- High school diploma or general equivalency diploma (GED)

The apprentice must:

- Wear proper safety equipment on the job
- Purchase and maintain tools of the trade as needed and required by the contractor
- Submit a monthly on-the-job training report to the committee
- Report to the committee if a change in employment status occurs
- Attend classroom-related instruction and adhere to all classroom regulations

4.3.0 Bureau of Apprenticeship and Training (BAT)

In 1934, Congress passed the Fitzgerald Act, which was the first step in forming a national system of apprenticeship training. Since then, the Bureau of Apprenticeship and Training (BAT) has worked closely with employer and labor groups, vocational schools, and other groups such as state agencies. The BAT does not conduct training programs; its main function is promotional and advisory.

Programs registered by BAT must provide that:

- Apprentices are no less than 16 years old.
- Opportunities to apply are available to all.
- There is a program incorporating both training and on-the-job experience.
- A minimum of 144 hours per year of related training is given.
- There is a progressively increasing schedule of wages.
- There is proper supervision with adequate facilities.
- Job performance and related instruction are periodically evaluated.
- Successful completions are formally recognized.

The apprenticeship system has grown up with America. Like America, it is still growing and changing. Today it serves a far-different nation from the one of pioneer days. Scientific discoveries, new teaching methods, expanding industry, and a large population are among the demands of our present-day technological and social systems to which apprenticeship is responding.

To meet the need for changes in production methods and products, apprenticeships have been set up in new trades, and apprenticeships in many of the older trades have been updated.

4.4.0 Apprenticeship Today

For those just starting out in the world of work, apprenticeship has important advantages. It offers an efficient way to learn skills because the training is planned and organized. The apprentices earn as they learn because they are already working.

Industry, too, benefits greatly. Out of apprenticeship programs come all-around craftworkers competent in all branches of their trades and able to work without close supervision because their training has enabled them to use imagination, ability, and knowledge of their work. When changes are made in production, these workers provide the versatility needed for quick adaptation of work components to suit the changing needs. An adequate supply of skilled workers with these qualities is vital to industrial progress.

Apprentice training is a systematic way of providing the skilled workers necessary to supply current and future employment demands. Related instruction is an integral part of a planned apprentice-training program and ranks in importance with the skills learned on the job. The related instruction provided to trainees helps them better understand the scaffolding trade and to know why things are done, as well as how they are done.

Workers are encouraged to advance in knowledge and skill so that they can become productive employees and thereby contribute to the economic success of the company. Individual initiative is doubly important: the workers gain as they move up the ranks to positions of greater responsibility, and the company gains as a competitive business.

1. Some 4,000 years ago, the Code of Hammurabi made provision for training _____.

   a. agricultural workers
   b. camel drivers
   c. new craftsmen
   d. public servants

2. In a traditional apprenticeship program, on-the-job training is usually measured in increments of _____.

   a. hours per year
   b. days per month
   c. hours per month
   d. weeks per year

3. The first step in forming a national system of apprenticeship training came in 1934 with the passage of the _____.

   a. Smith-Fields Act
   b. Fitzgerald Act
   c. Taft-Hartley Act
   d. Wilson Act

4. Apprentice training is a systematic way of providing the skilled workers necessary to supply current and future employment demands.

   a. True
   b. False

SUMMARY

A scaffold is a temporary platform supported above the ground for the purpose of performing work or supporting someone at a certain height. Scaffolds have been around for thousands of years. Modern scaffolds are constructed from uniform components made of steel or aluminum. Modern scaffolds fall into three basic categories: supported, mobile, and suspension.

The role of a scaffold builder is to erect safe and secure scaffolds according to the latest local and national OSHA regulations. The erection of scaffolds requires that the scaffold builder work at heights while handling the various scaffold components. Each scaffold builder should become familiar with the terms used in the trade, and be able to use them to communicate with other scaffold builders and workers at the job site.

Scaffold builders must be familiar with, and be capable of using, the tools necessary to erect each type of scaffold. They must also be able to use basic math functions to determine the proper type of scaffold to erect for a given application. Scaffolds are classified according to the weight they can safely support.

The types of supported scaffolds are tubular welded-frame, tube-and-clamp, and system (modular) scaffolds. Other supported scaffolds include outrigger and pump-jack scaffolds. Other types of mobile scaffolds include scissors lifts, aerial lifts, and mast-climbing platforms. Suspension scaffolds are designed to be hung from cables so that they can travel up and down the face of a structure. Scaffold builders should become familiar with each type of scaffold and the correct uses of each type.

1. The first metal scaffold frames, designed in 1935, were made of _____.
   a. aluminum tubing
   b. iron pipe
   c. angle iron
   d. seamless steel tubing

2. On a heavy-duty frame, a 5-foot by 7-foot scaffold platform can safely support a load of up to _____.
   a. 875 pounds
   b. 1,255 pounds
   c. 1,566 pounds
   d. 2,625 pounds

3. OSHA regulations covering scaffold safety are divided into sections, with each section devoted to a _____.
   a. particular industry
   b. type of scaffold
   c. geographic region
   d. specific type of hazard

4. The Occupational Safety and Health Act was passed by the US Congress in _____.
   a. 1925
   b. 1960
   c. 1970
   d. 1985

5. When erecting scaffolds, compensate for uneven ground by using _____.
   a. concrete blocks
   b. leveling jacks
   c. boxes
   d. railroad ties

6. On tubular welded-frame scaffolds, the vertical members of the erected scaffold are automatically squared and vertically aligned by the _____.
   a. braces
   b. bearers
   c. clamps
   d. cam locks

7. The wooden planks placed under the base plates of a scaffold on dirt or similar surfaces are referred to as _____.
   a. mudsills
   b. floaters
   c. ground anchors
   d. bearers

8. A platform of scaffold boards supported by beams secured to the building structure and extending out a window is called a(n) _____.
   a. suspension scaffold
   b. cantilevered scaffold
   c. outrigger scaffold
   d. supported scaffold

9. A type of scaffold that can be raised manually by stepping on a foot lever is a _____.
   a. bootstrap lift
   b. pump-jack scaffold
   c. mast-climbing platform
   d. scissors lift

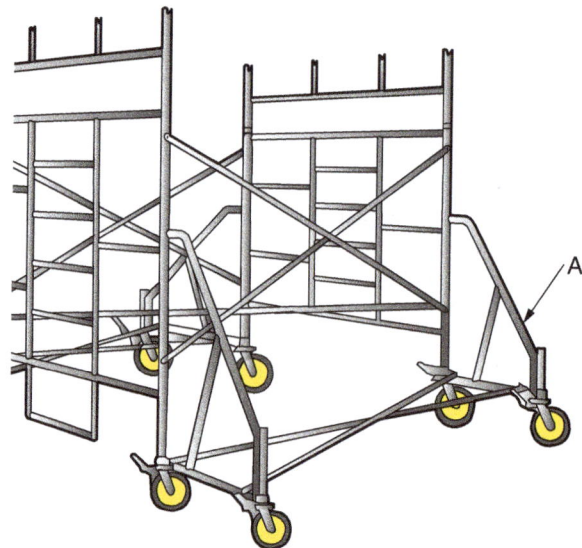

Figure 1

31101-14_RQ01.EPS

10. The component labeled A on the mobile scaffold shown in Review Question *Figure 1* is called a(n) _____.
    a. diagonal brace
    b. cantilever
    c. strut
    d. outrigger

11. The lift height on a scissors lift may be as much as _____.

    a. 22 feet
    b. 37 feet
    c. 41 feet
    d. 52 feet

12. A type of scaffold that can have a maximum height of more than 300 feet is the _____.

    a. mast-climbing platform
    b. beam-suspended scaffold
    c. tube-and-clamp scaffold
    d. aerial lift

13. Another name for the two-point suspension scaffold is the _____.

    a. float scaffold
    b. platform hoist
    b. swing scaffold
    d. boatswain's chair

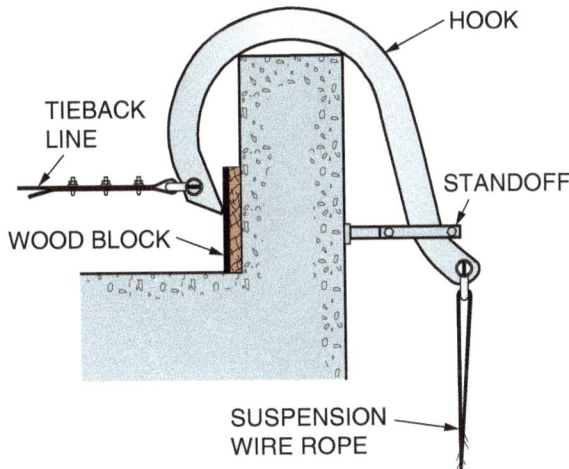

Figure 2

14. The support method for a suspension scaffold shown in Review Question *Figure 2* is called _____.

    a. cornice rigging
    b. parapet rigging
    c. roof rigging
    d. tieback rigging

15. OSHA requires that the strength of swing-scaffold cables or ropes exceed the rated load of the hoist by at least _____.

    a. three times
    b. four times
    c. five times
    d. six times

16. A type of scaffold often used on bridges or other industrial steel structures is the _____.

    a. outrigger scaffold
    b. float scaffold
    c. beam-suspended scaffold
    d. mobile scaffold

17. Fall protection must be provided separately if the platform of a catenary scaffold is above the next lowest level by more than _____.

    a. 6 feet
    b. 10 feet
    c. 12 feet
    d. 15 feet

18. If you can't come to work on a given day, you should _____.

    a. ask a fellow worker to tell your supervisor
    b. try to find a substitute
    c. phone your supervisor as early in the day as possible
    d. explain your absence the next time you come to work

19. A positive attitude is _____.

    a. essential to safety on the job
    b. not an important part of the job
    c. always wearing a smile
    d. less important than getting work done

20. During the Middle Ages, apprentices worked for and learned from master craftsmen in a system called _____.

    a. tutelage
    b. involuntary servitude
    c. bonding
    d. indenture

Fill in the blank with the correct term that you learned from your study of this module.

1. A(n) _____ is an individual who spends a specific amount of time learning a craft from someone who has mastered the craft.

2. A person with extensive knowledge, training, and experience who has successfully demonstrated an ability to solve or resolve problems related to a task or project is a(n) _____.

3. Classroom instruction that contributes to training received on the job site is _____.

4. Uneven ground on which a scaffold is being erected is compensated for by using adjusting screws or _____.

5. A person who is capable of identifying existing and predictable hazards on a project and has the authorization to take corrective actions to immediately eliminate the hazards is known as a(n) _____.

Trade Terms

Apprentice
Competent person
Leveling jack

Qualified person
Related instruction

Trade Terms Introduced in This Module

Apprentice: A person that agrees to work for another for a specific amount of time in return for instruction in a trade, art, or business.

Competent person: Defined by OSHA as one who is capable of identifying existing and predictable hazards in the surroundings or working conditions that are unsanitary, hazardous, or dangerous to employees, and who has authorization to take prompt corrective measures to eliminate them.

Leveling jack: A threaded, adjustable screw jack inserted between scaffold legs and their base plates or caster wheels, used to bring scaffolds into a level position when being erected on uneven surfaces.

Qualified person: Defined by OSHA as one who, by possession of a recognized degree, certificate, or professional standing, or who by extensive knowledge, training, and experience, has successfully demonstrated the ability to solve or resolve problems related to the subject matter, the work, or the project.

Related instruction: Classroom instruction that contributes to training received on the job site.

Additional Resources

This module presents thorough resources for task training. The following resource material is suggested for further study.

"Scaffolding." OSHA. **www.osha.gov/SLTC/scaffolding/index.html**

Fall Protection and Scaffolding Safety: An Illustrated Guide. 2000. Grace Drennan Ganget. Government Institutes.

International Building Code®, Latest Edition. Falls Church, VA: International Code Council.

International Residential Code®, Latest Edition. Falls Church, VA: International Code Council.

Figure Credits

Section Review Answer Key

Answer	Section Reference	Objective
Section One		
1. c	1.1.1	1a
2. b	1.2.2	1b
3. b	1.3.0	1c
Section Two		
1. d	2.0.0	2
2. c	2.1.2	2a
3. b	2.3.1	2c
Section Three		
1. a	3.1.3	3a
2. d	3.2.0	3b
Section Four		
1. c	4.0.0	4
2. a	4.2.0	4b
3. b	4.3.0	4c
4. a	4.4.0	4d

NCCER CURRICULA — USER UPDATE

NCCER makes every effort to keep its textbooks up-to-date and free of technical errors. We appreciate your help in this process. If you find an error, a typographical mistake, or an inaccuracy in NCCER's curricula, please fill out this form (or a photocopy), or complete the online form at **www.nccer.org/olf**. Be sure to include the exact module ID number, page number, a detailed description, and your recommended correction. Your input will be brought to the attention of the Authoring Team. Thank you for your assistance.

Instructors – If you have an idea for improving this textbook, or have found that additional materials were necessary to teach this module effectively, please let us know so that we may present your suggestions to the Authoring Team.

NCCER Product Development and Revision
13614 Progress Blvd., Alachua, FL 32615

Email: curriculum@nccer.org
Online: www.nccer.org/olf

❏ Trainee Guide ❏ Lesson Plans ❏ Exam ❏ PowerPoints Other _____

Craft / Level: _____ Copyright Date: _____

Module ID Number / Title: _____

Section Number(s): _____

Description: _____

_____

_____

_____

Recommended Correction: _____

_____

_____

_____

Your Name: _____

Address: _____

_____

Email: _____ Phone: _____

31102-15

Trade Safety

OVERVIEW

The scaffolding industry, including manufacturers, assemblers, and users, is guided by regulations, codes, and standards. To fully understand the relationship between the industry and this information, it is necessary to understand the meaning of codes and regulations. This module deals with trade safety and the codes and regulations that apply to scaffolds.

Module Two

Trainees with successful module completions may be eligible for credentialing through the NCCER Registry. To learn more, go to **www.nccer.org** or contact us at **1.888.622.3720**. Our website has information on the latest product releases and training, as well as online versions of our *Cornerstone* magazine and Pearson's product catalog.

Your feedback is welcome. You may email your comments to **curriculum@nccer.org**, send general comments and inquiries to **info@nccer.org**, or fill in the User Update form at the back of this module.

This information is general in nature and intended for training purposes only. Actual performance of activities described in this manual requires compliance with all applicable operating, service, maintenance, and safety procedures under the direction of qualified personnel. References in this manual to patented or proprietary devices do not constitute a recommendation of their use.

Objectives

When you have completed this module, you will be able to do the following:

1. Identify the reasons for the Occupational Safety and Health Act (OSHA) regulations that govern the scaffolding industry.
   a. Explain the development and intent of the regulations and standards.
   b. Describe common safety practices used in the scaffolding industry.
2. Explain the basic guidelines for planning, erecting, and using scaffolds.
   a. Explain the basic guidelines for planning a scaffold project.
   b. Explain the basic guidelines for erecting a scaffold.
   c. Explain the basic guidelines for using a scaffold.
3. Identify the equipment and tasks required for safe scaffold erection.
   a. List the personal protective equipment required for safe scaffold erection.
4. Identify the fall protection and lifesaving measures employed in the scaffolding trade.
   a. Identify the appropriate fall protection and lifesaving equipment, and describe their proper use.
   b. Describe proper rescue procedures after a fall.
5. Identify common electrical hazards and sources when working with scaffolds.
   a. Identify common electrical hazards when working with scaffolds.
   b. Identify common electrical sources when working with scaffolds.

Performance Tasks

When you have completed this module, you will be able to do the following:

1. Inspect the work site and identify potential safety hazards.
2. Plan and sketch a scaffold for a specific application specified by the instructor.
3. Examine and identify the soil type at the work site, and determine the proper mudsills to use.
4. Demonstrate the proper method for wearing and rigging a body harness.

Trade Terms

Decibels (dBs)
Galvanic action
Laminated
Hazard analysis

Pounds per square foot (psf)
Putlog
Tensile strength

Industry-Recognized Credentials

If you're training through an NCCER-accredited sponsor, you may be eligible for credentials from NCCER's Registry. The ID number for this module is 31102-14. Note that this module may have been used in other NCCER curricula and may apply to other level completions. Contact NCCER's Registry at 888.622.3720 or go to **www.nccer.org** for more information.

Code Note

Codes vary among jurisdictions. Because of the variations in code, consult the applicable code whenever regulations are in question. Referring to an incorrect set of codes can cause as much trouble as failing to reference codes altogether. Obtain, review, and familiarize yourself with your local adopted code.

Contents

Topics to be presented in this module include:

Figures and Tables

SECTION ONE

1.0.0 REGULATIONS AND STANDARDS

Objective

Identify the reasons for the Occupational Safety and Health Act (OSHA) regulations that govern the scaffolding industry.

a. Explain the development and intent of the regulations and standards.
b. Describe common safety practices used in the scaffolding industry.

Performance Task

Inspect the work site and identify potential safety hazards.

A code is a systematically arranged and comprehensive collection of laws and rules. A regulation is a principle, rule, or law designed to govern behavior. A standard is a degree or level of performance. Standards are developed and maintained by industry groups, while codes and regulations are developed by government agencies.

1.1.0 The Development and Intent of Regulations and Standards

The primary code, or regulation, guiding the scaffolding industry is the Occupational Safety and Health Administration (OSHA) code. This code is federal law. Violations of the code are punishable by fines or other similar measures. Individual states have their own codes, which are also enforceable by law. The intent of all of these laws is to provide a safe workplace. These laws apply to workers and employers equally, and should be used as a basis for all rules, policies, and behaviors.

The OSHA regulations form the basic set of rules governing the scaffolding industry. These rules cover all aspects of scaffold use and misuse. The OSHA regulations are divided into sections. Each industry is covered in a separate section, as follows:

- General Industry Standard (29 *CFR* [*Code of Federal Regulations*] 1910.28)
- Maritime Standard (29 *CFR* 1915.71, section k)
- Construction Standard (29 *CFR* 1926.451–454)

Each regulation is designed to have jurisdiction over an industry that uses scaffolds.

In addition to federal, state, and local regulations, organizations working with scaffolds have developed standards that are useful to the scaffold builder. The main sources of these standards are:

- Army Corps of Engineers (ACE) (EM-385-1.1)
- Mine Safety & Health Administration (MSHA)
- American National Standards Institute (ANSI)
- Associated Builders and Contractors (ABC)
- Scaffold and Access Industry Association (SAIA)
- Scaffold, Shoring, and Forming Institute (SSFI)
- Association of General Contractors (AGC)
- Various scaffold manufacturers' company manuals

Copies of the standards published by each of the industry associations listed are available from the individual association and are not included in this module. General industry regulations are applicable to all types of scaffolds. Safety recommendations for the planning, erection, use, and disassembly of individual scaffold types will be covered in other modules in this curriculum.

1.1.1 General Safety and Health Provisions

OSHA provides general safety and health regulations that apply to the construction industry as a whole. A few excerpts from relevant regulations are as follows. The actual text from the regulations is in *italics*, and the comments and explanations are in regular print.

Accident prevention responsibilities include the following:

- *It shall be the responsibility of the employer to initiate such programs as may be necessary to comply with this part.*
- *Such programs shall provide for frequent and regular inspections of the job sites, materials, and equipment to be made by competent persons designated by the employers.*

This subpart requires the employer to develop and implement a safety program. The goal of the safety program should be to safely erect, use, and dismantle scaffold in compliance with general industry regulations. The employer is also required to designate a competent person to make regular inspections of the job site and the materials and equipment used at the site.

1.1.2 Safety Training and Education

Employer responsibilities for safety training include the following:

- *The employer shall have each employee who performs work while on a scaffold trained by a person qualified in the subject matter to recognize the hazards associated with the type of scaffold being used and to understand the procedures to control or minimize those hazards. (OSHA 29 CFR 1926.454-a)*
- *The employer shall instruct each employee who is involved in erecting, disassembling, moving, operating, repairing, maintaining, or inspecting a scaffold trained by a competent person to recognize any hazards associated with the work in question. (OSHA 29 CFR 1926.54-b)*

Employers are responsible for training all of their employees in safe work practices and the regulations that apply to the work they are performing.

The regulations use the terms *should* and *shall*. When dealing with regulations, *should* means that it is recommended, while *shall* means that it is required. The employer is required to provide the necessary training, and is free to designate as many competent people, as defined in the next section, as necessary.

1.1.3 Define and Explain Hazard Analysis

Two key definitions presented in the supplemental document are repeated below:

- *A competent person is someone who is capable of identifying existing and predictable hazards in the surroundings or working conditions which are unsanitary, hazardous, or dangerous to employees, and who has authorization to take prompt corrective measures to eliminate them.*

The first part of this definition requires that the competent person has the ability to recognize hazards associated with the activity. OSHA has identified the "Fatal Four" job-site hazards: electrocution hazards, fall hazards, struck-by hazards, and caught-in or-between hazards. To recognize these hazards, the competent person must be thoroughly trained in the safety procedures, OSHA regulations, and safety practices associated with the activity. This person must know the safe way to do a job in order to recognize an unsafe practice. The competent person must be given authority by the employer to eliminate the hazard. This may require stopping work until the hazard can be corrected.

- *A qualified person is someone who, by possession of a recognized degree, certificate, or professional standing, or who by extensive knowledge, training, and experience, has successfully demonstrated the ability to solve or resolve problems related to the subject matter, the work, or the project.*

When specifying that all scaffolds must be designed by a qualified person, the definition above applies. The qualified person must design and plan a scaffold project, and then pass the project to a competent person to supervise the erection. It is possible for one person to fulfill both roles. Notice that the qualified person is not required to have a degree, but must have adequate expertise for the particular project.

Regulations require that a registered engineer design any scaffold over 125 feet tall and determine the proper number of working levels. The number of planked and working levels is limited, based on the type of scaffold erected (see *Table 1*).

1.2.0 Common Scaffold Industry Safety Practices

Scaffold users are not required to be trained in scaffold construction. However, training should cover the hazards that might be encountered, such as electrical hazards, fall hazards, falling-object hazards, and caught-in or -between hazards. The training of scaffold builders and any employees who might modify, repair, maintain, or inspect scaffolds is required, not just recommended. Employees must be trained when conditions change, and must be retrained as necessary. Industry practices state the following:

- *The employer shall have each employee who performs work while on a scaffold trained by a person qualified in the subject matter to recognize the hazards associated with the type of scaffold being used and to understand the procedures to control or minimize those hazards. The training shall include the following areas as applicable:*

Table 1 Maximum Number of Planked Levels for Each Scaffold Rating

Number Of Working Levels	Maximum Number Of Additional Planked Levels			Maximum Height Of Scaffold (In Feet)
	Light Duty	Medium Duty	Heavy Duty	
1	16	11	6	125
2	11	1	0	125
3	6	0	0	125
4	1	0	0	125

- *The nature of any electrical hazards, fall hazards, and falling-object hazards in the work area;*
- *The correct procedures for dealing with electrical hazards for erecting, maintaining, and disassembling the fall protection system and falling-object protection system being used;*
- *The proper use of the scaffold, and the proper handling of materials on the scaffold;*
- *The maximum intended load and the load-carrying capacity of the scaffold used; and*
- *Any other pertinent requirements of this subpart.*

- *The employer shall have each employee who is involved in erecting, disassembling, moving, operating, repairing, maintaining, or inspecting a scaffold trained by a competent person to recognize any hazards associated with the work in question. The training shall include the following topics as applicable:*
  - *The nature of scaffold hazards;*
  - *The correct procedures for erecting, disassembling, moving, operating, repairing, inspecting, and maintaining the type of scaffolds in question;*

- *The design-criteria maximum intended load-carrying capacity and use of the scaffold;*
- *Any other pertinent requirements of this subpart.*

- *When the employer has reason to believe that an employee lacks the skill or understanding needed for safe work involving the erection, use, or dismantling of scaffolds, the employer shall retrain each employee so that the requisite proficiency is regained. Retraining is required in at least the following situations:*
  - *Where changes at the work site present a hazard about which the employee has not been previously trained; or*
  - *Where changes in the types of scaffolds, fall protection, falling-object protection, or other equipment present a hazard about which an employee has not been previously trained; or*
  - *Where inadequacies in an affected employee's work involving scaffolds indicate that the employee has not retained the requisite proficiency.*

Additional Resources

"Scaffolding." OSHA. www.osha.gov/SLTC/scaffolding/index.html

Fall Protection and Scaffolding Safety: An Illustrated Guide. 2000. Grace Drennan Ganget. Government Institutes.

1.0.0 Section Review

1. OSHA regulations require an employer to develop and implement a safety program.

   a. True
   b. False

2. Scaffold safety training should include electrical hazards, falling-object hazards, and _____.

   a. vehicle hazards
   b. respiratory hazards
   c. fall hazards
   d. water hazards

SECTION TWO

2.0.0 SCAFFOLD PLANNING, ERECTING, AND USE

Objective

Explain the basic guidelines for planning, erecting, and using scaffolds.

a. Explain the basic guidelines for planning a scaffold project.
b. Explain the basic guidelines for erecting a scaffold.
c. Explain the basic guidelines for using a scaffold.

Performance Tasks

Plan and sketch a scaffold for a specific application specified by the instructor.

Examine and identify the soil type at the work site, and determine the proper mudsills to use.

Trade Terms

Galvanic action: Corrosion caused at the point of contact of dissimilar metals such as steel and aluminum.

Hazard analysis: Condensed safety/planning meeting that is most often conducted prior to each day's activity; also called a prejob meeting or toolbox talk.

Laminated: Layers of wood glued together with the grain parallel.

Pounds per square foot (psf): The weight present for each square foot of surface area.

Putlog: A horizontal scaffold member on which the scaffold platform rests; also known as a truss.

NOTE	It is not uncommon for trade terms to vary from one region of the country to another. For example, what might be called a putlog in the Midwest is known as a truss elsewhere.

This section covers basic guidelines for the planning, erecting, and use of scaffolds. All scaffold builders should consult their supervisor for any job-site rules or guidelines that may be more specific and more stringent.

Many of these guidelines are discussed daily during lunch-box talks, and documented on a daily job-site hazard analysis form. These discussions commonly cover such subjects as tasks to be performed, fall protection, emergency actions, rescue plans, possible hazard exposures, and other related topics.

2.1.0 Explain the Need and Reason for Careful Planning

All scaffold builders are responsible for complying with all applicable codes, standards, and regulations as well as commonsense practices designed for safely erecting, using, or dismantling scaffolds. General industry regulations require that scaffold builders and users be trained in the safe performance of their duties on or around the scaffolds and that this training be performed under the supervision of a competent person. To help scaffold builders and users better understand the codes, standards, and regulations that apply to scaffolds, the following paragraphs cover some of the planning involved in scaffold building.

2.1.1 Planning by Qualified Person

Each scaffold job must be evaluated, designed, and planned by a qualified person. Scaffold design must comply with all site, local, state, and federal safety requirements. The qualified person should use common sense, good judgment, and sound reasoning to evaluate the following:

- The proximity of the completed scaffold to power lines, process piping, and any overhead obstructions. *Table 2* shows the minimum distance allowed from electrical power lines.

WARNING!	To avoid the possibility of electrical shock while erecting or working from a scaffold, maintain a safe distance from electrical lines. Electrical shock could cause serious injury or death.

Table 2 Minimum Distance Allowed from Electrical Power Lines

Insulated Lines Voltage	Minimum Distance	Alternatives
Less than 300 Volts	3 Feet	
300 Volts to 50 KV	10 Feet	
More than 50 KV	10 Feet plus .4 inch for each 1 KV over 50 KV	Two times thelength of the line insulator, but never less than 10 feet
Uninsulated Lines Voltage	**Minimum Distance**	**Alternatives**
Less than 50 KV	10 Feet	
More than 50 KV	10 Feet plus .4 inch for each 1 KV over 50 KV	Two times the length of the line insulator, but never less than 10 feet

- Adequate access to the platforms to be used. (Is more than one access necessary?)
- Atmospheric and weather conditions. (Is wind protection required? Is protection from other work in the area required?)
- Ground or foundation conditions. (Are there excavations or buried utilities in the area? Is the ground suitable for the expected loads?)
- Interference with other jobs or workers during or after erection.
- Proper bracing so that the scaffold is rigid in all directions.
- Proper guardrails and toeboards.
- Adequate decking and overhead protection for the working levels.
- Protection for anyone walking or working near or underneath the scaffold.

2.1.2 Classifications of Scaffolds

Depending on their design, scaffolds are classified for light-duty, medium-duty, or heavy-duty use, meaning that they can safely support loads up to 25, 50, or 75 pounds per square foot (psf), respectively. When deciding if the scaffold will be safe for the intended use, the weight of all personnel, equipment, and materials must be considered.

The most common scaffold is 5 feet wide by 7 feet long with a light-duty scaffold frame. Using the following formula, this scaffold can safely support up to 875 pounds.

Light-duty scaffold = 25 psf
5 feet × 7 feet = 35 square feet
35 square feet × 25 psf = 875 pounds

By the same formula, a medium-duty scaffold (which can support up to 50 psf) of the same size could support up to 1,750 pounds. The same 5-foot × 7-foot scaffold using a heavy-duty frame (75 psf) could support up to 2,625 pounds.

Most scaffold applications can be accommodated with light-duty scaffolds; remember, the average worker with a tool belt and tools weighs 250 pounds. Some crafts, such as masonry, at times require the placement of heavy pallets of brick, stone, or other materials on the platform, so plan on loads of 50 psf to accomodate this material.

Properly assembled scaffold legs can carry between 2,000 and 3,000 pounds per leg. On low scaffolds, the legs are strong enough to support light-, medium-, or heavy-duty loads. Higher scaffolds require special consideration. The total weight of the scaffold as well as the load on the platforms must be considered for total leg loading.

2.1.3 Mixing of Scaffold Materials

Do not mix scaffold materials from different manufacturers, unless determined appropriate under the supervision of a competent person. Different manufacturers may use different sizes of support members, different leg spacing, or different bracing attachments. Improperly mixing scaffold materials could result in an unsafe scaffold.

2.1.4 Inspection of Scaffold Materials and Equipment

Carefully inspect all scaffold material before use to ensure that it is in good condition. Scaffold materials should be inspected when removed from storage, just before use, during disassembly, and before storage. Any damaged equipment must be removed from use immediately. The following is a partial checklist for inspecting scaffold components:

- Rust or corrosion inside or outside tubing
- Bent components
- Crushed, kinked, or flattened tubing
- Cracks around welds, joints, or around the circumference of the tubing

- Free movement of all latches and locks
- Deformation of attachment points, joints, and brace holes
- Manufactured planking: missing locks, hooks, rivets; damaged surfaces or bent side rails
- Paint blisters that might indicate rust or other hidden damage
- Casters, for free movement and damaged rollers, brakes, axles, or stems

2.1.5 Scaffold Plank Types and Markings

Scaffold planks (*Figure 1*) may be made of any of the following materials, depending on the application:

- Solid sawn wood, approved National Lumber Graders Authority (NLGA)-graded and marked scaffold grade
- Manmade laminated wood planks specifically manufactured as scaffold planks
- Fiberglass planks
- Aluminum or steel hook-on planks manufactured as scaffold planks

Figure 2 shows examples of scaffold plank grade stamps.

2.1.6 Possible Changes to Scaffold Installations

Consult with the qualified person to accommodate any changes that may arise during the time the scaffold is in place.

2.1.7 Scaffold Erection and Dismantling Safety

Examine the availability of fall protection for use during erection and dismantling. Due to the requirements of fall arrest system tie-off points, scaffold materials are not suitable for this use unless approved by the manufacturer.

2.2.0 Safety Concerns during Scaffold Erection

The primary consideration during the erection or dismantling of a scaffold system is safety. A safe job site is a result of proper training and use of proper procedures. During any type of dismantling or modification, the safety status of scaffolds must be properly identified using tags until deemed completely safe for use. Common scaffold identification tags are colored red (danger, keep off), yellow (does not meet OSHA standards), and green (safe for use).

WARNING!

Serious injury or death can result from improper use, erection, or dismantling of scaffolds.

2.2.1 Erection Requirements

Scaffolds must be erected by a properly trained crew under the supervision of a competent person. Appropriate personal protective equipment, such as hard hats, safety glasses, gloves, and appropriate footwear should be worn. Personal fall arrest equipment must be used according to the guidelines of the latest regulations.

2.2.2 Use of Mudsills

Mudsills are used to distribute the scaffold loads to the ground or supporting structure. The type of soil will determine the mudsills used. A competent person should provide the results of soil testing for the loocation.

- Type A soil refers to cohesive soil with a compressive strength of at least 1.5 tons per square foot. Cohesive soils are soils that do not crumble, are plastic when moist, and are hard to break up when dry. Examples of cohesive soils are clay, silty clay, sandy clay, and clay loam. Cemented soils, such as caliche and hardpan, are considered Type A. No soil can be considered Type A if any of the following conditions exist:
  - The soil is fissured. Fissured means a soil that has a tendency to break along definite planes of fracture with little resistance, or a material that has cracks (such as tension cracks) in an exposed surface.
  - The soil is subject to vibration from heavy traffic, pile driving, or other similar effects.
  - The soil has been previously disturbed.
  - The material is subject to any other factors that would cause it to be less stable.

  For a 3,000-pound leg load on Type A soils, a scaffold-grade 2 × 10 at least 18 inches long is recommended as a mudsill.
- Type B soil refers to cohesive soils with a compressive strength of greater than 0.5 tons per square foot but less than 1.5 tons per square foot. Type B soils include granular soils, including angular gravel, which are similar to crushed rock, silt, sandy loam, unstable rock, and any unstable or fissured Type A soils. Type B soils also include previously disturbed soils except those that would fall into the Type C classification. For a 3,000-pound leg load on Type B soils, use an 18-inch square mudsill.

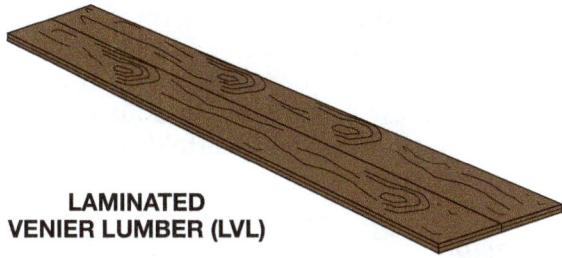

LAMINATED VENIER LUMBER (LVL)

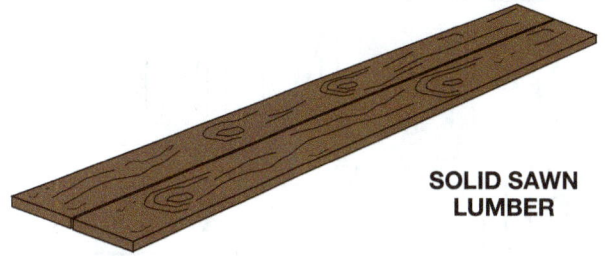

SOLID SAWN LUMBER

FABRICATED SCAFFOLD DECK

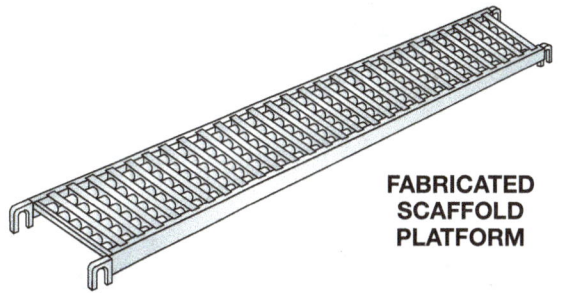

FABRICATED SCAFFOLD PLATFORM

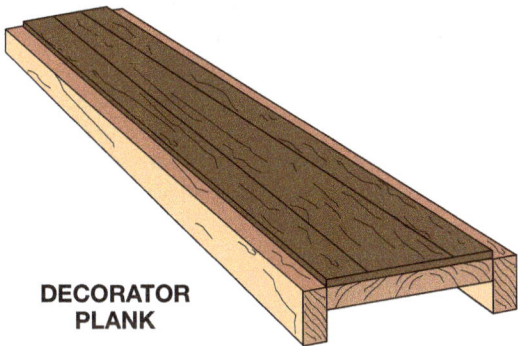

DECORATOR PLANK

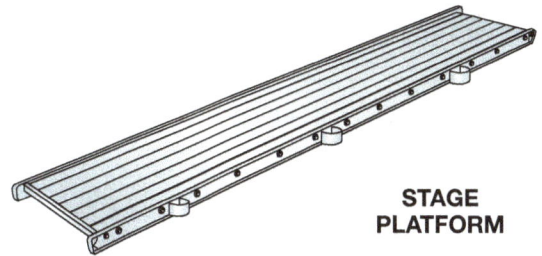

STAGE PLATFORM

WOOD SCAFFOLD PLATFORM

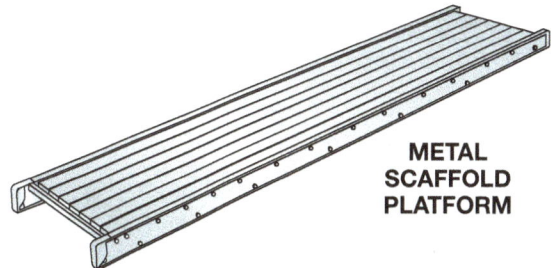

METAL SCAFFOLD PLATFORM

31102-14_F01.EPS

Figure 1 Scaffold planks.

SPIB₋ ONS INO 65
KO19 S-DRY ⑦

SCAFFOLD PLANK

MILL 10
SEL STR
WC
LB
SCAF PLK
D. FIR S. DRY

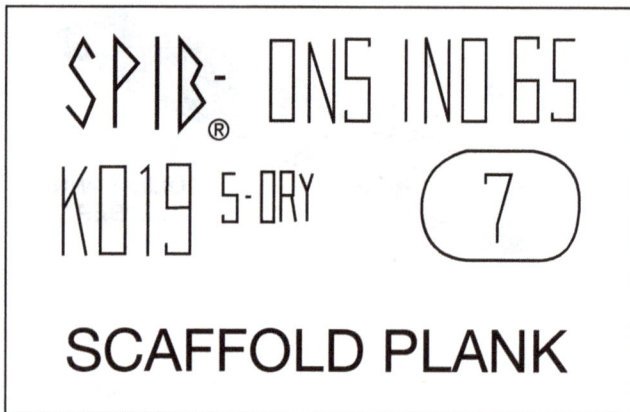

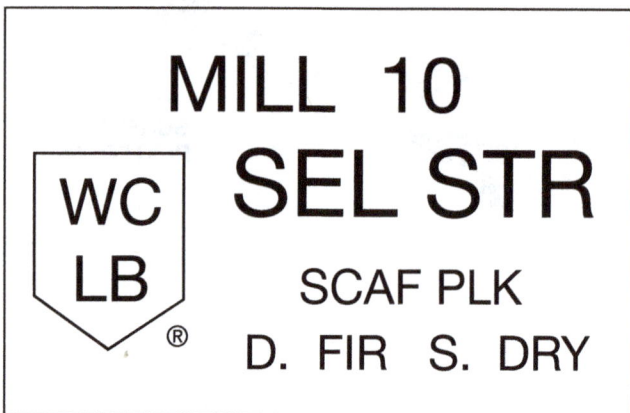

31102-14_F02.EPS

Figure 2 Scaffold plank grade stamps.

- Type C soil is the most unstable soil type. Type C soil refers to cohesive soil with a compressive strength of 0.5 tons per square foot or less. Gravel, loamy soil, sand, any submerged soil, soil from which water is freely seeping, and unstable submerged rock are considered Type C soils. As best practice, consider all soil as class C.

2.2.3 Base Plates and Screw Jacks

Use base plates with screw jacks to provide a means to make leveling adjustments.

- Start erecting the scaffold at the highest point to keep the screw jack extension to a minimum.
- Set the screw jack handles in firm contact with the frame leg.

> **NOTE**
> The maximum extension allowed for a screw jack varies by manufacturer.

Figure 3 shows the maximum screw extension (12 inches) for one manufacturer's scaffold.

Plumb, level, and square the scaffold at the base. Use the screw jacks to continuously plumb and level the scaffold as new scaffold sections are added.

2.2.4 Bracing

Brace the scaffold to prevent it from tipping. The following are bracing recommendations:

- Limit the height of the scaffolds to four times the base width in the narrow direction (refer to state and local regulations).
- Increase the base width with outriggers.
- Tie off to the adjacent structure.

  Vertical ties – The first tie-off must be no higher than four times the minimum base width, and the upper tie-offs no more than 20 feet apart vertically for scaffolds 3 feet wide or narrower, and 26 feet vertically on scaffolds over 3 feet wide. For example: if the scaffold is 5 feet wide and 50 feet high, the first tie should be no higher than 20 feet and the second tie between 40 and 46 feet. Ties should be placed close to the frame header or bearer.

  Horizontal ties – These ties should be placed at both ends and every 30 feet horizontally. For example: if the scaffold is 150 feet long, ties should be at 30 feet, 60 feet, 90 feet, 120 feet, and at each end.

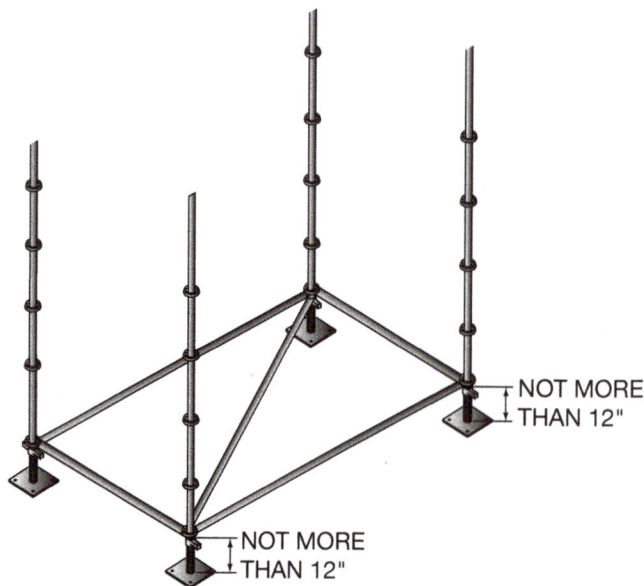

NOT MORE THAN 12"

NOT MORE THAN 12"

31102-14_F03.EPS

Figure 3 Maximum screw extension.

- All ties should be installed during the erection process and left in place until the scaffold is dismantled in that location.
- Prior to starting the dismantling process, check to see if any tie-offs, braces, etc., have been compromised during construction. Best practice is to have a predismantle checklist. When dismantling scaffolds, start at the top, removing all components except braces. Never remove all braces from all levels.
- Anchors, ties, and guys should be attached to structurally sound components. The attachment points must be able to carry the loads. (See the module, *Supported Scaffolds*, for details and proper tie procedures.) Guying systems must be designed by a professional engineer.
- The specifications for securing listed above are a minimum. Wind, side brackets, cantilevered platforms, hoist arms, or work performed from the platforms may require additional ties and uplift protection.
- Completely circular scaffolds inside or outside a structure may be braced by installing buttress ties.

2.2.5 Guying

Guying may be used for stability if there is no structure available for bracing. Guy cables may be used for lateral stability.

- Guying cannot be used as a substitute for horizontal or vertical bracing.
- Use proper-size wire cable for guying.
- Use a minimum of three cable clips for each cable connection.

> **CAUTION**
> For optimal, safe location, always communicate with a qualified person before and during the installation of guy wires.

- Remove all slack from the cables as they are installed, but do not overtighten them.

> **WARNING!**
> Overtightening the cables can preload the legs to the point of failure. Cable failure can cause a scaffold to collapse, causing serious injury or death.

2.2.6 Access

Access must be provided to each platform that is more than 2 feet above the next lower level. Access can be provided by a scaffold ladder, portable ladder, stair tower, ramp, walkway, prefabricated rungs in the frame, or direct access from another structure.

- Cross braces should not be used as a ladder.
- Install scaffold ladders according to manufacturers' specifications.
- Access ladders must:
  - Be positioned so that they do not cause the scaffold to tip
  - Have the lowest rung of attached or built-in ladders within 24 inches of the ground
  - Have rest levels provided at least every 35 feet
  - Have rungs at least 11½ inches wide
  - Have rung spacing 16¾ inches or less
  - Have evenly spaced rungs
  - Attachable ladders should extend at least 3 feet above the top platform
- Access ladders must be installed when a scaffold is one lift high. *Figure 4* shows a ladder-and-landing-platform arrangement.
- Stairway-type ladders must meet the criteria specified.
- Stair towers must meet the criteria specified.
- Ramps must meet the criteria specified.
- Prefabricated rungs in the scaffold frame must:
  - Be designed as ladder rungs
  - Have a rung length of least 8 inches
  - Be uniformly spaced
  - Not be used as a work platform
  - Have a rest platform at least every 35 feet (Army Corps of Engineers EM-385 regulations call for platforms at intervals of no greater than 20 feet).
  - Have a rung spacing of 16¾ inches or less

2.2.7 Platform Planking

Work platforms should be fully planked. The following are guidelines for platform planking:

- All planking must be scaffold grade or manufactured specifically for planking.
- All platforms must be at least 18 inches wide unless it is not physically possible.
- Planks shall extend at least 6 inches beyond their support unless cleated or secured. Planks may be secured by tying the toeboard to the scaffold frame. This will also help to prevent planking from becoming dislodged by wind.

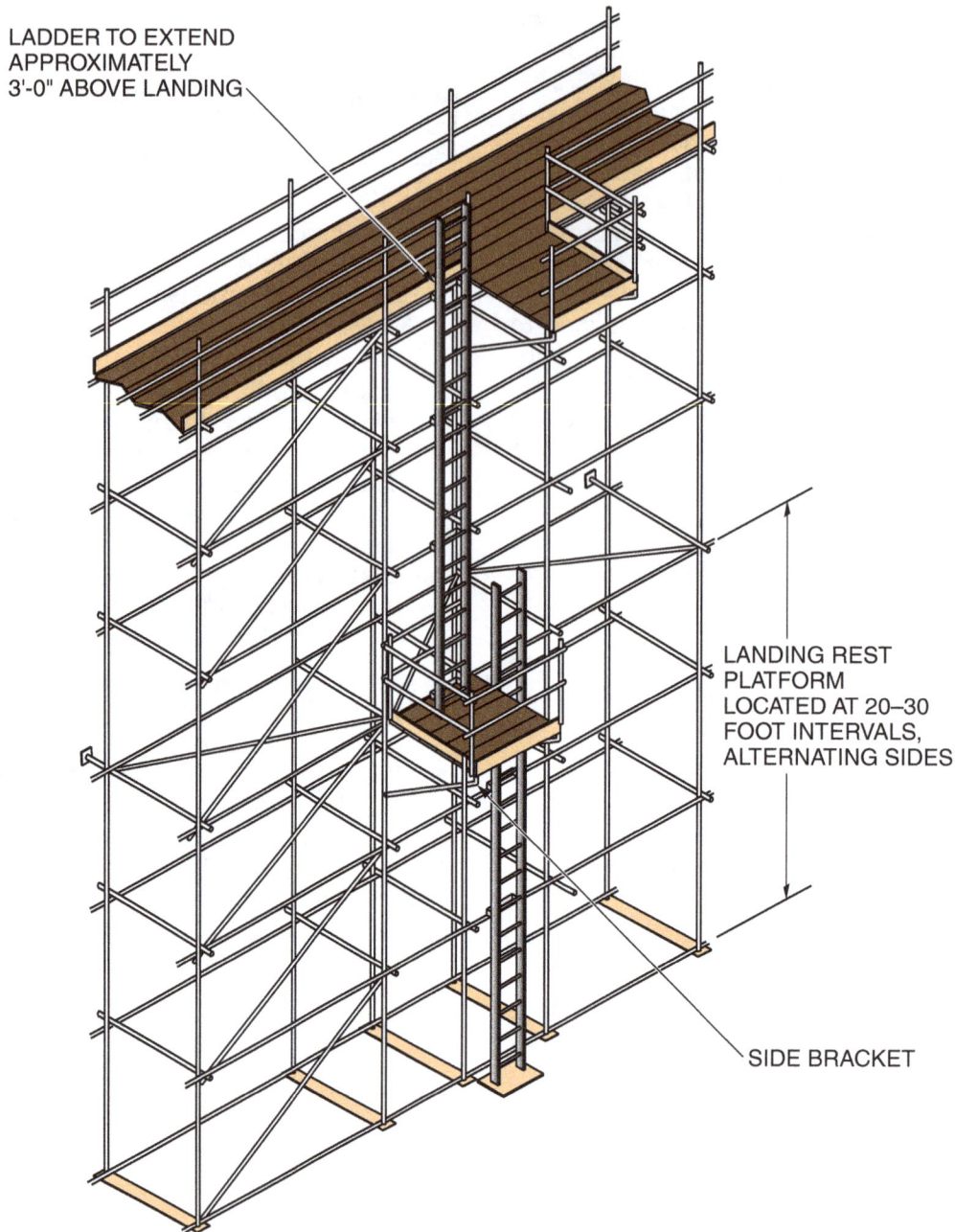

LADDER TO EXTEND
APPROXIMATELY
3'-0" ABOVE LANDING

LANDING REST
PLATFORM
LOCATED AT 20–30
FOOT INTERVALS,
ALTERNATING SIDES

SIDE BRACKET

31102-14_F04.EPS

Figure 4 Ladder and landing-platform arrangement.

- Planks 10 feet long or less must not extend more than 12 inches beyond the last support unless the cantilevered area is guardrailed to prevent access.
- Planks 10 feet long or longer must not extend more than 18 inches beyond the last support unless the cantilevered area is guardrailed to prevent access.
- Planking runs must overlap at least 12 inches.
- Any damaged or weakened planks must be removed and replaced immediately.
- Any spills must be removed as soon as possible.

> **WARNING!**
> To eliminate slipping hazards, remove Ice, snow, water, oil, or dirt before using a scaffold. Slipping hazards can cause serious injury or death.

- Scaffold platforms or planking must not be painted.
- When planking changes direction, such as at a corner, the planks that would lie across the bearers at less than a right angle should be laid first. Then the right-angle planking can be laid on top.

- Planking spans should be in accordance with the manufacturer's specifications when using nominal-thickness planking.
- Do not plank between towers unless designed by a qualified person.

2.2.8 Protection from Falling Objects

Objects may fall onto personnel working on or below scaffolds. Protective devices must be installed to protect against all falling objects.

- Barricade the area below the scaffold to prevent access under the work zone.
- Provide overhead protection if an overhead hazard exists.
- Do not toss items up or down on scaffolds, and do not jump from one level to another.
- Falling-object protection should be provided to meet the following conditions:
  - Toeboards should be installed on all sides of a platform.
  - Toeboards must be at least 3½ inches in height.
  - Toeboards must be secured to withstand a 50-pound force applied downward or horizontally.
  - If materials are piled higher than the toeboards, netting must be provided from the platform to the guardrail. Do not stack materials above the guardrail.
  - Debris nets, canopies, or catch platforms may be acceptable in some areas.

2.2.9 Guardrails Systems

On each platform more than 10 feet above the ground or the next lower level, guardrails (consisting of toprails and midrails) are required on all open sides that are more than 14 inches from a solid-faced structure. Guardrail requirements can vary from state to state.

- Guardrails must be able to withstand a force of 200 pounds applied downward or horizontally.
- Minimum height for the toprail is between 38 inches and 45 inches. (Always check regional and state requirements.)
- A midrail must be placed halfway between the platform and the toprail.
- Wire rope may be used as guardrail but must not deflect below the minimum guardrail height.
- Guardrails must be smooth to prevent cuts. Steel or plastic banding may not be used as guardrails.

- The cross bracing may be considered as part of the toprail or midrail, depending on the height of the cross bracing.

2.2.10 Sway Prevention

After the scaffold is erected, measures must be taken to ensure that the scaffold will not sway or become displaced.

- Legs, poles, or uprights must be plumbed, secured, and rigidly braced to prevent swaying or displacement.
- Putlogs must satisfy the following:
  - Putlogs are for support of personnel only. Do not place materials on platforms supported by putlogs, unless designed by a qualified person.
  - They must overhang the support points by at least 6 inches.
  - Lengths greater than 10 feet long require knee bracing.
  - Depending on the expected loading, putlogs may require close spacing.
  - Follow the recommendations of the manufacturer.

2.2.11 Dissimilar Metal Components

Do not mix components made of dissimilar metals (steel clamps on aluminum bracing) unless a competent person, capable of identifying existing and predictable hazards, has determined that the galvanic action will not be a problem, and the scaffold will be erected only for a short time.

2.2.12 Final Inspection and Tagging

If the erected scaffold has been modified, or its structural integrity has been compromised, it must be inspected by a competent person before use, and re-inspected by a competent person during each shift. Placing a tag on the scaffold to indicate the condition of the scaffold is not required by OSHA, but it is recommended. However, tagging scaffolds is mandatory in EM-385 regulations.

- *Green tag* — indicates scaffold is 100-percent OSHA compliant.
- *Yellow tag* — indicates hazards exist and have been identified.
- *Red tag* — indicates the scaffold is not available for use.

All hazard recognition and tagging should be well documented. Documentation of these types of safety issues are commonly recorded on job hazard analysis (JHA) forms, task safety analysis (TSA) forms, or Army Corps of Engineers (EM) forms.

2.3.0 Scaffold Use

A brief list of precautions to be considered while working on scaffolds is presented here. It is not intended to be a comprehensive guide to proper scaffold use.

Scaffold loading must be constantly monitored. In addition to the final inspection by a competent person, scaffold users should always perform a visual inspection prior to use.

- Do not use a scaffold that has not been inspected during your work shift. Check for a tag indicating that the scaffold is safe and when the inspection was done.
- Do not use a scaffold that does not have proper access.

> **WARNING!**
> Do not climb guardrails or bracing. Use only approved ladders, stairs, or access from an existing structure. Guardrails and braces could fail, causing personal injury or death.

- Do not work from a platform that is not fully planked, unless fall protection is used.
- Do not use the scaffold if the planking is not scaffold grade, was not made specifically for decking, or is not in good condition.
- Do not use a scaffold if the planking is bowing more than ¹⁄₆₀ of the span (2 inches on a 10-foot span).
- Do not use a scaffold that is not rigid, plumb, and square.
- Do not use a scaffold that is taller than four times the minimum base dimension unless it is properly braced.
- Do not work on a scaffold if you feel sick, dizzy, or weak.

- Do not climb scaffold ladders carrying tools or materials. Use both hands on the side-rails.
- Do not jump on the planking.
- Do not use a ladder to extend the height of the scaffold.
- Do not use a scaffold if you see damaged parts.
- Do not allow material, debris, or tools to create a hazard on a platform.

> **WARNING!**
> Due to fire hazard, do not use gas-powered tools, equipment, or hoists when working from a suspended scaffold. If a fire should occur there is no escape.

- Do not work on a platform that is covered with snow or ice, unless you are involved in removing it.

> **WARNING!**
> Snow and ice contribute to overloading and cause slipping hazards. Always remove snow and ice from scaffolds to prevent personal injury.

- Do not work on a scaffold during high winds or storms.
- Do not use a scaffold as a platform for a hoist unless it has been designed and built for it.
- Do not load a platform beyond the designed capacity.
- Do not concentrate loads in a single spot on a frame or platform.
- Do not violate clearances from electrical or process lines.
- Do not use the scaffold unless proper falling-object protection is in place.
- Do not ride on rolling scaffold towers.
- Do not use flame-producing equipment unless proper safeguards are protecting the flammable portions of the scaffolds.
- Do not attach fall arrest equipment to any guardrail system unless it is recommended by the scaffold manufacturer.

Additional Resources

"Scaffolding." OSHA. **www.osha.gov/SLTC/scaffolding/index.html**

Fall Protection and Scaffolding Safety: An Illustrated Guide. 2000. Grace Drennan Ganget. Government Institutes.

2.0.0 Section Review

1. Heavy-duty scaffolds can support a per-square-foot load of _____.

   a. 100 pounds
   b. 75 pounds
   c. 50 pounds
   d. 25 pounds

2. The maximum extension allowed for a screw jack is _____.

   a. 6 inches
   b. 8 inches
   c. 12 inches
   d. 18 inches

3. An unacceptable means of access to scaffolds is _____.

   a. stairs
   b. bracing
   c. ladder
   d. from an existing structure

3.0.0 SAFE SCAFFOLD ERECTION

Objective

Identify the equipment and tasks required for safe scaffold erection.

a. List the personal protective equipment required for safe scaffold erection.

Trade Term

Decibels (dBs): The intensity of sound.

Depending on the specific job and conditions, appropriate personal protective equipment is often neded. Scaffold builders should check the local regulations to determine the specific PPE required for each situation.

3.1.0 Personal Protective Equipment

The following types of common personal protective equipment are discussed in the sections that follow:

- Foot protection
- Hand protection
- Eye and face protection
- Head protection
- Hearing protection
- Respiratory protection

3.1.1 Foot Protection

Protective footwear should be worn when working where falling, rolling, or sharp objects pose a danger of foot injuries, and where feet are exposed to electrical hazards. Leather footwear with reinforced soles or innersoles of flexible metal is recommended for any construction site. Steel-toed safety shoes should be worn for operations involving the handling of heavy materials. Protective footwear must comply with American National Standards Institute (ANSI) standard Z41-1991. Sneakers, tennis shoes, or similar types of footwear should never be worn at the job site.

3.1.2 Hand Protection

A scaffold builder's hands may be exposed to a variety of hazards (temperature extremes, abrasive materials, and so on) that may cause inflam-

mation of the skin, known as dermatitis. Gloves are the primary type of hand protection. They may be made of leather, rubber, cotton, or a variety of plastics or synthetics. There is no all-purpose glove. Gloves must be selected on the basis of the hazards involved in doing the work. *Figure 5* shows a pair of work gloves.

> **WARNING!**
> Do not wear loose-fitting gloves around machinery. The moving parts of the machine can snag the gloves and pull your hands in, causing injury or death.

3.1.3 Eye and Face Protection

It is extremely important to take care of your eyes. Lost vision cannot be restored. Proper eyewear and face protection must be worn for the job at hand. There are tinted safety glasses for working outside, and prescription safety glasses of all types. Adding eye shields to prescription eyewear does not make them safety glasses. Make sure eyewear fits properly and wear head straps if necessary. When operating rotating or high-pressure equipment, a face shield must be worn. Eye protection is required, even when using a face shield (see ANSI Z87).

31102-14_F05.EPS

Figure 5 Gloves.

Figure 6 shows examples of proper eyewear and face protection.

3.1.4 Head Protection

An approved hard hat must be worn on the job site in all posted areas and other areas where there is any danger of head injury due to falling or flying objects. Nonconductive hard hats must be worn when working in the vicinity of equipment or electrical lines with potential high-voltage electrical shock hazards. Hard hats used for protection against falling and/or flying objects must meet the American National Standards Institute (ANSI) Z89.1-1969 requirements. Those used to protect against electrical shock must meet ANSI standard Z89.2-1971. The ANSI standard for a hard hat is stamped or labeled on the inside of the hat by the manufacturer. Only hard hats stamped or labeled with the correct ANSI classification should be used. *Figure 7* shows a hard hat.

The correct way to wear a hard hat is to adjust its inner suspension so that the hat is slightly raised off the head. A suspension that is too flexible will permit contact with the head upon impact and can result in a skull fracture or concussion. A suspension that is too rigid can transmit the shock impact and fracture the neck vertebrae. Never remove the hard hat's inner suspension for any reason or keep anything under your hard hat. Both actions will prevent the suspension from acting as a shock absorber. Other practices that should be followed are:

- Clean your hard hat and suspension at least every 30 days.
- Immediately replace a broken or punctured hard hat.
- Never drill holes into your hard hat for ventilation.
- Never leave your hard hat in the rear window of your car or truck. The sunlight may adversely affect its protective quality. Also, an emergency stop can turn your hard hat into a dangerous projectile.
- For the protection of your eyes, always wear your hard hat with the brim facing forward.

3.1.5 Hearing Protection

Construction-site activities and/or the use of some equipment can sometimes create noise at unhealthy levels. If you find yourself asking questions such as "What?" "Pardon?" or "What did you say?" because you cannot hear what another person is saying, then you are probably in an area that contains a noise level that can cause permanent hearing loss.

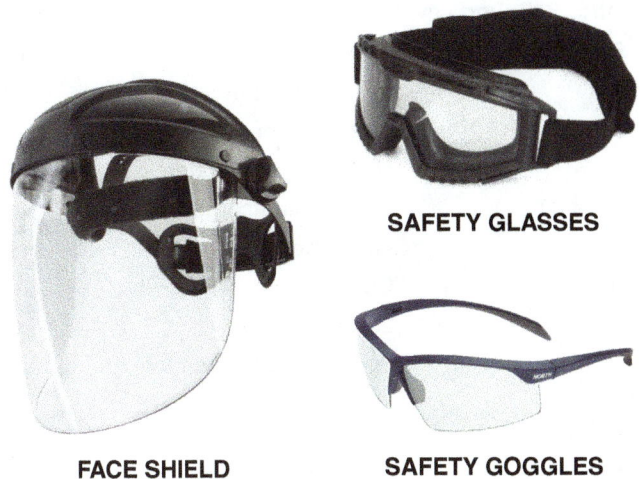

SAFETY GLASSES

FACE SHIELD

SAFETY GOGGLES

31102-14_F06.EPS

Figure 6 Examples of proper eyewear and face protection.

31102-14_F07.EPS

Figure 7 Hard hat.

Not all sounds have the same effect on hearing. Sound or noise has three different characteristics: intensity, pitch, and length of exposure. Intensity means loudness of sound and is measured in decibels (dBs). Pitch refers to the frequency of sound waves. A high-frequency (high-pitched) whistle, even though it is no louder than a low-pitched thud, is usually more harmful than the low-frequency sound. Length of exposure refers to the amount of time a person is subjected to a noise. Continual exposure to certain noises can be more harmful than occasional bursts of offensive sound. Exposure levels can range from 0 dB (the faintest sound a human ear can hear) to 140 dB (the threshold of pain) and higher. *Table 3* shows the relative sound levels for some common noises.

Table 3 Sound Levels of Some Common Noises

Sound Level (dB)	Source
0	Threshold of human hearing
30	Soft whisper
60	Normal conversation
60 to 80	Printers and copiers
90	Lawn mower
100	Chainsaw
110	Loud nightclub
115	Portable stereo headset on high setting
140	Gunshot
180	Rocket pad during launch

Table 4 OSHA-Permissible Noise Levels

Duration Per Day (Hours)	Sound Level (Db)
8	90
6	92
4	95
3	97
2	100
1½	102
1	105
½	110
¼ or less	115

OSHA regulations allow workers to be exposed to no more than an average of 85 dB of continuous noise (as opposed to impact noise, such as a shotgun blast) over an eight-hour period. When exposed to sound levels that exceed allowable levels, OSHA requires that approved earplugs or earmuffs be used to provide hearing protection. *Table 4* shows OSHA-permissible noise levels.

Figure 8 shows examples of earplugs and earmuffs.

Disposable earplugs are the most common form of hearing protection. These devices usually have an outer layer of pliable foam and a core layer of acoustical fiber that filters out harmful noise, yet allows you to hear normal conversation. To use disposable earplugs, simply roll one into a cylinder and insert the tapered end into the ear canal. The earplugs will expand, filling the ear canal and making a proper fit. Reusable earplugs are also common.

Earmuffs are usually worn in severe noise environments. They offer greater protection from all noise, including low frequencies. Most earmuffs have adjustable headbands that can be worn over the head, behind the neck, or under the chin. To use earmuffs, adjust the tension on the headband and ear cushion pads to obtain the best possible seal. Check the shell of the earmuff for cracks and the ear cushion pads for tears before each use. Any damaged, cracked, or torn part must be repaired or replaced.

3.1.6 Respiratory Protection

Scaffold builders may have to erect scaffolds in atmospheric conditions that are hazardous to their health. These environments may include petrochemical plants, sandblasting areas, or forest-product plants. Enclosed containers present additional respiratory hazards.

When working in these areas, respirators are required to provide proper protection for the employee's safety and health. A complete respiratory protection plan must be established that includes proper selection, maintenance, use, evaluation, training, and inspection of respirator equipment. Scaffold builders must be aware of their responsibilities concerning respiratory protection equipment.

There are two basic types of respirators: the air-supplying respirator and the air-purifying respirator. Air-supplying respirators provide a supply of breathable air different from the workplace air.

DISPOSABLE EARPLUGS **REUSABLE EARPLUGS**

EARMUFFS

31102-14_F08.EPS

Figure 8 Earplugs and earmuffs.

Air-purifying respirators purify the workplace air through a mask and filters designed for the hazard presented.

A walk-through survey of the workplace must be performed to identify work conditions and processes that could require the use of respiratory protective equipment. There are instruments designed to determine the concentrations of airborne contaminants. Oxygen-deficient atmospheres may require special analysis.

WARNING!

Only qualified personnel should use atmospheric testing equipment. If the facility does not have in-house qualified personnel, outside consultation will be required to do sampling and testing to determine the hazard levels. Do not enter any suspect area unless it has been tested and you have been advised that it is safe, or unless you are provided with proper respiratory protection equipment.

Respiratory equipment shall be visually inspected for defects prior to use, during use, and after use. The respirator shall be cleaned after each use and stored in a clean plastic bag or other such container, to prevent contamination.

NOTE

Specific inspection, cleaning, and maintenance of respiratory equipment will be covered in the company respiratory protection plan.

If you are required to use a respirator, you must receive specific training on the use, care, and responsibilities associated with the respirator to be used. OSHA regulations require a medical clearance and a fit test prior to wearing any respirator. When conditions change and a different type of respirator is needed, additional training is required.

Additional Resources

"Scaffolding." OSHA. **www.osha.gov/SLTC/scaffolding/index.html**

Fall Protection and Scaffolding Safety: An Illustrated Guide. 2000. Grace Drennan Ganget. Government Institutes.

3.0.0 Section Review

1. Dermatitis is a physical condition involving _____.

   a. eye irritation
   b. repiratory difficulty
   c. skin inflmation
   d. hearing loss

2. Over an eight-hour period, a worker must not be exposed to continuous noise averaging more than _____.

   a. 40 dB
   b. 55 dB
   c. 70 dB
   d. 85 dB

4.0.0 FALL PROTECTION AND LIFESAVING EQUIPMENT

Objective

Identify the fall protection and lifesaving measures employed in the scaffolding trade.

a. Identify the appropriate fall protection and lifesaving equipment, and describe their proper use.

b. Describe proper rescue procedures after a fall.

Performance Task

Demonstrate the proper method for wearing and rigging a body harness.

Trade Term

Tensile strength: The resistance of a material to a force tending to tear it apart.

Scaffold builders spend much of their time working in elevated work areas. Falls from high places can cause serious injury or death when the wrong kinds of fall protection equipment are used, or when the right equipment is used improperly. Often the scaffold is the highest structure in the area, and most scaffolds do not meet the requirements for a tie-off point, so the use of fall protection must be balanced against the additional hazards caused by its use. A qualified person must make the decision.

4.1.0 Use of Fall Protection and Lifesaving Equipment

There are three common types of fall protection equipment:

- Personal fall arrest systems
- Personal fall restraint systems
- Safety-net systems

4.1.1 Personal Fall Arrest Systems

Personal fall arrest systems catch workers after they have fallen. They are designed to activate only if a fall occurs. They are designed and rigged to prevent a worker from free-falling a distance of more than 6 feet (1.8 meters) and hitting the ground or a lower work area.

> **WARNING!**
>
> Putting on personal fall arrest systems requires instruction from a qualified person. Failure to properly wear and install various types of fall arrest equipment can result in a personal injury or death from falls.

Equipment used in personal fall arrest systems includes:

- Body harnesses
- Lanyards
- Deceleration devices
- Lifelines
- Anchoring devices and equipment connectors
- Safety nets

4.1.2 Body Harnesses

Full-body harnesses with sliding back D-rings are used in personal fall arrest systems. They are made of straps that are designed to be worn securely around the user's body. This allows the arresting force to be distributed throughout the body (shoulders, legs, torso, and buttocks) by the harness straps. This distribution decreases the chance of injury. If a fall occurs, the sliding D-ring moves to the nape of the neck, keeping the worker in an upright position and helping to distribute the arresting force. This keeps the worker in a relatively comfortable position while waiting for rescue. *Figure 9* shows a full-body harness.

Selecting the right full-body harness depends on a combination of job requirements and personal preference. Harness manufacturers normally provide selection guidelines in their product literature. Other types of full-body harnesses are equipped with front chest D-rings, side D-rings, or shoulder D-rings. Harnesses with front chest D-rings are typically used in ladder climbing and personal positioning systems. Those with side D-rings are also used in personal positioning systems.

Personal positioning systems are systems that allow workers to hold themselves in place, keeping their hands free to accomplish a task. Per OSHA regulations, a personal positioning system should not allow a worker to free-fall more than 2 feet (0.61 meter), and the anchorage to which it is attached should be able to support at least twice the impact load of a worker's fall, or 3,000 pounds (13.3 kilonewtons), whichever is greater. Harnesses equipped with shoulder D-rings are typically used with a spreader bar or rope yoke for entry into and retrieval from confined spaces.

SHOULDER STRAPS

CHEST STRAP

LEG STRAPS

FULL-BODY HARNESS (FRONT)

SLIDING BACK D-RING

FULL-BODY HARNESS (BACK)

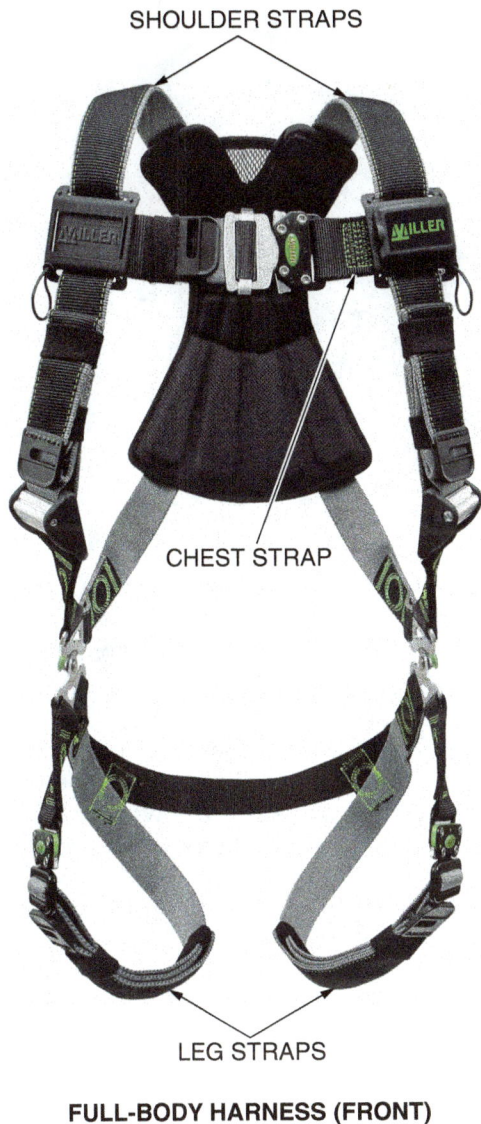

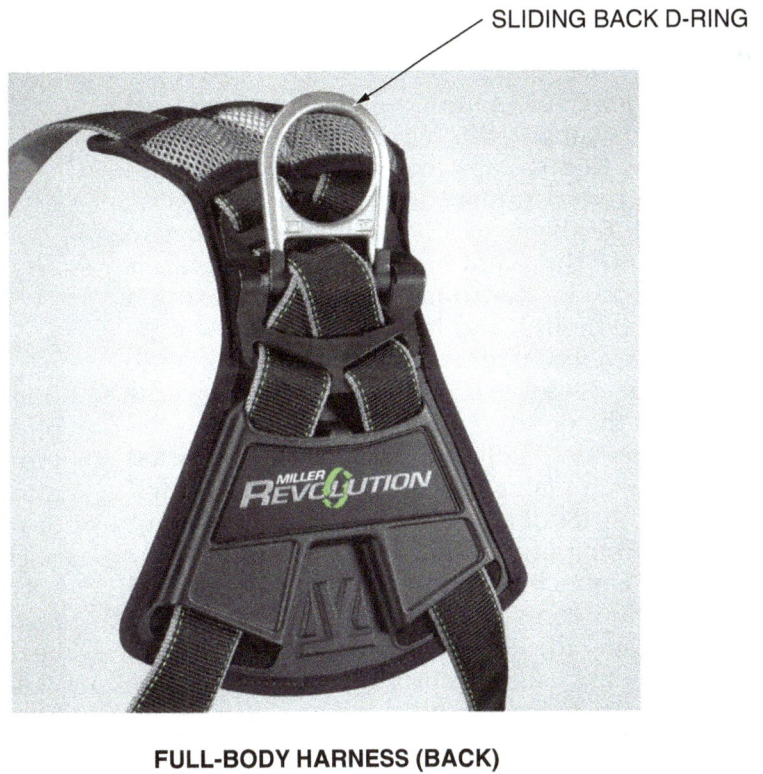

31102-14_F09.EPS

Figure 9 Full-body harness.

4.1.3 Lanyards

Lanyards are short, flexible lines with connectors on each end. They are used to connect a body harness or body belt to a lifeline, deceleration device, or anchorage point. There are many kinds of lanyards made for different uses and climbing situations. All must have a minimum breaking strength of 5,000 pounds (22.2 kilonewtons). Lanyards come in both fixed lengths and adjustable lengths, and are made out of steel, rope, or nylon webbing. Some have a shock absorber that absorbs up to 80 percent of the arresting force when a fall is being stopped. When choosing a lanyard for a particular job, always follow the manufacturer's recommendations. *Figure 10* shows a lanyard.

31102-14_F10.EPS

Figure 10 Lanyard.

When activated during the fall-arresting process, a shock-absorbing lanyard stretches as it acts to reduce the fall-arresting force. This potential increase in length must always be taken into consideration when determining the total free-fall distance from an anchor point. Failure to properly determine the free-fall distance can cause injury or death if a fall occurs.

4.1.4 Deceleration Devices

Deceleration devices limit the arresting force that a worker is subjected to when the fall is stopped suddenly. Rope grabs and self-retracting lifelines are two common deceleration devices. A rope grab connects to a lanyard and attaches to a lifeline. In the event of a fall, the rope grab is pulled down by the attached lanyard, causing it to grip the lifeline and lock in place. Some rope grabs have a mechanism that allows the worker to unlock the device and slowly descend down the lifeline to the ground or surface below.

Self-retracting lifelines provide for unrestricted movement and fall protection while climbing and descending ladders and similar equipment, or when working on multiple levels. Typically, they have a 25-foot to 100-foot galvanized steel cable that automatically takes up the slack in the attached lanyard, keeping the lanyard out of the worker's way. In the event of a fall, a centrifugal braking mechanism engages to limit the worker's free fall.

Per OSHA requirements, self-retracting lifelines and lanyards that limit the free-fall distance to 2 feet (0.61 meter) or less must be able to support a minimum tensile load of 3,000 pounds (13.3 kilonewtons). Those that do not limit the free fall to 2 feet (0.61 meter) or less must be able to hold a tensile load of at least 5,000 pounds (22.2 kilonewtons). *Figure 11* shows a rope grab and a self-retracting lifeline.

4.1.5 Lifelines

Lifelines are normally flexible steel cables that are attached to an anchorage. They provide a means for tying off personal fall protection equipment. Vertical lifelines are suspended vertically from a fixed anchorage at the upper end to which a fall arrest device, such as a rope grab, is attached. Vertical lifelines must have a minimum breaking strength of 5,000 pounds (22.2 kilonewtons). Each worker must use his or her own line. This is because if one worker falls, the movement of the

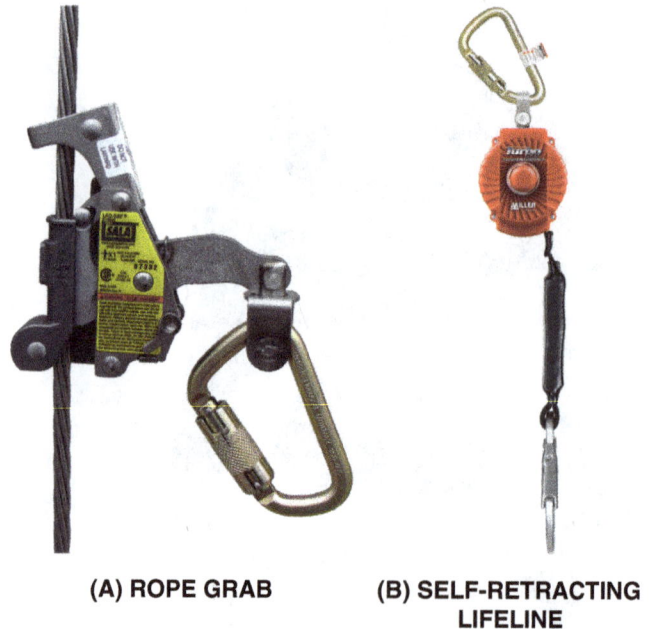

(A) ROPE GRAB **(B) SELF-RETRACTING LIFELINE**

31102-14_F11.EPS

Figure 11 Rope grab and self-retracting lifeline.

lifeline during the fall arrest may also cause the other workers to fall. Vertical lifelines must be terminated in a way that will keep the worker from moving past its end, or must extend to the ground or the next lower working level.

Horizontal lifelines are connected horizontally between two fixed anchorages to which a fall arrest device is attached. Horizontal lifelines must be designed, installed, and used under the supervision of a qualified and competent person. The line must be capable of supporting 5,000 pounds for each worker attached to it.

4.1.6 Anchoring Devices and Equipment Connectors

Anchoring devices, commonly called tie-off points, support the entire weight of the fall arrest system. The anchorage must be capable of supporting 5,000 pounds (22.2 kilonewtons) for each worker attached. Eyebolts, overhead beams, and integral parts of building structures are all types of anchorage points.

The D-rings, buckles, snap hooks, and other equipment that fasten and/or connect the parts of a personal fall arrest system are called connectors. Regulations specify how they are to be made, and require D-rings and snap hooks to have a minimum tensile strength of 5,000 pounds. All such components should be designed for use with the attached hardware. Only locking-type snap hooks are permitted for use in personal fall arrest systems.

4.1.7 Safety-Net Systems

Safety nets are used for fall protection on bridges and similar projects. They must be installed as close as possible beneath the work area, and not more than 30 feet (9.14 meters) below the work area. There should be enough clearance under a safety net to prevent a worker who falls into it from hitting the surface below. Also, there must be no obstruction between the work area and the net. Depending on the actual vertical distance between the net and the work area, the net must extend at least 8 to 13 feet (2.44 to 3.96 meters) beyond the edge of the work area. Mesh openings in the net are limited to 36 square inches (91.44 square centimeters) and 6 inches (15.24 centimeters) on the side. The border rope must have a 5,000-pound minimum breaking strength, and connections between net panels must be as strong as the nets themselves. Safety nets must be inspected at least once a week and after any event that might have damaged or weakened them. Worn or damaged nets must be removed from service.

4.2.0 Rescue Procedures after a Fall

Every elevated job site shall have a rescue and retrieval plan in case it is necessary to rescue a fallen worker. Planning is especially important in remote areas that are not readily accessible to rescue crews. Before there is a risk of a fall, make sure that you know what your employer's rescue plan requires you to do. Find out what rescue equipment is available and where it is located. Learn how to use equipment for self-rescue and the rescue of others.

If a fall occurs, any employee hanging from the fall arrest system must be rescued safely and quickly. Your employer should have previously determined the method of rescue for fall victims, which may include equipment that lets the victim rescue themselves, a system of rescue by co-workers, or a way to alert a trained rescue squad. If fall rescue depends on calling for outside help, such as the fire department or rescue squad, all phone numbers must be posted in plain view at the work site. In the event a co-worker falls, follow your employer's rescue plan. Call any special rescue service needed. Communicate with the victim and monitor the victim constantly during the rescue.

Never allow a worker that has been suspended to lie down, as this can cause a sudden rush of de-oxygenated blood to the heart, triggering a cardiac arrest. This condition is known as suspension trauma. Proper procedure requires placing the person in a sitting position, with the person's knees pulled up tight to the chest.

Additional Resources

"Scaffolding." OSHA. www.osha.gov/SLTC/scaffolding/index.html

Fall Protection and Scaffolding Safety: An Illustrated Guide. 2000. Grace Drennan Ganget. Government Institutes.

4.0.0 Section Review

1. D-rings and snap hooks must have a minimum tensile strength of _____.

   a. 1,000 pounds
   b. 1,500 pounds
   c. 3,000 pounds
   d. 5,000 pounds

2. Allowing a rescued worker who has been suspended to lie down can cause a dangerous condition called _____.

   a. suspension trauma
   b. deceleration shock
   c. vertigo
   d. prone seizure

SECTION FIVE

5.0.0 POTENTIAL ELECTRICAL HAZARDS AND SOURCES

Objective

Identify common electrical hazards and sources when working with scaffolds.

a. Identify common electrical hazards when working with scaffolds.
b. Identify common electrical sources when working with scaffolds.

There are three possible results when electrical current shocks a worker: bodily harm from the shock itself, which can range from minor to severe; injury or death resulting from an accident caused by involuntary muscle response to shock, such as falling from a scaffold; and death from electrocution.

5.1.0 Electrical Hazards

An awareness of electrical hazards and the ability to react properly to varying circumstances in a timely manner can mean the difference between life or death for an injured co-worker.

5.1.1 Electrical Shock

The amount of current that passes through the human body determines the outcome of an electrical shock. The higher the voltage is, the greater the chance for a fatal shock. Electrical current flows along the path of least resistance to return to its source. If you come in contact with a live conductor, the shock could result in electrocution.

Figure 12 shows how much resistance the human body presents under various circumstances, and how this converts to amps or milliamps when the voltage is 110V. Note that the potential for shock increases dramatically if the skin is damp. A cut will also reduce your resistance. *Table 5* shows the effects of current on the human body.

WARNING!

High voltage, defined as 600 volts or more, is almost 10 times as likely to kill as low voltage. On the job, most time is spent working on or near lower voltages. However, lower voltages kill more people than high voltages because they are exposed to low voltages more often. Currents of less than 1 amp can severely injure and even kill a person.

5.1.2 Fire and Explosion

In addition to shock, electricity poses the hazard of fire from short-circuiting or overheating, and the hazard of explosion from arcing in an atmosphere that contains combustible dust or flammable vapors. Even the low-energy discharge of static electricity from the operation of electrical equipment can cause a disaster. The severity of these hazards makes it necessary to take careful precautions to avoid them. Under these circumstances, nonsparking tools should be employed by scaffold builders.

5.2.0 Electrical Hazard Sources

There are two areas where electrical hazards must be considered: the general use of electrical tools and equipment; and high voltages, which are encountered most often in industrial plants and in transmission, substation, and power generation equipment.

5.2.1 Power Tools and Equipment

The following precautions can be taken to minimize the hazards of electricity when using power tools and equipment:

- Always inspect all cords for cuts, flat areas, and any other damage prior to use.
- When available, use double-insulated power tools (ones that are encased entirely in nonconductive, shatterproof materials, and that have a nonconductive switch).
- Do not use electrical tools in a wet environment.
- Inspect the cords and plugs of tools before use to make sure they are serviceable.

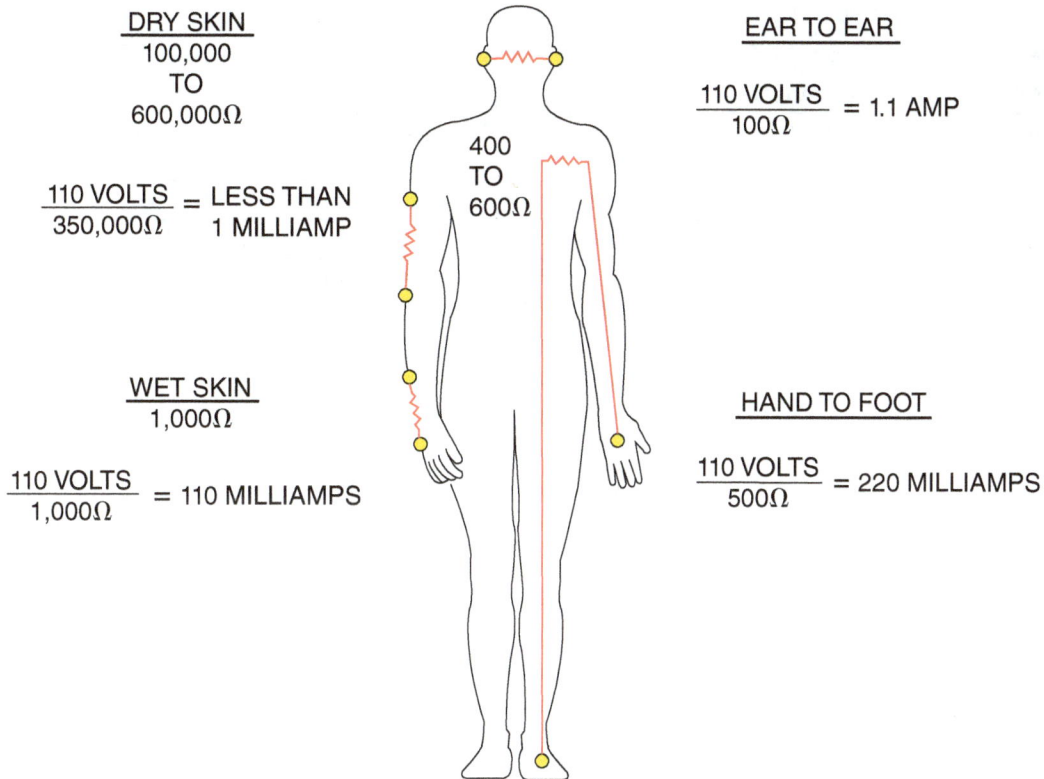

DRY SKIN
100,000
TO
600,000Ω

$$\frac{110 \text{ VOLTS}}{350,000\Omega} = \text{LESS THAN } 1 \text{ MILLIAMP}$$

400
TO
600Ω

EAR TO EAR

$$\frac{110 \text{ VOLTS}}{100\Omega} = 1.1 \text{ AMP}$$

WET SKIN
1,000Ω

$$\frac{110 \text{ VOLTS}}{1,000\Omega} = 110 \text{ MILLIAMPS}$$

HAND TO FOOT

$$\frac{110 \text{ VOLTS}}{500\Omega} = 220 \text{ MILLIAMPS}$$

31102-14_F12.EPS

Figure 12 Typical body resistances and current.

- Do not trim the larger, polarized prong of a three-pronged plug to make it fit into an outlet. This prong is for safety.
- When using an adapter for a three-pronged plug, secure the grounding wire on the adapter to the faceplate of a grounded outlet box.
- Wherever possible, use a ground fault circuit interrupter (GFCI box), which is designed to shut down a circuit when current leakage is detected.
- Use properly grounded, three-wire extension cords.
- Use jacketed extension cords with power tools, and use specially rated cords with heating devices that require large amounts of current.
- Protect extension cords from sharp surfaces. Tape them down in areas of foot traffic.

When working around high voltages, workers must have special training, use standard operating procedures, and follow a site safety plan.

Safety measures must be taken to avoid high voltages, which can lead to death. Chief among these safety measures are locking and tagging out equipment to avoid accidental energizing. Lockout/tagout procedures involve placing locks and warning tags on each device to de-energize circuits or equipment in order to prevent their use.

5.2.2 From Other Sources

Scaffolds may need to be installed and used near power lines, process piping, and other overhead obstructions. See *Table 2*, which shows the minimum distance allowed from electrical power lines.

Since most scaffolds are made of metal, and metal conducts electricity, care must also be taken when operating electrically powered tools or equipment on the scaffold. Damaged or pinched electrical power cords can allow electricity to transfer from the cord(s) into the scaffold itself.

Table 5 Effects of Current on the Human Body

Current Value	Typical Effects
Less Than 1 Milliamp	No sensation.
1 To 20 Milliamps	Sensation of shock, possibly painful. Between 10 and 20 milliamps, may lose some muscular control.
20 To 50 Milliamps	Painful shock, severe muscular contractions, breathing difficulties.
50 To 200 Milliamps	Up to 100 milliamps same symptoms as above, only more severe.
	Between 100 and 200 milliamps, ventricular fibrillation may occur. This typically results in almost immediate death unless special medical equipment and treatment are available.
Over 200 Milliamps	Severe burns and muscular contractions. The chest muscles contract and stop the heart for the duration of the shock.

Additional Resources

"Scaffolding." OSHA. **www.osha.gov/SLTC/scaffolding/index.html**

Fall Protection and Scaffolding Safety: An Illustrated Guide. 2000. Grace Drennan Ganget. Government Institutes.

5.0.0 Section Review

1. High voltage is defined as being _____.

   a. 300 volts or more
   b. 600 volts or more
   c. 900 volts or more
   d. 1,200 volts or more

2. When necessary to fit into an outlet, the larger, polarized prong of a three-prong plug can be trimmed.

   a. True
   b. False

SUMMARY

This module covered the industry requirements for scaffold design, assembly, and use. The requirements for training were identified, and general safety guidelines for the scaffold builder were presented.

Scaffold builders are responsible complying with regulations, codes, and standards, as well as commonsense practices, while they are performing their duties. Employers are required to implement and maintain a training program to ensure that scaffold builders are properly trained and kept up to date. Refresher training or new training may be needed as conditions change at the work site. All training must be documented. Without this training and it's proper application, serious injury or death can result from improper erection or use of scaffolds.

A scaffold job must be evaluated, designed and planned by a qualified person. The erection of the scaffold must be done by trained personnel under the supervision of a competent person. A job hazard analysis should be conducted with each member of the crew prior to erection, dismantling, or modification of any scaffold.

1. The agency with primary responsibility for regulating the scaffolding industry is the _____.

   a. Scaffold Access Industry Association
   b. Occupational Safety and Health Administration
   c. National Safety Council
   d. American National Standards Institute

2. Someone who, by possession of a recognized degree, certificate, or professional standing, or who by extensive knowledge, training, and experience, has successfully demonstrated the ability to solve or resolve problems related to the subject matter, the work, or the project is a(n) _____.

   a. designated person
   b. competent person
   c. authorized person
   d. qualified person

3. The most common scaffold is 5 feet wide and has a length of _____.

   a. 5 feet
   b. 6 feet
   c. 7 feet
   d. 8 feet

4. Damaged scaffold equipment must be _____.

   a. reported to the scaffold supervisor
   b. braced
   c. used until it fails
   d. removed from service

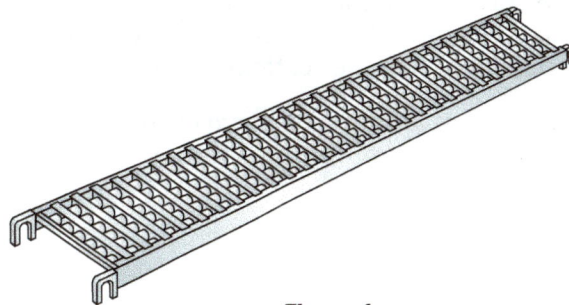

Figure 1

31102-14_RQ01.EPS

5. The scaffold plank in Review Question *Figure 1* is a _____.

   a. fabricated scaffold platform
   b. stage platform
   c. decorator plank
   d. fabricated scaffold deck

6. The primary consideration during the erection of a scaffold system is _____.

   a. the contractor's schedule and deadlines
   b. cost
   c. safety
   d. avoiding interference with other contractors

7. Scaffolds must be erected by a properly trained crew under the supervision of a _____.

   a. professional engineer
   b. competent person
   c. safety technician
   d. qualified person

8. With a compressive strength of 0.5 tons per square foot or less, the most unstable soil is Type _____.

   a. Type D soil
   b. Type C soil
   c. Type B soil
   d. Type A soil

9. To make leveling adjustments, screw jacks are used in combination with _____.

   a. base plates
   b. ground planks
   c. trimmer blocks
   d. mudsills

10. When tying off a scaffold to an adjacent structure, the first vertical tie-off must be no higher than _____.

   a. twice the scaffold's base width
   b. one-eighth the scaffold's final height
   c. four times the scaffold's base width
   d. one-fourth the scaffold's final height

11. An access ladder must have evenly spaced rungs with a minimum width of _____.

   a. 11.5 inches
   b. 12 inches
   c. 14.5 inches
   d. 16 inches

12. Work platforms must be fully planked, with planking runs overlapped by at least _____.

   a. 8 inches
   b. 1 foot
   c. 18 inches
   d. 2 feet

13. To protect against falling objects, all sides of scaffold platforms must be provided with _____.

   a. guardrails
   b. lifelines
   c. toeboards
   d. overhead deflectors

14. If an inspection indicates that hazards exist and have been identified, the scaffold should be marked with a tag colored _____.

   a. yellow
   b. red
   c. green
   d. black

15. Unless properly braced, a scaffold should not be used if its height is greater than its minimum base dimension by _____.

   a. three times
   b. four times
   c. six times
   d. eight times

16. For use on a construction site, footwear should be made of leather with _____.

   a. rubber soles
   b. ankle support
   c. reinforced soles or flexible metal inner-soles
   d. waterproof seams and uppers

17. The most common form of hearing protection worn on construction sites is _____.

   a. earmuffs
   b. disposable earplugs
   c. cotton wads
   d. reusable earplugs

18. According to OSHA regulations, a personal positioning system (used with a full-body harness) should not allow a worker to free-fall more than _____.

   a. 2 feet
   b. 4 feet
   c. 6 feet
   d. 8 feet

19. Depending on its distance below the work area, a safety net must extend beyond the work-area edge by at least _____.

   a. 6 to 10 feet
   b. 8 to 13 feet
   c. 11 to 15 feet
   d. 14 to 20 feet

20. Power tools with a nonconductive switch and a shatterproof, nonconductive casing are described as _____.

   a. safety rated
   b. shockproof
   c. double insulated
   d. nonsparking

Trade Terms Quiz

Fill in the blank with the correct term that you learned from your study of this module.

1. A _____ is a horizontal scaffold member on which the scaffold platform rests.

2. Corrosion that occurs at the point of contact of dissimilar materials is caused by _____.

3. Layers of wood that are glued together parallel with the grain is considered _____.

4. The intensity of sound is measured in _____.

5. The resistance of a material to a force tending to tear it apart is its _____.

6. The weight present for each square foot of surface area is measured in _____.

Trade Terms

Decibels (dBs)
Galvanic action
Laminated

Pounds per square foot (psf)
Putlog
Tensile strength

Trade Terms Introduced in This Module

Decibels (dBs): The intensity of a sound.

Galvanic action: Corrosion caused at the point of contact of dissimilar metals such as steel and aluminum.

Laminated: Layers of wood glued together with the grain parallel.

Pounds per square foot (psf): The weight present for each square foot of surface area.

Putlog: A horizontal scaffold member on which the scaffold platform rests.

Tensile strength: The resistance of a material to a force tending to tear it apart.

Additional Resources

This module presents thorough resources for task training. The following resource material is suggested for further study.

"Scaffolding." OSHA. **www.osha.gov/SLTC/scaffolding/index.html**

Fall Protection and Scaffolding Safety: An Illustrated Guide. 2000. Grace Drennan Ganget. Government Institutes.

International Building Code®, Latest Edition. Falls Church, VA: International Code Council.

International Residential Code®, Latest Edition. Falls Church, VA: International Code Council.

Figure Credits

Courtesy of PERI Formwork Systems, Inc., Module Opener

Irwin Tools, Figure 5
Honeywell International, Figures 6, 7, and 8

Section Review Answer Key

Answer	Section Reference	Objective
Section One		
1. a	1.1.1	1a
2. c	1.2.0	1b
Section Two		
1. b	2.1.2	2a
2. c	2.2.3	2b
3. b	2.3.0	2c
Section Three		
1. c	3.1.2	3a
2. d	3.1.5	3a
Section Four		
1. d	4.1.6	4a
2. a	4.2.0	4b
Section Five		
1. b	5.1.1	5a
2. b	5.2.1	5b

NCCER CURRICULA — USER UPDATE

NCCER makes every effort to keep its textbooks up-to-date and free of technical errors. We appreciate your help in this process. If you find an error, a typographical mistake, or an inaccuracy in NCCER's curricula, please fill out this form (or a photocopy), or complete the online form at **www.nccer.org/olf**. Be sure to include the exact module ID number, page number, a detailed description, and your recommended correction. Your input will be brought to the attention of the Authoring Team. Thank you for your assistance.

Instructors – If you have an idea for improving this textbook, or have found that additional materials were necessary to teach this module effectively, please let us know so that we may present your suggestions to the Authoring Team.

NCCER Product Development and Revision

13614 Progress Blvd., Alachua, FL 32615

Email: curriculum@nccer.org
Online: www.nccer.org/olf

❏ Trainee Guide   ❏ Lesson Plans   ❏ Exam   ❏ PowerPoints   Other _____

Craft / Level: _____   Copyright Date: _____

Module ID Number / Title: _____

Section Number(s): _____

Description: _____

_____

_____

_____

_____

Recommended Correction: _____

_____

_____

_____

_____

Your Name: _____

Address: _____

_____

Email: _____   Phone: _____

31103-15

Trade Tools and Equipment

OVERVIEW

As a scaffold builder, you must know how to properly store, inspect, and possibly repair scaffold equipment. You will also be required to use many types of hand and power tools. You will need to recognize these tools and be familiar with their safe and proper use. This module provides basic guidelines for the storage, handling, and inspection of scaffold equipment, and identifies some basic hand and power tools used in the scaffolding trade. Information on personal fall arrest systems is also presented.

Module Three

Trainees with successful module completions may be eligible for credentialing through the NCCER Registry. To learn more, go to **www.nccer.org** or contact us at **1.888.622.3720**. Our website has information on the latest product releases and training, as well as online versions of our *Cornerstone* magazine and Pearson's product catalog.

Your feedback is welcome. You may email your comments to **curriculum@nccer.org**, send general comments and inquiries to **info@nccer.org**, or fill in the User Update form at the back of this module.

This information is general in nature and intended for training purposes only. Actual performance of activities described in this manual requires compliance with all applicable operating, service, maintenance, and safety procedures under the direction of qualified personnel. References in this manual to patented or proprietary devices do not constitute a recommendation of their use.

Objectives

When you have completed this module, you will be able to do the following:

1. Explain the proper methods of storing, handling, and inspecting scaffold equipment.
   a. Explain the proper methods for storing scaffold equipment at the laydown area.
   b. Explain the proper methods for handling scaffold equipment.
   c. Explain the general inspection procedures for equipment of various types of scaffold systems.
   d. Explain the procedures for the repair of damaged equipment.
2. Identify the hand and power tools commonly used by scaffold builders, and describe their proper use.
   a. Identify the hand tools commonly used by scaffold builders, and describe their proper use.
   b. Identify the power tools commonly used by scaffold builders, and describe their proper use.
3. Identify the components used to level scaffolds, and how to properly use these items.
   a. Describe the purpose and proper use of screw jacks.
   b. Describe the general guidelines for properly using jacks.
4. Describe the proper use of personal fall arrest equipment used in the scaffolding industry.
   a. Describe the proper use of vertical and horizontal lifeline systems.
   b. Describe the proper use of self-retracting fall arrest devices.

Performance Tasks

Under the supervision of your instructor, you should be able to do the following:

1. Perform inspection of available scaffold equipment.
2. Demonstrate the safe and effective use of available hand tools.
3. Demonstrate the safe and effective use of power tools.
4. Demonstrate the proper rigging of available fall arrest devices.

Trade Terms

Columns
Level
Plumb

Ratcheting
Reciprocating

Industry-Recognized Credentials

If you're training through an NCCER-accredited sponsor, you may be eligible for credentials from NCCER's Registry. The ID number for this module is 31103-14. Note that this module may have been used in other NCCER curricula and may apply to other level completions. Contact NCCER's Registry at 888.622.3720 or go to **www.nccer.org** for more information.

Code Note

Codes vary among jurisdictions. Because of the variations in code, consult the applicable code whenever regulations are in question. Referring to an incorrect set of codes can cause as much trouble as failing to reference codes altogether. Obtain, review, and familiarize yourself with your local adopted code.

Contents

Topics to be presented in this module include:

Figures

1.0.0 STORAGE, HANDLING, AND INSPECTION OF SCAFFOLD EQUIPMENT

Objective

Explain the proper methods of storing, handling, and inspecting scaffold equipment.

  a. Explain the proper methods for storing scaffold equipment at the laydown area.

  b. Explain the proper methods for handling scaffold equipment.

  c. Explain the general inspection procedures for equipment of various types of scaffold systems.

  d. Explain the procedures for the repair of damaged equipment.

Performance Task

Perform inspection of available scaffold equipment.

In order to prevent scaffold accidents, there must be a program in place for storage, inspection, and preventive maintenance of scaffold equipment. All reputable scaffold suppliers have recommendations and guidelines for the storage, inspection, and repair of their equipment. All scaffold erectors and users are encouraged to ask the supplier for this information. The following sections serve as a basic guideline for scaffold equipment, regardless of manufacturer.

1.1.0 Proper Storage of Scaffold Equipment

Proper storage of scaffold equipment is one step to ensure that scaffolds are safe. Improperly stored equipment can result in damage, loss of equipment strength, and loss of safe operation.

1.1.1 Storage at Laydown Area

Whether the storage is in a rental building, a scaffold builder's warehouse, or at the job site, the following conditions should always be met:

- Develop a plan for storing equipment. The effort used to maintain a clean and orderly storage area will result in a safer, more efficient work environment.
- Store equipment so that it is easily accessible.
- A 12-foot alley is needed for the movement of trucks, forklifts, and other handling equipment. Give attention to water that may be trapped under ladder rungs.
- Provide dunnage to keep the equipment off the ground.
- Provide support between layers so that suitable materials can be handled by the forklifts.
- Store different-sized parts in separate racks.
- Provide baskets or racks for small parts.
- Always keep inventory and equipment in need of repair in separate, clearly marked areas. Prior to repair, this equipment should be tagged (out of service).
- Red-tagged parts should be removed from the job site.
- Do not stack equipment high enough to become unstable. This requires even more diligent care when racks must be stored on unstable ground.
- Provide blocking to prevent scaffold tube from rolling while being handled.

1.2.0 Proper Handling of Scaffold Equipment

Handling scaffold equipment can be hazardous. Scaffold planking is heavy, scaffold frames can be awkward, and braces can pinch. Appropriate personal protective equipment can reduce the risk of injury, but the following guidelines should also be followed:

- Never throw any type of equipment off of the scaffold to the ground.
- Do not climb on stacks of equipment.
- Be careful when stacking or removing scaffold tube from storage. It is round and will roll if not properly blocked.
- Use caution when loading/unloading any materials from scaffold racks, giving particular attention to crush points.
- Do not try to catch falling, sliding, or slipping equipment.
- Use extreme caution when handling wet equipment; it can be very slippery.
- You must be properly trained in the use of banding equipment. Manufacturers' instructions must be considered.

- When cutting banding, stand clear of the ends of the bands as well as any area where the equipment may fall when released. (Banding is very sharp, so appropriate gloves should be worn.)
- Respect the physical dangers inherent when handling scaffold equipment. Remember to warm up, stretch, and use your legs when lifting. Develop a plan when using multiple people to pass materials or heavier items.
- Throwing equipment from one person to another is discouraged.

1.2.1 General Inspections

Visual inspection is essential to provide a safe and reliable scaffold inventory. Inspections should be made by the supplier, erector, and the user. The ultimate responsibility of avoiding the use of damaged equipment is with the scaffold erector and user. Inspections of individual components should be performed when the component is received, when it is removed from storage, as it is installed in a scaffold system, when it is removed from the system, and when it is returned to storage. Occupational Safety and Health Administration (OSHA) regulations require that assembled scaffolds be inspected prior to each work shift, or after any occurrence that affects the structural integrity of the scaffold. Inspection of the equipment includes checking for:

- Broken welds
- Split scaffold tube
- Bent scaffold tube
- Cracks around the scaffold tube
- Crushed scaffold tube
- Excessive rust (inside and outside)

> **NOTE**
> Use a flashlight to inspect the inside of the scaffold tube.

- Corrosion and pitting
- Properly working brace locks
- Evidence of exposure to extreme heat or fire
- Straightness, especially for vertical legs of frames
- Damaged threads on screw jacks
- Excessively worn rivets or bolts on braces
- Split or bent ends on cross braces
- Properly working caster brakes and swivels
- Bent or broken clamp parts

- Any additional conditions specified by the individual manufacturer
- Defects and damage on all scaffold boards and lumber
- Signs of exposure to chemicals

1.3.0 Inspection Procedures

Different types of scaffolds require different inspection methods. The following are a few scaffold items that require specific inspections:

- Frames
- Approved mudsills
- Scaffold gates and safety stop (gates swing inward)
- Ladder installation and proper clearance
- Tube-and-clamp scaffolds
- System scaffolds
- Scaffold braces and guardrails
- Scaffold brackets
- All metal, synthetic, and wood planks (check to see if they have are OSHA approved; make sure that any nails that were placed in the planking in a prior setting have been removed)

1.3.1 Inspection Points for Frames

The following are points to inspect or check on scaffold frames:

- Verify that the coupling pins are held securely, and that they are not bent or missing.
- Verify that the legs are not dented, bent, twisted, or cracked.
- Verify that the Speedlocks™ are not twisted, missing, or jammed.
- Verify that the ledger and rungs are not twisted, dented, cracked, or missing.
- Verify that the weld areas are not cracked or rusted.

Figure 1 shows the points to inspect on a scaffold frame.

1.3.2 Inspection Points for Braces and Guardrails

The following are points to inspect or check on scaffold braces and guardrails:

- Verify that the ends are not twisted.
- Verify that the rivets are not pitted, loose, or rusty.
- Verify that the angles are correct, and that the tubes are not bent or cracked.

Figure 2 shows the points to inspect on scaffold braces and guardrails.

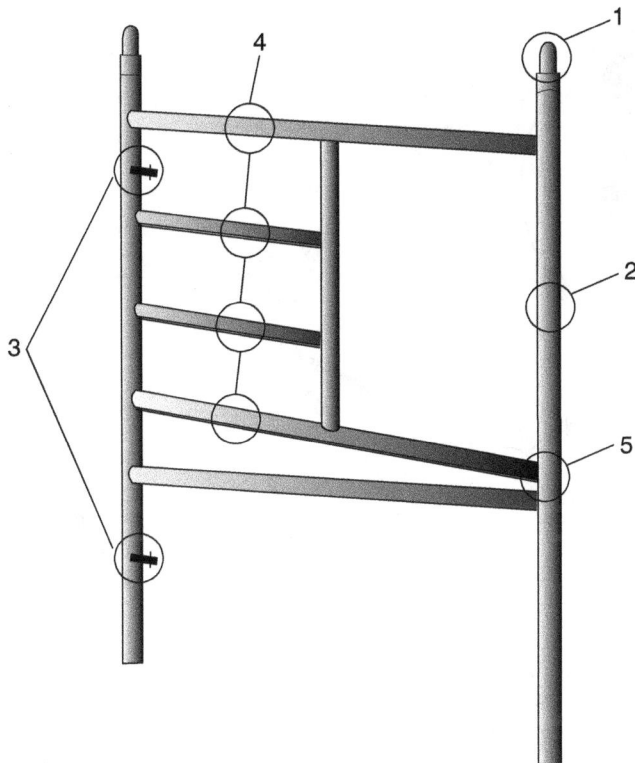

1 - CHECK COUPLING PINS: BENT? MISSING?
   NOT HELD SECURELY?

2 - CHECK LEGS: DENTED? BENT? TWISTED?
   CRACKED?

3 - CHECK SPEEDLOCKS™: TWISTED?
   MISSING? JAMMED?

4 - CHECK LEDGER AND RUNGS: TWISTED?
   DENTED? CRACKED? MISSING?

5 - CHECK WELD AREAS: CRACKS? RUSTED?

31103-14_F01.EPS

Figure 1 Frame inspection points.

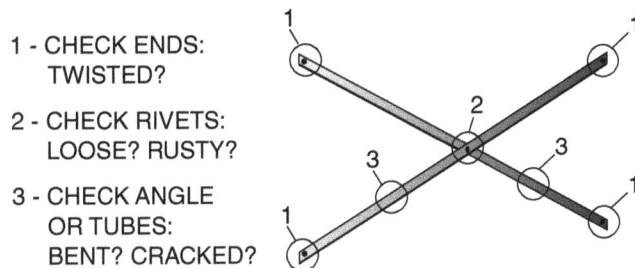

1 - CHECK ENDS:
   TWISTED?

2 - CHECK RIVETS:
   LOOSE? RUSTY?

3 - CHECK ANGLE
   OR TUBES:
   BENT? CRACKED?

31103-14_F02.EPS

Figure 2 Inspection points for scaffold braces and
   guardrails.

1.3.3 Inspection Points for Tube-and-Clamp Scaffolds

The following are points to inspect or check on tube-and-clamp scaffolds:

- Verify that the end fittings hold securely, and that they are not bent or missing.
- Verify that the verticals (tubes) are not dented, bent, or cracked.
- Verify that the clamps are not twisted, broken, or jammed.
- Verify that the horizontals (tubes) are not cracked or dented.
- Verify that the nut on the clamp bolt is operational and will turn on the threads.

Figure 3 shows the points to inspect on tube-and-clamp scaffolds.

1.3.4 Inspection Points for System Scaffolds

The following are points to inspect or check on system scaffolds:

- Verify that the coupling pins are held securely, and that they are not bent or missing.
- Verify that the legs and uprights are not dented, bent, twisted, or cracked.
- Verify that the wedges are not twisted, missing, or jammed.
- Verify that the bearer ledger and runners are not twisted, dented, cracked, or bent.
- Verify that the ring sets are not bent, cracked, or missing.
- Verify that the diagonals are not dented, bent, twisted, swiveled, or cracked.

Figure 4 shows the points to inspect on system scaffolds.

1.3.5 Inspection Points for Brackets

The following are points to inspect or check on scaffold brackets:

- Verify that the wedges, hooks, and U's are not missing, bent, straight, or damaged.
- Verify that the latches are working.
- Verify that the ledger, brace, and back are not bent or broken.
- Verify that the weld areas are not cracked, broken, or rusted.
- Verify that the guardrail posts and attaching members are not missing, bent, or unusable.

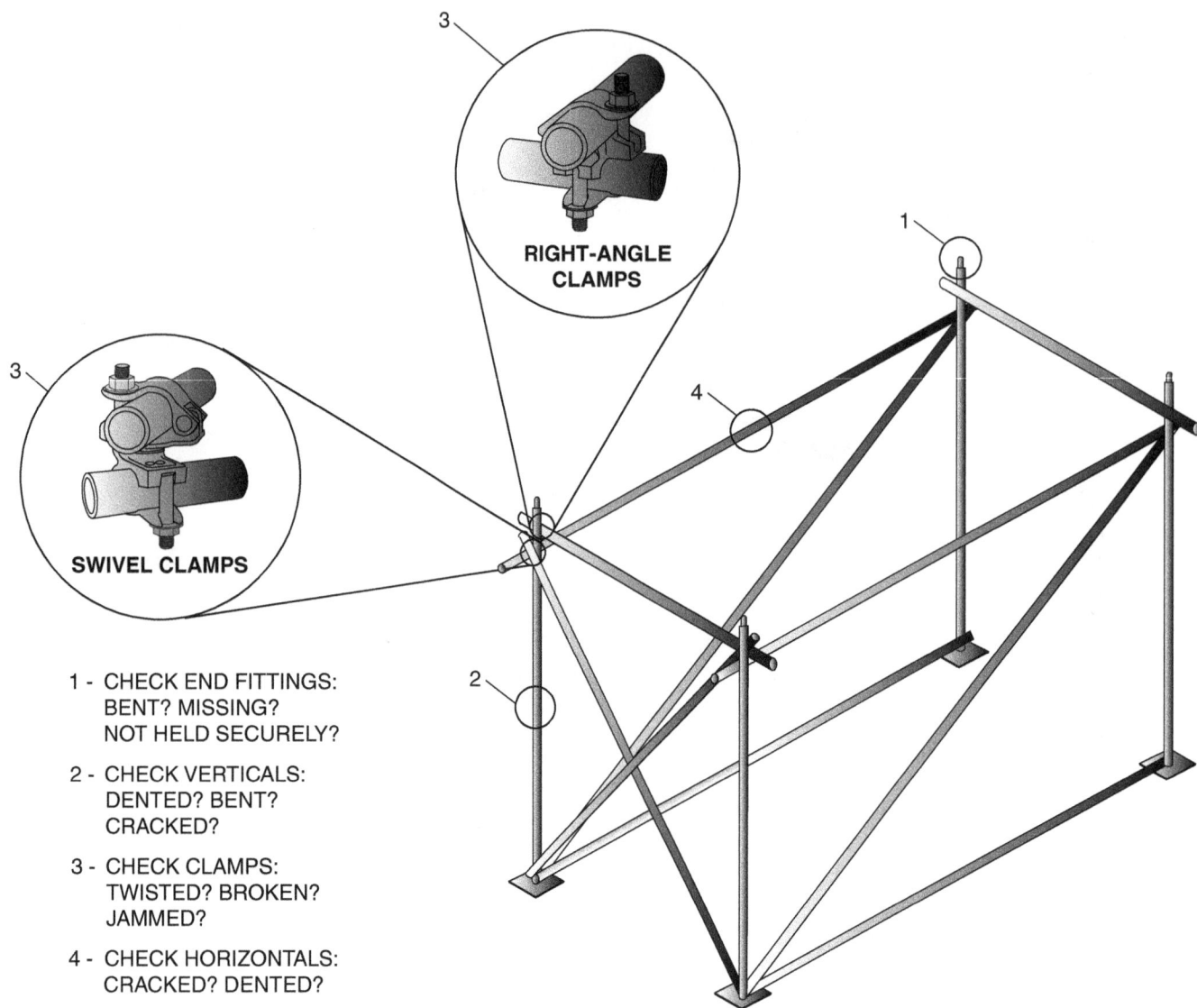

RIGHT-ANGLE
CLAMPS

SWIVEL CLAMPS

1 - CHECK END FITTINGS:
     BENT? MISSING?
     NOT HELD SECURELY?

2 - CHECK VERTICALS:
     DENTED? BENT?
     CRACKED?

3 - CHECK CLAMPS:
     TWISTED? BROKEN?
     JAMMED?

4 - CHECK HORIZONTALS:
     CRACKED? DENTED?

31103-14_F03.EPS

Figure 3 Tube-and-clamp inspection points.

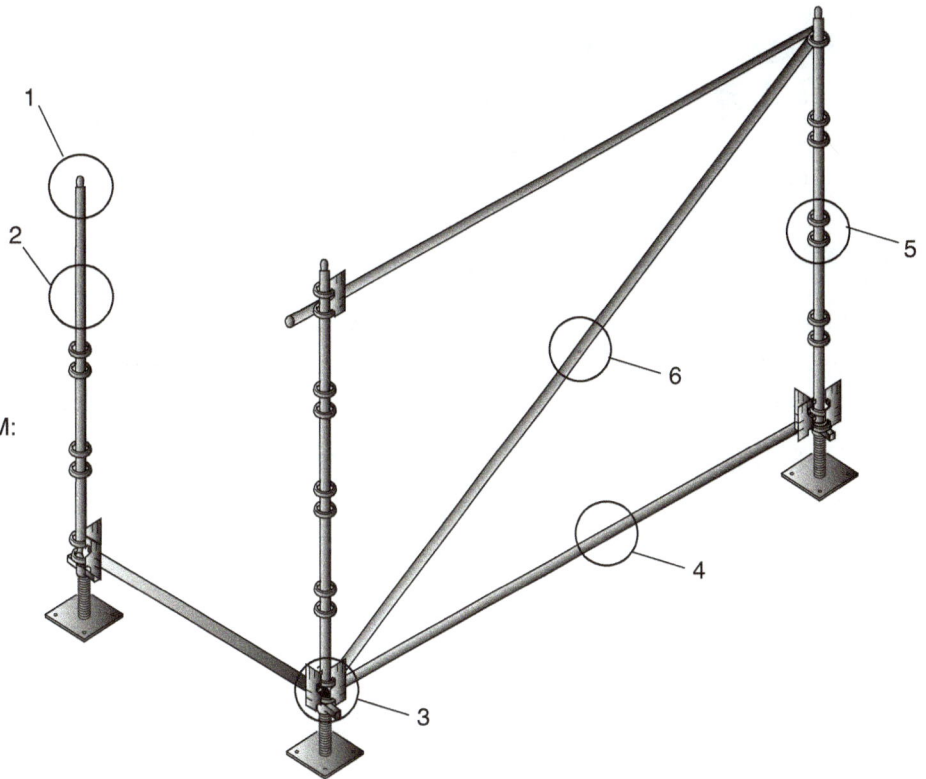

1 - CHECK COUPLING PINS:
   BENT? MISSING?
   NOT HELD SECURELY?

2 - CHECK STANDARD:
   DENTED? BENT?
   TWISTED? CRACKED?

3 - CHECK WEDGES:
   TWISTED? MISSING?
   JAMMED?

4 - CHECK LEDGER AND TRANSOM:
   TWISTED? DENTED?
   CRACKED? BENT?

5 - CHECK CLUSTERS:
   BENT? CRACKED?
   MISSING?

6 - CHECK DIAGONALS:
   DENTED? BENT? TWISTED?
   SWIVEL? CRACKED?

31103-14_F03.EPS

Figure 4 Inspection points for system scaffolds.

Figure 5 shows the points to inspect on scaffold brackets.

1.4.0 Equipment Repair

Repairs on scaffold equipment by the scaffold builder are very limited. All repairs must be performed by qualified persons who have received instructions specific to the equipment being repaired. Some points to consider for equipment repair include the following:

- A qualified person must determine the repairability of scaffold components.
- Only qualified persons should repair components.
- Check with the manufacturer before attempting any repairs. There are many things that cannot be done to repair scaffolds.
- When in doubt, discard the equipment rather than risk an accident.

Never alter or modify a scaffold component in the field, unless the qualified person representing the manufacturer determines that the structural integrity of the scaffold will not be adversely affected by the modification.

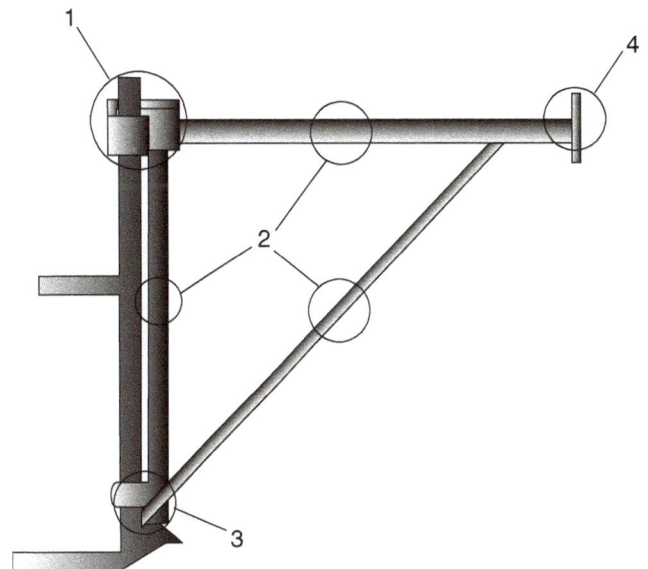

1 - CHECK HOOKS OR U's: MISSING? BENT?
   STRAIGHT OR DAMAGED? LATCHES NOT WORKING?

2 - CHECK LEDGER, BRACE, AND BACK: BENT? BROKEN?

3 - CHECK WELD AREAS: CRACKS? BROKEN? RUSTED?

4 - CHECK GUARDRAIL POST AND ATTACHING MEMBER:
   MISSING? BENT? UNUSABLE?

31103-14_F05.EPS

Figure 5 Scaffold bracket inspection points.

Additional Resources

"Scaffolding." OSHA. **www.osha.gov/SLTC/scaffolding/index.html**

Fall Protection and Scaffolding Safety: An Illustrated Guide. 2000. Grace Drennan Ganget. Government Institutes.

1.0.0 Section Review

1. To prevent scaffold tube from rolling while being handled, provide _____.

   a. dunnage
   b. rigging
   c. banding
   d. blocking

2. Cut-resistant gloves should be worn when working with banding.

   a. True
   b. False

3. Scaffolds should be inspected for all the following, *except* _____.

   a. cracking
   b. cleanliness
   c. corrosion
   d. straightness

4. When inspecting tube-and-clamp scaffolds, verify that there are no twisted, broken, or jammed _____.

   a. clamps
   b. verticals
   c. end fittings
   d. horizontals

2.0.0 COMMON TOOLS USED BY SCAFFOLD BUILDERS

Objective

Identify the hand and power tools commonly used by scaffold builders, and describe their proper use.

a. Identify the hand tools commonly used by scaffold builders, and describe their proper use.
b. Identify the power tools commonly used by scaffold builders, and describe their proper use.

Performance Tasks

Demonstrate the safe and effective use of available hand tools.

Demonstrate the safe and effective use of power tools.

Trade Terms

Level: Completely horizontal; at right angles to plumb.

Plumb: Completely vertical, or straight up and down; at right angles to level.

Ratcheting: To lock when moving in one direction and release when moved in the other.

Reciprocating: Moving back and forth.

This section provides information about the hand tools and power tools commonly used in the erection and dismantling of scaffolds. Some of these tools were covered in the *Core Curriculum* and are presented here as a refresher.

WARNING!	Tools can easily fall out of even the best tool belts. For the safety of anyone passing below, tools should be tethered to the tool belt using lanyards.

2.1.0 Types of Hand Tools

A scaffold builder must understand the proper usage of various types of hand and power tools. Although other tools may be used at times, the following hand tools will be necessary in the erection and dismantling of most scaffold systems.

2.1.1 Plumb Bobs

A plumb bob (*Figure 6*) is a pointed weight attached to a string or wire. It uses the force of gravity to make a vertical or plumb line. When the weight is suspended by the string and allowed to hang freely, the string is plumb. If the string of a hanging plumb bob is attached a specific distance from the top of a scaffold leg and the string is that same distance from the bottom of the leg, then the leg is plumb.

2.1.2 Rules and Tapes

Tapes and rules are used to accurately measure the distance between points. The tape is wound on a reel and enclosed in a housing. The tape can be made of steel, plastic, or fiberglass. The tape may be marked in feet and inches or in metric gradations. Pocket tapes are available in various lengths, the most common being 25 feet. Pocket tapes are fitted with a spring inside the housing to rewind the tape after use. For measuring longer distances, scaffold runs, long walls, etc., a steel tape is used. These tapes must be manually rewound. If the tape has picked up dirt or oil while extended, wipe it with a cloth as it is rewound. *Figure 7* shows steel tapes.

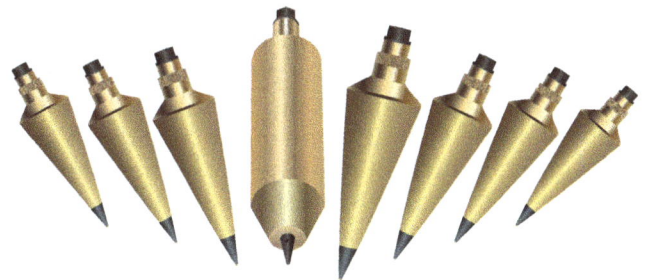

31103-14_F06.EPS

Figure 6 Selection of plumb bobs.

Figure 7 Steel tapes.

(A) LEVEL

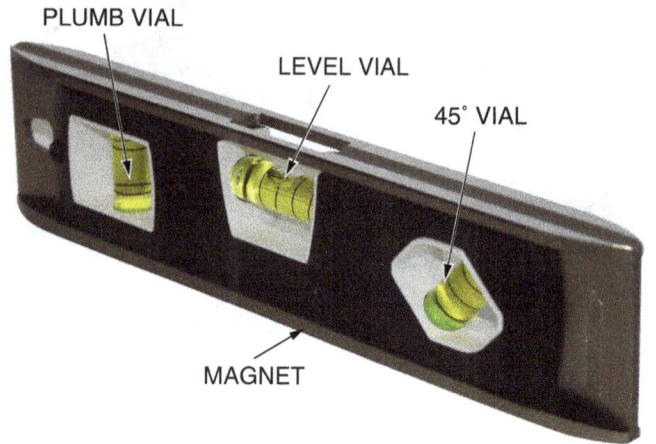

(B) TORPEDO LEVEL

Figure 8 Two styles of levels.

2.1.3 Levels

Spirit levels and torpedo levels (*Figure 8*) are the two most common levels used by scaffold builders. They consist of a rectangular frame containing a series of bubble tubes. Depending on their orientation, the bubble tubes can be used to check for plumb or level. When the bubble in the tube is centered between the two etched lines, the surface supporting the level is plumb or level. Common sizes are 2 or 4 feet long. Some levels contain a bubble tube that allows you to check a 45-degree angle, which may be helpful when setting bracing. The smaller versions of spirit levels are referred to as torpedo levels.

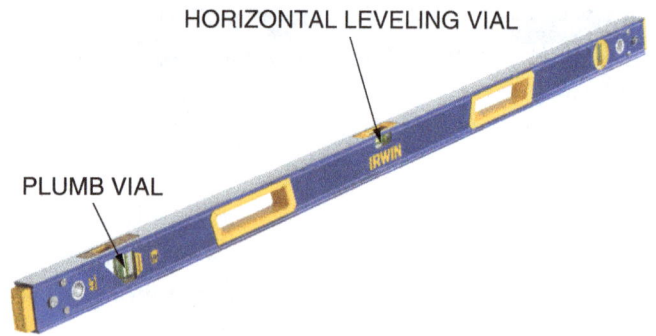

2.1.4 String Lines

The string line (*Figure 9*) is used to establish baselines in a scaffold layout and to ensure that scaffolds are straight. The most common material for this type of use is nylon. The string is wound on a spool to allow for ease of use and to prevent tangling. Woven varieties of string are stronger than braided versions.

2.1.5 Handsaws

Handsaws (*Figure 10*) are very useful for small cutting jobs, such as cutting toeboards. It is often quicker to grab a handsaw and make the cut rather than stringing out an extension cord and using a circular saw. Handsaws are designed to

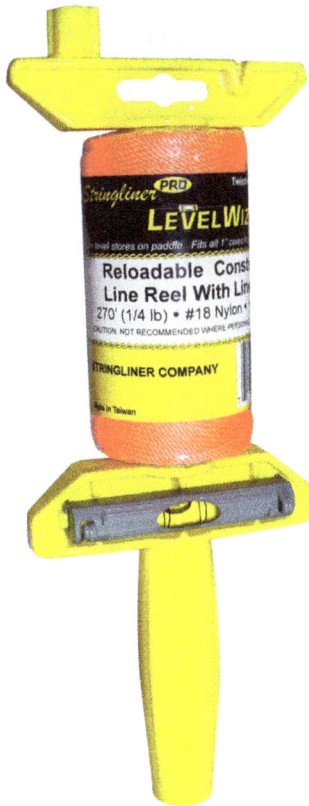

31103-14_F09.EPS

Figure 9 String line.

cut across the grain of a plank. If a lengthy cut is needed along the grain, a ripsaw should be used.

> **WARNING!**
> Handsaw blades are sharp. Wear gloves when using a handsaw to protect your hands form cuts.

2.1.6 Hammers

Straight-claw hammers (*Figure 11*) are commonly used by scaffold builders. Hammers with handles made of metal should be used. The straight-claw hammer is preferred by scaffold builders due to the design of the claw. The straight claw is better suited to prying boards apart and is still adequate for nail pulling. The heavier head weights are not recommended for scaffold use because of the risk of damaging equipment.

> **NOTE**
> Brass tools should be used in hazardous environments, Class 1, 2, and 3. (29 *CFR* 1910.146)

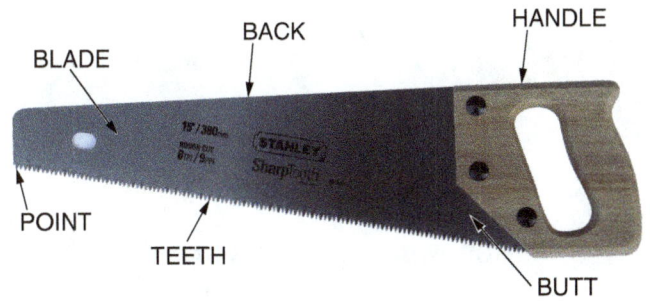

31103-14_F10.EPS

Figure 10 Handsaw.

> **WARNING!**
> It is recommended that you use a hammer that weighs 23 ounces or less. Larger sizes can cause elbow damage. Do not use hammers to notch wood. Flying wood can cause personal injury.

2.1.7 Types of Crowbars and Nail Pullers

Scaffold builders use crowbars and various types of nail pullers to dismantle wood pole scaffolds and to remove planking boards from other types of scaffolds. These tools are used to pry nailed boards apart and to remove nails from the boards.

The crowbar is sometimes called a wrecking bar or gooseneck. It is made from six-sided steel. The curved end is notched to allow it to be used to pull nails, and the other end is flattened to allow it to fit between boards and pry them apart. Crowbars come in a variety of lengths from 18 to 48 inches.

The Wonder Bar™ is a flat prying and nail-pulling bar. They are relatively small (12 to 18 inches long) and have notches in each end to assist with nail pulling. There is a teardrop-shaped hole near the end that allows a better bite on a nail than the end notches. This hole can sometimes be used to pull nails that have had the heads broken off. A little more force is required when using this method, due to the loss of leverage.

The cat's paw is used along with a hammer. The claw end can be driven under a nail head. The nail is then pried out enough to remove it with a hammer or another type of bar. *Figure 12* shows bars and nail pullers.

31103-14_F11.EPS

Figure 11 Straight-claw hammer.

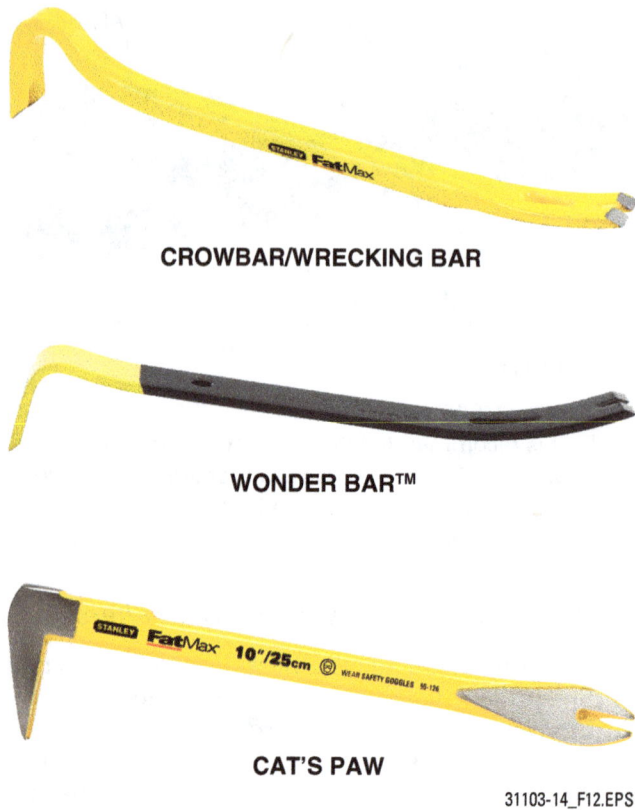

CROWBAR/WRECKING BAR

WONDER BAR™

CAT'S PAW

31103-14_F12.EPS

Figure 12 Bars and nail pullers.

31103-14_F13.EPS

Figure 13 Lineman's pliers.

2.1.8 Lineman's Pliers

Lineman's pliers (*Figure 13*) are used by scaffold builders to pull, twist, and cut a wire. These pliers are often needed when using wire to secure toeboards to the scaffold support members. The tie wire can be positioned, pulled tight, twisted, and cut with a single tool. When twisting wire, always tuck the sharp, cut ends of the wire so they do not pose a hand safety hazard. Lineman's pliers are known by many as "Kleins." Never remove the rubber protectors on the handle grips, as it removes the insulating properties of the tool, and the uninsulated grips could pose a shock hazard.

2.1.9 Various Wrenches

Scaffold builders have a choice of a variety of wrenches to use when assembling scaffolds. When assembling tube-and-clamp scaffolds, the clamps must be tightened well to ensure that the scaffold is rigid. Each clamp must be checked periodically to ensure that it has not worked loose. To tighten a nut, pull the wrench toward you from right to left, and to loosen a nut, pull from left to right. Any of the following wrenches will work in this application.

Box-end wrenches are preferred to open-end wrenches because the grip on the bolt or nut is much better. Box-end wrenches usually have different sizes on each end. The wrench must be lifted off the bolt or nut and repositioned to continue tightening or loosening.

A combination wrench is a both a box-end wrench and an open-end wrench. Both ends are the same size. The open end is used when it is impossible to fit the box end around the bolt or nut.

The jaws of an adjustable wrench can be opened or closed so that it can be made to fit a variety of bolt or nut sizes. Be careful when using this wrench—it has a tendency to slip, and can round off the shoulders of the nut or bolt.

Figure 14 shows commonly used wrenches.

Scaffold wrenches are designed specifically for tightening and loosening the nuts on the clamps used in tube-and-clamp scaffold systems. The scaffold wrench fits over and completely around a nut like a box-end wrench and has a ratcheting action like a socket wrench. The scaffold wrench must be turned over to change the turning direction. A swing-over wrench uses a socket that fits over the nut. The swing-over wrench does not ratchet. A ratchet wrench for scaffolds works just like a typical ratchet wrench used for other purposes.

BOX-END WRENCH

COMBINATION WRENCH

ADJUSTABLE WRENCH

31103-14_F14.EPS

Figure 14 Commonly used wrenches.

2.1.10 Snips and Shears

Snips and shears of various types are useful for cutting thin sheets of brass, aluminum, copper, black and galvanized iron, and stainless steel. Hand snips cannot be used for cutting tempered steel. Snips and shears are separated into two general classifications: those designed to make straight cuts, and those that are designed to make circular cuts. Both types are available in right-hand or left-hand styles. Snips and shears come in various types and sizes. With the exception of compound lever shears, these hand-operated snips and shears can only cut metals up to 20-gauge thickness. Compound lever shears can be used to cut metal up to 12-gauge thickness. Heavier-gauge metals must be cut with either a chisel or power tools. *Figure 15* shows some common hand-operated snips and shears.

> **WARNING!**
> The unpolished newly cut edges of metal are extremely sharp, and are in close proximity to your hands as cuts are made. Caution is needed, and gloves should be used to avoid hand injury.

The following are the snips or shears used most often in scaffold activities:

RIGHT-CUT OFFSET SNIPS

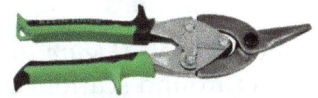

RIGHT-CUT AVIATION SNIPS

LEFT-CUT OFFSET SNIPS

LEFT-CUT AVIATION SNIPS

STRAIGHT-CUT OFFSET SNIPS

STRAIGHT-CUT AVIATION SNIPS

31103-14_F15.EPS

Figure 15 Snips and shears.

- *Tin snips* – Tin snips are used like scissors to cut thin, soft metals. Tin-snip blades are of two basic types: straight blades or combination blades. The straight blade has the face of the blade running up from the cutting edges, and the combination blade is curved back from the cutting edge. This curved surface allows the metal to slip over the top blade when cutting curves. Straight-blade snips have greater strength and can be constructed with longer blades.
- *General-purpose snips* – General-purpose snips may be either straight-blade or combination-blade design. They are used for most general-purpose cutting of 26-gauge or thinner metals.
- *Aviation snips* – Aviation snips have a compound leverage design that allows them to cut thicker metal than the general-purpose snips. The blade design allows them to be used for cutting small, irregular curves and even inside 90-degree corners. Aviation snips are available in straight, left-hand, or right-hand cutting models. The blades are serrated, which allows for better gripping of the metal being cut. Aviation snips usually come with insulated (though not dielectric strength) handle grips. The color of the handle grips denotes whether the snips are straight-line (yellow); left-handed (red); or right-handed (green) cutting models. If the handle grip color is not identifiable, the right-hand and left-hand snips can be distinguished by the positioning of the upper blade. If the upper blade is on the right, the snips are right-handed snips. If the upper blade is on the left, the snips are left-handed snips.

2.1.11 Banders

Banding tools (*Figure 16*) are used to secure metal bands around scaffold components that are being stored or shipped. These bands keep components from different types or brands of scaffolds from being intermixed. The bands also make the components easier to load when shipping, and make it easier to secure the components safely.

> **CAUTION**
>
> Do not use the claw end of a hammer to break steel straps or banding. Using any tool in a manner for which it is not designed poses a safety hazard.

2.2.0 Types of Power Tools

Portable power tools are often used by scaffold builders in the construction of wood pole scaffolds. These tools allow more efficient production than hand tools. It is important that the scaffold builder be familiar with the tools that are available and their basic use. Unless battery operated, all power tools must be plugged in to a ground fault circuit interrupter (GFCI) receptacle.

The circular saw, jigsaw, and reciprocating saw are the most common power tools used by scaffold builders. They are all needed due to the variety of functions that they perform. Each has unique functions or a design that makes it useful for different tasks and in different locations during scaffold erection and dismantling.

2.2.1 Circular Saws

The circular saw is the most commonly used power tool. There are two basic styles: the standard circular saw and a worm drive. These saws are used for cutting lumber, planking, and plywood. Special blades are available for cutting plastic and metal.

The depth and angle of the cut can be adjusted. The blade should be set to extend about $\frac{1}{4}$ inch below the material being cut. These changes in depth, or cutting angle, should be made each time a material of a different thickness is being cut.

31103-14_F16.EPS

Figure 16 Banding tools.

Before starting a cut, check the material to make sure that it does not contain any foreign objects, such as staples or nails. Then line up the saw, but make sure the blade is not in contact with the material to be cut prior to pulling the trigger switch. Allow the saw to come to full speed before the blade contacts the material.

When making a cut, make sure that the material is properly supported. When cutting plywood, it is good practice to support the material so that it remains supported after the cut is made. When making plunge cuts, make sure that the toe of the saw is firmly against the material before setting the blade into the material. Allow the saw to come up to full speed before the blade contacts the plywood.

> **WARNING!**
>
> The main cause of accidents with a circular saw is the failure of the blade guard to snap back into place when the cut is completed. This is often caused by a small piece of material jamming the guard. Always ensure that the guard has returned to the proper place before putting the saw down. Never intentionally disable the snapback feature of the guard.

NCCER – *Scaffolding* 31103-15

It is important to replace the blade when it becomes dull. A dull blade makes the cut more difficult and can cause the saw to kick back, which can cause damage to the material being cut, or serious injury to the operator. When replacing the blade, always unplug the saw and refer to the operator's manual for the proper blade replacement procedure. *Figure 17* shows commonly used circular saws.

2.2.2 Saber Saws

A saber saw, or jigsaw, is made to cut irregular shapes, curves, or circles. A scaffold builder may use one to cut plywood to fit close to an irregular-shaped wall or around a pole or pipe. Some of the larger models are capable of cutting planking, but that type of cut is best done by a circular saw. *Figure 18* shows a saber saw.

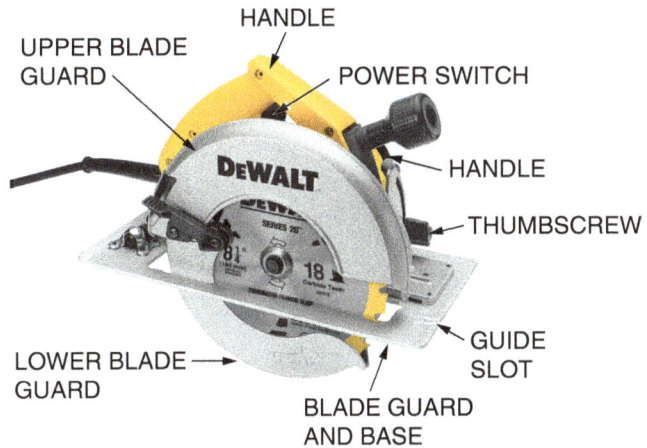

Figure 17 Types of circular saws.

2.2.3 Reciprocating Saws

Reciprocating saws (*Figure 19*) operate much the same as the saber saw. This type of saw is often referred to as a Sawsall™ or chop saw. These saws are suited to fairly heavy work, and can easily be used in areas where it is difficult or impossible to use a circular saw.

31103-14_F18.EPS

Figure 18 Saber saw.

31103-14_F19.EPS

Figure 19 Battery-powered reciprocating saws.

Additional Resources

"Scaffolding." OSHA. **www.osha.gov/SLTC/scaffolding/index.html**

Fall Protection and Scaffolding Safety: An Illustrated Guide. 2000. Grace Drennan Ganget. Government Institutes.

2.0.0 Section Review

1. To minimize the possibility of elbow damage, use a hammer that weighs no more than _____.

    a. 23 ounces
    b. 28 ounces
    c. 32 ounces
    d. 36 ounces

2. A circular-saw blade should be in contact with the work before the cut is started.

    a. True
    b. False

3.0.0 MANUAL LIFTING TOOLS (JACKS)

Objective

Identify the components used to level scaffolds, and how to properly use these items.

    a. Describe the purpose and proper use of screw jacks.

    b. Describe the general guidelines for properly using jacks.

31103-14_F20.EPS

Figure 20 Screw jack.

Manually operated lifting tools called jacks are commonly used to lift or reposition scaffolds.

3.1.0 Screw Jacks

Scaffolds must be erected plumb and level so as to safely carry the load applied by equipment, materials, and workmen. Scaffold builders use screw jacks in order to achieve a plumb and level scaffold on uneven ground.

Screw jacks (*Figure 20*) are used to lift loads and level scaffolds. There are two general types of screw jacks: regular and inverted. The screw jack uses the screw-and-nut principle. A simple lever will apply enough power to turn the screw.

3.2.0 General Guidelines for Using Screw Jacks

Using screw jacks to align scaffolds involves placing the jack between a foundation or base and the scaffold. As the jack is extended, force from the jack presses against the scaffold. As the weight of the scaffold starts to resist the force being created by the jack, the resistance is transmitted back through the jack to the foundation where the jack is positioned. As the jacking actions continue, the increasing force causes one of the following situations to occur:

- The scaffold moves.
- The foundation moves.
- The jack fails.

There are three key components involved in using jacks: the foundation, the jack, and the scaffold. The jack is used to transmit force from the foundation to the object being moved. The force must be transmitted directly from the foundation to the object. The two weak points are the point where the bottom of the jack meets the foundation and the foundation itself. To prevent a collapse, ensure that all the components stay aligned to the center line between the foundation and the scaffold.

When the reach of a jack needs to be extended, additional blocking may be installed between the foundation and the bottom of the jack. However when more components are involved in the jacking action, the number of contact points is also increased. When the number of contact points increases, so does the possibility of collapse. Using fewer jacking components or pieces results in a safer installation.

> **CAUTION**
>
> Never modify a screw jack to try to gain additional scaffold height. A scaffold collapse due to an overextended screw jack can cause injury to the scaffold builder, other workers in the area, and possible damage to the equipment.

Follow these guidelines when using jacks:

- Inspect jacks before using them. Ensure that they are not damaged in any way.
- To prevent the scaffold from shifting or falling, position the jacks properly and raise the load evenly.
- Never place jacks directly on the ground when lifting. Use a solid footing.
- Never exceed the manufacturer's rated lifting capacity of the jack.

Additional Resources

"Scaffolding." OSHA. **www.osha.gov/SLTC/scaffolding/index.html**

Fall Protection and Scaffolding Safety: An Illustrated Guide. 2000. Grace Drennan Ganget. Government Institutes.

International Building Code®, Latest Edition. Falls Church, VA: International Code Council.

International Residential Code®, Latest Edition. Falls Church, VA: International Code Council.

3.0.0 Section Review

1. A manually operated lifting tool used to level and reposition scaffolds is a _____.

   a. scaffold lever
   b. shifter
   c. jack
   d. gandy bar

2. When using a jack to lift an object, force must be transmitted directly from the _____.

   a. jack to the object
   b. foundation to the object
   c. jack to the foundation
   d. object to the jack

4.0.0 FALL ARREST EQUIPMENT

Objective

Describe the proper use of personal fall arrest equipment used in the scaffolding industry.

   a. Describe the proper use of vertical and horizontal lifeline systems.

   b. Describe the proper use of self-retracting fall arrest devices.

Performance Task

Demonstrate the proper rigging of available fall arrest devices.

Trade Term

Columns: Upright building supports.

There are many types of fall arrest equipment available in the market today. The use of any of these systems must be designed by, and its use overseen by, a qualified person in accordance with OSHA standards.

> **WARNING!**
> Proper training is critical for authorized persons, qualified persons, and competent persons per OSHA and equipment manufacturer requirements.

One of the newer fall prevention systems is called the beamer. The beamer can be used as a singular anchor point or between two points as part of a pre-engineered horizontal anchoring device system for personnel working on high steel I-beams and columns. The beamer equipment includes a lifeline cable extended between two stationary clamps mounted onto the I- or H-beams of a structure. The lifeline should be connected with a certain amount of slack as recommended by the manufacturer. *Figure 21* shows the beamer clamps and cable installed in a variety of situations.

On columns, one clamp is installed on the column at a point below the work area, and the second clamp is installed on the column at a point above the work area. A cable is connected to each clamp, and the two clamps are separated enough to provide the cable sag specified by the manu-

facturer. After the clamps and cable are installed, workers can attach their safety lanyards to the horizontal cable, and then move around the work area near the vertical column.

4.1.0 Components of Lifeline Systems

The following devices or systems are used at various times when erecting or dismantling scaffolds:

- Self-retracting lifeline devices (SRD)s (sometimes called yo-yos), the primary lifeline used when working 20 feet or less above the ground
- Self-retracting device leading edge (SRD-LE), which has a built-in connector for twin-leg SRL direct attachment to a harness
- Ropes and rope grabs
- Beam straps/anchor slings (cheaters)
- Beamers or beam clamps
- Removable anchors and anchor plates
- Double-Y lanyards (ineffective for falls less than 20 feet)

> **WARNING!**
> A qualified person must ensure compatibility of fall protection devices such as harnesses, lanyards, and anchorage.

4.1.1 Horizontal Lifeline System

For horizontal beams, the two beamer clamps are installed on the vertical columns supporting the horizontal beam. The clamps are positioned a few feet above the horizontal beam and separated enough to provide the cable sag specified by the manufacturer. This type of scaffold installation must be handled by a qualified person. After the beamer clamps and cable are in place, workers can attach their safety lanyards to the horizontal cable and move freely in the area above the horizontal beam. The horizontally mounted cable also serves as a handrail for walking across the horizontal beam. If workers need to work below a horizontally mounted beam, the beamer clamps and cable can be lowered so that the cable extends horizontally across the space below the beam. After the beamer clamps and cable are installed, the workers can connect lanyards to the beamer cable and then move freely below the lifeline cable.

4.2.0 Self-Retracting Devices

Self-retracting devices provide for unrestricted movement and fall protection while climbing and descending ladders and similar equipment, or when working on multiple levels. Typically, they have a

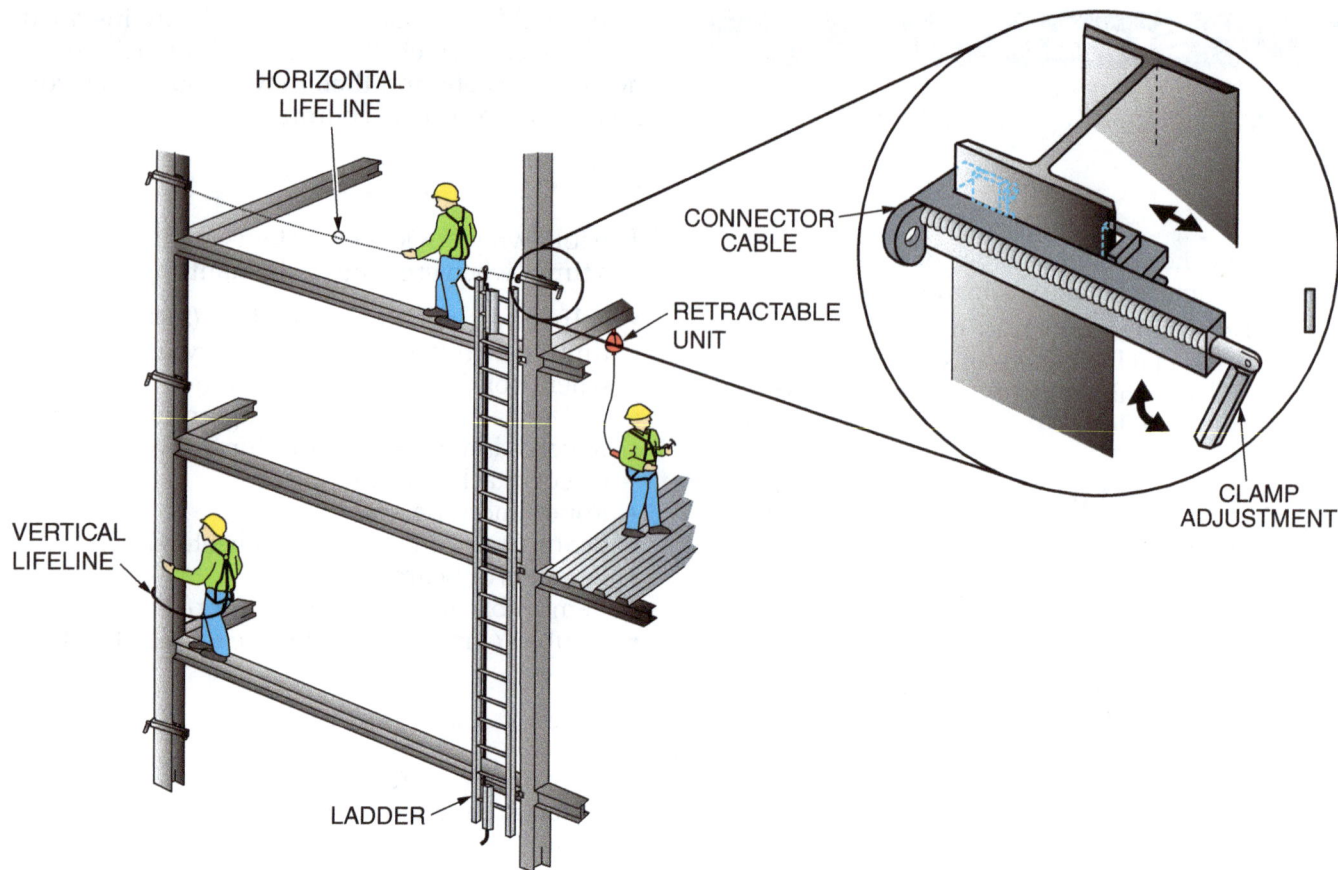

Figure 21 Variations of beamer equipment installations and uses.

25-foot to 100-foot galvanized steel cable that automatically takes up the slack in the attached lanyard, keeping the lanyard out of the worker's way. In the event of a fall, a centrifugal braking mechanism engages to limit the worker's free fall. Per OSHA requirements, self-retracting lifelines and lanyards that limit the free-fall distance to 2 feet (0.61 meter) or less must be able to support a minimum tensile load of 3,000 pounds (13.3 kilonewtons). Those that do not limit the free fall to 2 feet (0.61 meter) or less must be able to hold a tensile load of at least 5,000 pounds (22.2 kilonewtons).

4.2.1 Self-Retracting Device Leading Edge (SRD-LE) Models

The self-retracting device leading edge (SRD-LE) models are tested in conditions similar to those present in steel erection. These products may or may not be suitable for attachment to anchorages that are not located overhead. Follow manufacturer's instructions closely, and be sure to contact the manufacturer if there are any questions.

SRD-LE self-retracting devices include leading edge self-retracting lanyards (SRL). SRLs should be installed overhead. At lower heights, they may

not be able to handle the force inputs caused by the free-fall. Ensure you are accounting for any free-fall distance in your clearance calculations.

> **WARNING!**
> Follow the manufacturer's instructions carefully. If the manufacturer of choice does not include instructional guidance on anchorages below the level of the back D-ring, stop use and contact the manufacturer to obtain free-fall limits.

SRL edges can be hazardous when falling due to line failure caused by edge abrasion. The LE model is used in fall-arrest systems that could potentially result in the line passing over edges of structures such as I-beams, angle irons, decking, and so on. The energy absorber part (constituent) is positioned at the harness end of the SRL. This attachment is usually permanent. It serves as a supplemental deceleration device when the line passes over an edge during a fall in order to minimize the cutting effects of the edge. SRL-LE models are recommended for all SRL's used where edges are expected during the work.

Additional Resources

"Scaffolding." OSHA. **www.osha.gov/SLTC/scaffolding/index.html**

Fall Protection and Scaffolding Safety: An Illustrated Guide. 2000. Grace Drennan Ganget. Government Institutes.

International Building Code®, Latest Edition. Falls Church, VA: International Code Council.

International Residential Code®, Latest Edition. Falls Church, VA: International Code Council.

4.0.0 Section Review

1. For a horizontal cable used with a beamer, the amount of sag is _____.

   a. as little as possible
   b. determined by the safety committee
   c. 1 inch per 10 feet of length
   d. specified by the manufacturer

2. A self-retracting lifeline is also known as a _____.

   a. yo-yo
   b. snapback
   c. snatch block
   d. pull-tight

SUMMARY

Scaffold builders use a number of hand and power tools. It is necessary to be able to recognize these tools and be familiar with their safe and proper use. Remember that you should always use a tool for what it was designed to do. Never use a tool in a manner not recommended by the manufacturer. You will be required to store, inspect, and sometimes repair scaffold equipment. This module identifies some basic hand and power tools and provides basic guidelines for the storage and inspection of scaffold equipment.

1. One step to ensure that scaffolds are safe is _____.

    a. annual inspections
    b. regular cleaning
    c. proper storage
    d. posting warning signs

2. When storing scaffold equipment, remove items from the job site that have a _____.

    a. red tag
    b. yellow tag
    c. green tag
    d. multicolored tag

3. Ultimate responsibility for avoiding the use of damaged equipment rests with the _____.

    a. manufacturer
    b. OSHA inspector
    c. supervisor
    d. scaffold erector and user

4. Verifying that rivets are not loose or rusty is a step involved when inspecting _____.

    a. tubing frame legs
    b. bracket latches
    c. braces and guardrails
    d. tube-and-clamp scaffolds

5. Any alteration or modification of scaffold equipment must be approved by a _____.

    a. licensed engineer
    b. qualified manufacturer's representative
    c. supervisor
    d. site safety officer

6. The most common length for a pocket tape is _____.

    a. 10 feet
    b. 15 feet
    c. 20 feet
    d. 25 feet

7. When setting bracing, a helpful tool is a level with a bubble tube that can be used to check an angle of _____.

    a. 33 degrees
    b. 45 degrees
    c. 66 degrees
    d. 90 degrees

8. String lines used to establish baselines when laying out scaffolds are usually made of _____.

    a. nylon
    b. cotton
    c. polyester
    d. braided fine wire

9. For cutting toeboards, the most convenient tool to use is the _____.

    a. circular saw
    b. saber saw
    c. handsaw
    d. reciprocating saw

10. Hammers with the heavier head weights are not recommended for scaffold work because they can _____.

    a. cause elbow damage
    b. be hard to control
    c. rapidly fatigue the user
    d. damage equipment

11. Another name for a crowbar is a _____.

    a. wrecking bar
    b. power bar
    c. long bar
    d. Wonder Bar™

12. A flat, 12"–18" bar used primarily for pulling nails is referred to as a _____.

    a. cat's paw
    b. Wonder Bar™
    c. short bar
    d. nail grabber

13. For twisting and cutting wire, the tool that is usually used is the _____.

    a. side cutter
    b. needle-nose pliers
    c. lineman's pliers
    d. end nipper

14. The box-end wrench _____.

    a. doesn't grip bolt heads as well as an open-end wrench
    b. has a ratcheting action
    c. can round off shoulders of a nut or bolt
    d. usually has a different size on each end

15. Compound lever shears can be used to cut metal as thick as _____.

    a. 12 gauge
    b. 16 gauge
    c. 18 gauge
    d. 20 gauge

16. The power tool most commonly used when erecting scaffolds is the _____.

    a. chain saw
    b. saber saw
    c. circular saw
    d. reciprocating saw

17. For cutting circles, curves, or irregular shapes, scaffold builders commonly use a _____.

    a. circular saw
    b. saber saw
    c. reciprocating saw
    d. handsaw

18. To plumb and level scaffolds on uneven ground, scaffold builders use _____.

    a. hydraulic cylinders
    b. mudsills
    c. screw jacks
    d. shims

19. The weak points in a jacking action are the point where the bottom of the jack meets the foundation and the _____.

    a. foundation itself
    b. jack screw
    c. jack lever
    d. scaffold frame

20. The self-retracting lifeline is used in situations where the distance to the ground is not more than _____.

    a. 10 feet
    b. 15 feet
    c. 20 feet
    d. 25 feet

Fill in the blank with the correct term that you learned from your study of this module.

1. A _____ action locks when moved in one direction and releases when moved in the other.

2. A surface is considered _____ when it is completely horizontal.

3. A surface is considered _____ when it is completely vertical.

4. Moving back and forth is a _____ action.

5. Upright building supports are known as _____.

Trade Terms

Columns
Level
Plumb

Ratcheting
Reciprocating

Trade Terms Introduced in This Module

Columns: Upright building supports.

Level: Completely horizontal; at right angles to plumb.

Plumb: Completely vertical, or straight up and down; at right angles to level.

Ratcheting: To lock when moving in one direction and release when moved in the other.

Reciprocating: Moving back and forth.

Additional Resources

This module presents thorough resources for task training. The following resource material is suggested for further study.

"Scaffolding." OSHA. **www.osha.gov/SLTC/scaffolding/index.html**

Fall Protection and Scaffolding Safety: An Illustrated Guide. 2000. Grace Drennan Ganget. Government Institutes.

International Building Code®, Latest Edition. Falls Church, VA: International Code Council.

International Residential Code®, Latest Edition. Falls Church, VA: International Code Council.

Figure Credits

Courtesy of PERI Formwork Systems, Inc., Module Opener

© Stanley Tools, Figures 7, 11, 12, 13, and E01

Courtesy of S4Carlisle Publishing Services, Figures 8B, 10, and 14

Stringliner by U.S. Tape Co., Figure 9

Klein Tools, Inc./**www.kleintools.com**, Figure 15

Kleton Manufacturing Inc., Figure 16

Stanley Black and Decker/DeWalt Industrial Co., Figures 17, 18, and 19

Section Review Answer Key

Answer	Section Reference	Objective
Section One		
1. d	1.1.1	1a
2. a	1.2.0	1b
3. b	1.3.3	1c
4. a	1.3.3	1c
Section Two		
1. a	2.1.6	2a
2. b	2.2.1	2b
Section Three		
1. c	3.0.0	3
2. b	3.2.0	3b
Section Four		
1. d	4.0.0	4a
2. a	4.1.0	4b

NCCER CURRICULA — USER UPDATE

NCCER makes every effort to keep its textbooks up-to-date and free of technical errors. We appreciate your help in this process. If you find an error, a typographical mistake, or an inaccuracy in NCCER's curricula, please fill out this form (or a photocopy), or complete the online form at **www.nccer.org/olf**. Be sure to include the exact module ID number, page number, a detailed description, and your recommended correction. Your input will be brought to the attention of the Authoring Team. Thank you for your assistance.

Instructors – If you have an idea for improving this textbook, or have found that additional materials were necessary to teach this module effectively, please let us know so that we may present your suggestions to the Authoring Team.

NCCER Product Development and Revision

13614 Progress Blvd., Alachua, FL 32615

Email: curriculum@nccer.org
Online: www.nccer.org/olf

❏ Trainee Guide   ❏ Lesson Plans   ❏ Exam   ❏ PowerPoints   Other _____

Craft / Level: _____   Copyright Date: _____

Module ID Number / Title: _____

Section Number(s): _____

Description: _____

_____

_____

_____

_____

Recommended Correction: _____

_____

_____

_____

_____

Your Name: _____

Address: _____

_____

Email: _____   Phone: _____

Trade Math

OVERVIEW

In order to certify that a scaffold is safe to use for a given task, or to select materials for a scaffold job, the scaffold builder must be able to calculate the actual or expected loads on the scaffold. The loads to be applied to the scaffold by workers, tools, equipment, and materials must be considered. The loads applied to the scaffold by scaffold decking, upper frames and decking, and the environmental conditions must also be taken into account. Each scaffold builder must be able to use basic mathematical calculations to determine the combined loads on a scaffold system.

Module Four

Trainees with successful module completions may be eligible for credentialing through the NCCER Registry. To learn more, go to **www.nccer.org** or contact us at **1.888.622.3720**. Our website has information on the latest product releases and training, as well as online versions of our *Cornerstone* magazine and Pearson's product catalog.

Your feedback is welcome. You may email your comments to **curriculum@nccer.org**, send general comments and inquiries to **info@nccer.org**, or fill in the User Update form at the back of this module.

This information is general in nature and intended for training purposes only. Actual performance of activities described in this manual requires compliance with all applicable operating, service, maintenance, and safety procedures under the direction of qualified personnel. References in this manual to patented or proprietary devices do not constitute a recommendation of their use.

Objectives

When you have completed this module, you will be able to do the following:

1. Explain how to calculate the area and linear dimensions of plane surfaces.
   a. Explain how to calculate the area of rectangles and circles.
   b. Explain how to calculate the perimeter or linear dimensions of structures.
2. Explain how to reference and use tables commonly used in the scaffolding trade to solve math problems.
   a. Explain how to reference and use comparative value tables.
   b. Explain how to reference and use mathematical tables.
3. Identify types of live and dead loads on scaffolds and explain how to calculate these loads.
   a. Identify common types of live and dead loads.
   b. Explain how to calculate equipment loads.
   c. Explain how to calculate human loads.
   d. Explain how to calculate material loads.
4. Explain how to calculate loads as to their placement on scaffold platforms.
   a. Explain how to calculate concentrated loads.
   b. Explain how to calculate distributed loads.
   c. Explain how to calculate cantilevered loads.

Performance Tasks

This is a knowledge-based module; there are no performance tasks.

Trade Terms

Allowable load
Concentrated load
Distributed load

Load
Safety factor

Industry-Recognized Credentials

If you're training through an NCCER-accredited sponsor, you may be eligible for credentials from NCCER's Registry. The ID number for this module is 31104-14. Note that this module may have been used in other NCCER curricula and may apply to other level completions. Contact NCCER's Registry at 888.622.3720 or go to **www.nccer.org** for more information.

Code Note

Codes vary among jurisdictions. Because of the variations in code, consult the applicable code whenever regulations are in question. Referring to an incorrect set of codes can cause as much trouble as failing to reference codes altogether. Obtain, review, and familiarize yourself with your local adopted code.

Contents

Topics to be presented in this module include:

Figures and Tables

SECTION ONE

1.0.0 CALCULATE THE AREA OF PLANE SURFACES

Objectives

Explain how to calculate the area and linear dimensions of plane surfaces.

a. Explain how to calculate the area of rectangles and circles.
b. Explain how to calculate the perimeter or linear dimensions of structures.

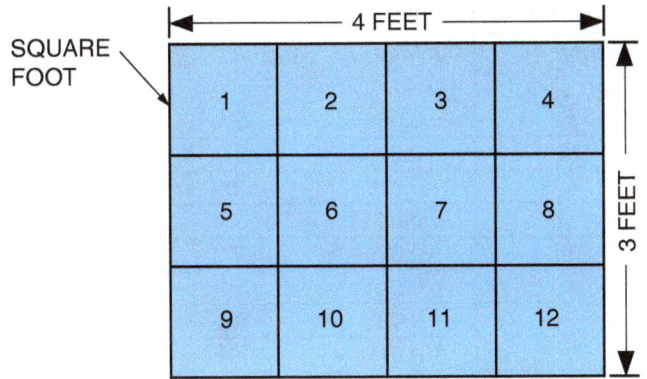

Figure 1 Rectangle.

Area is the amount of plane surface in a defined space. The amount of surface for any given area can be calculated for practical applications, such as the following:

- Room layout
- Material estimates
- Cost estimates
- Sizes of stock or parts

1.1.0 Various Shapes

Area is measured by using standard areas of smaller units, such as the square inch or square foot. A square inch is a surface enclosed by a square that is one inch on each side. A square foot is enclosed by a square that is one foot on each side. There are 144 square inches in one square foot (12 × 12 = 144). These units can be used to find the area of rectangles, squares, and circles.

1.1.1 Rectangles

A rectangle is a four-sided figure. The length of its opposite sides are equal, and all sides are joined at right angles. To find the area of a rectangle, it is necessary to find the number of surface units it contains. For example, if a rectangle contains three rows of square feet and there are four square feet in each row, the rectangle contains 12 square feet. *Figure 1* shows this rectangle.

The area of a rectangle can be found by multiplying the length by the width, or A = lw. Area is always expressed in square units.

In an area formula, all measurements must be expressed in (or converted to) the same type of units. For example, if a room is 20 feet, 6 inches long and 15 feet, 9 inches wide, the area of the floor is found as follows:

$$A = lw$$
$$A = 20 \text{ feet, 6 inches} \times 15 \text{ feet, 9 inches}$$

Convert inches into decimal feet (see *Table 1*):

$$A = 20.5 \text{ feet} \times 15.75 \text{ feet}$$
$$A = 322.875 \text{ or } 322.9 \text{ square feet}$$

A square is a rectangle with four equal sides. A square that measures 6 inches on each side is called a six-inch square. The formula for finding the area of a rectangle can be used to find the area of a square. However, since each side is the same length, the length of any side can be multiplied by itself or A = S². This is read as "area equals the sides squared." For example, the area of a 6-inch square is found as follows:

$$A = S^2$$
$$A = 6 \times 6$$
$$A = 36 \text{ square inches}$$

1.1.2 Circles

The area of a circle is calculated using the formula A = πr². The letter A represents the area of the circle. The symbol π, called pi (pronounced *"pie"*) represents a number that has been calculated by mathematicians to use when working with circles. The rounded-off value of π is 3.14. The small letter "r" represents the radius of the circle. The radius of a circle is the distance from the center of the circle to the side of the circle. The radius can also be calculated as one-half the diameter. The diameter of a circle is the distance from one side to the other measured through the center of the circle. To find the area of a circle, the radius is multiplied by itself, or squared.

Table 1 Inches Converted to Decimals of a Foot

Inches	Decimals of a Foot	Inches	Decimals of a Foot	Inches	Decimals of a Foot	Inches	Decimals of a Foot
1/16	0.005	3 1/16	0.255	6 1/16	0.505	9 1/16	0.755
1/8	0.010	3 1/8	0.260	6 1/8	0.510	9 1/8	0.760
3/16	0.016	3 3/16	0.266	6 3/16	0.516	9 3/16	0.766
1/4	0.021	3 1/4	0.271	6 1/4	0.521	9 1/4	0.771
5/16	0.026	3 5/16	0.276	6 5/16	0.526	9 5/16	0.776
3/8	0.031	3 3/8	0.281	6 3/8	0.531	9 3/8	0.781
7/16	0.036	3 7/16	0.286	6 7/16	0.536	9 7/16	0.786
1/2	0.042	3 1/2	0.292	6 1/2	0.542	9 1/2	0.792
9/16	0.047	3 9/16	0.297	6 9/16	0.547	9 9/16	0.797
5/8	0.052	3 5/8	0.302	6 5/8	0.552	9 5/8	0.802
11/16	0.057	3 11/16	0.307	6 11/16	0.557	9 11/16	0.807
3/4	0.063	3 3/4	0.313	6 3/4	0.563	9 3/4	0.813
13/16	0.068	3 13/16	0.318	6 13/16	0.568	9 13/16	0.818
7/8	0.073	3 7/8	0.323	6 7/8	0.573	9 7/8	0.823
15/16	0.078	3 15/16	0.328	6 15/16	0.578	9 15/16	0.828
1	0.083	4	0.333	7	0.583	10	0.833
1 1/16	0.089	4 1/16	0.339	7 1/16	0.589	10 1/16	0.839
1 1/8	0.094	4 1/8	0.344	7 1/8	0.594	10 1/8	0.844
1 3/16	0.099	4 3/16	0.349	7 3/16	0.599	10 3/16	0.849
1 1/4	0.104	4 1/4	0.354	7 1/4	0.604	10 1/4	0.854
1 5/16	0.109	4 5/16	0.359	7 5/16	0.609	10 5/16	0.859
1 3/8	0.115	4 3/8	0.365	7 3/8	0.615	10 3/8	0.865
1 7/16	0.120	4 7/16	0.370	7 7/16	0.620	10 7/16	0.870
1 1/2	0.125	4 1/2	0.374	7 1/2	0.625	10 1/2	0.875
1 9/16	0.130	4 9/16	0.380	7 9/16	0.630	10 9/16	0.880
1 5/8	0.135	4 5/8	0.385	7 5/8	0.635	10 5/8	0.885
1 11/16	0.141	4 11/16	0.391	7 11/16	0.641	10 11/16	0.891
1 3/4	0.146	4 3/4	0.396	7 3/4	0.646	10 3/4	0.896
1 13/16	0.151	4 13/16	0.401	7 13/16	0.651	10 13/16	0.901
1 7/8	0.156	4 7/8	0.406	7 7/8	0.656	10 7/8	0.906
1 15/16	0.161	4 15/16	0.411	7 15/16	0.661	10 15/16	0.911
2	0.167	5	0.417	8	0.667	11	0.917
2 1/16	0.172	5 1/16	0.422	8 1/16	0.672	11 1/16	0.922
2 1/8	0.177	5 1/8	0.427	8 1/8	0.677	11 1/8	0.927
2 3/16	0.182	5 3/16	0.432	8 3/16	0.682	11 3/16	0.932
2 1/4	0.188	5 1/4	0.438	8 1/4	0.688	11 1/4	0.938
2 5/16	0.193	5 5/16	0.443	8 5/16	0.693	11 5/16	0.943
2 3/8	0.198	5 3/8	0.448	8 3/8	0.698	11 3/8	0.948
2 7/16	0.203	5 7/16	0.453	8 7/16	0.703	1 17/16	0.953
2 1/2	0.208	5 1/2	0.458	8 1/2	0.708	11 1/2	0.958
2 9/16	0.214	5 9/16	0.464	8 9/16	0.714	11 9/16	0.964
2 5/8	0.219	5 5/8	0.469	8 5/8	0.719	11 5/8	0.969
2 11/16	0.224	5 11/16	0.474	8 11/16	0.724	11 11/16	0.974
2 3/4	0.229	5 3/4	0.479	8 3/4	0.729	11 3/4	0.979
2 13/16	0.234	5 13/16	0.484	8 13/16	0.734	11 13/16	0.984
2 7/8	0.240	5 7/8	0.490	8 7/8	0.740	11 7/8	0.990
2 15/16	0.245	5 15/16	0.495	8 15/16	0.745	11 15/16	0.995
3	0.250	6	0.500	9	0.750	12	1.000

Example 1: Calculate the area of a circle that is 4 feet in diameter.

Step 1 Calculate the radius. Remember, the radius is half the diameter.

$$4 \text{ feet} \div 2 = 2 \text{ feet}$$

Step 2 Square the radius (multiply 2 by itself).

$$2 \times 2 = 4$$

Step 3 Multiply 4 by π (3.14), which equals 12.56 square feet.

In this case, $A = \pi r^2$ works out like this: $A = 3.14 \times 2 \times 2$.

Example 2: Calculate the area of a circle that has a radius of 15 feet.

$$A = \pi r^2$$
$$A = 3.14 \times 15 \text{ feet} \times 15 \text{ feet}$$
$$A = 3.14 \times 225 \text{ square feet}$$
$$A = 706.5 \text{ square feet}$$

Figure 2 illustrates these examples for calculating the area of a circle.

1.2.0 Calculating Perimeters

A scaffold builder will often need to calculate the perimeter or the distance around the outside of a structure in order to calculate the total length of scaffold needed to wrap around the structure. There are two ways of determining the perimeter of a building or other structure: measure around it with a tape measure or calculate it. If the building is square (all four sides are equal) measure one side and multiply the distance by 4. The formula is ($P = S \times 4$), where P is perimeter and S is the length of one side. If the structure is rectangular, measure one short side and one long side, add these distances, and then double the sum. $P = (S1 + S2) \times 2$, where P is the perimeter and S1 and S2 are the length of the sides.

The perimeter of a circle is called the circumference. The formula for calculating the circumference of a circle is $C = \pi d$. The letter C is the circumference, π (pi) is 3.14, and the letter D is the diameter of the circle.

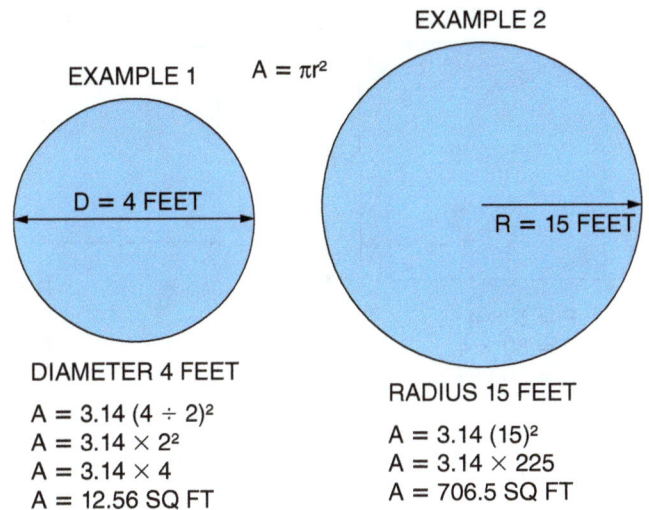

EXAMPLE 1 $A = \pi r^2$ EXAMPLE 2

D = 4 FEET

R = 15 FEET

DIAMETER 4 FEET

$A = 3.14 (4 \div 2)^2$
$A = 3.14 \times 2^2$
$A = 3.14 \times 4$
$A = 12.56$ SQ FT

RADIUS 15 FEET

$A = 3.14 (15)^2$
$A = 3.14 \times 225$
$A = 706.5$ SQ FT

31104-14_F02.EPS

Figure 2 Examples of calculating the area of a circle.

Figure 3 shows three examples of calculating perimeters.

Calculate the perimeter of each of the shapes shown in *Figure 3* for some practice at calculating perimeters. An explanation of each calculation follows the figure.

The first example shows a square that measures 50 feet on each side.

$$P = S \times 4$$
$$P = 50 \times 4$$
$$P = 200 \text{ feet}$$

The second example shows a rectangle that is 25 feet wide and 65 feet long.

$$P = (S1 + S2) \times 2$$
$$P = (25 + 65) \times 2$$
$$P = 90 \times 2$$
$$P = 180 \text{ feet}$$

The third example shows a circle that is 15 feet in diameter.

$$C = \pi d$$
$$C = 3.14 \times 15$$
$$C = 47.1 \text{ feet}$$

EXAMPLE 1 – SQUARE

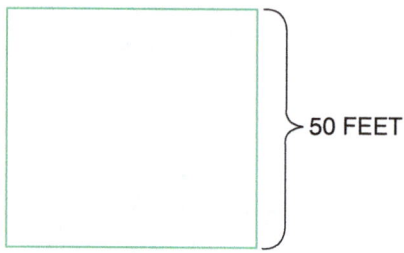

50 FEET

P = S × 4
P = 50 × 4
P = 200 FT

EXAMPLE 2 – RECTANGLE

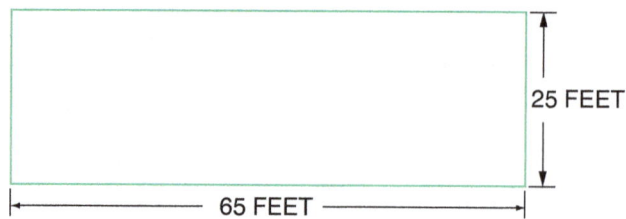

25 FEET

65 FEET

P = (S1 + S2) × 2
P = (25 + 65) × 2
P = 90 × 2
P = 180 FT

EXAMPLE 3 – CIRCLE

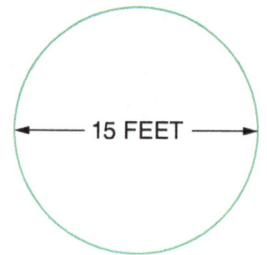

15 FEET

C = πD
C = 3.14 × 15
C = 47.1 FT

31104-14_F03.EPS

Figure 3 Calculating perimeters.

1.0.0 Section Review

1. An area of 144 square inches is equal to _____.

   a. 1 cubic foot
   b. 0.39 square yard
   c. 1.4 square feet
   d. 1 square foot

2. The formula $A = S^2$ is used to calculate the area of a _____.

   a. triangle
   b. parallelogram
   c. square
   d. circle

SECTION TWO

2.0.0 USING TABLES

Objectives

Explain how to reference and use tables commonly used in the scaffolding trade to solve math problems.

a. Explain how to reference and use comparative value tables.
b. Explain how to reference and use mathematical tables.

Tables consist of two or more parallel columns of data. They can be read quickly and can present large amounts of data clearly and concisely. Handbooks of tables are frequently useful as information references and for solving mathematical problems. While tables vary in form, they are read following the same basic steps.

2.1.0 Comparative Value Tables

Tables are a quick source of data and information. The following sections explain comparative and mathematical tables.

The simplest types of tables provide comparative values of related quantities. These values come from the definitions of quantities in the tables. Comparative value tables include the following:

- Tables of measure
- Tables of weight
- Multiplication tables
- Tables of money

For example, one type of comparative value table is a table of linear measures. *Table 2* shows equivalent linear measures.

When solving mathematical problems, it is sometimes necessary to know the decimal equivalent of a fraction. *Table 3* shows decimal equivalents of some common fractions. In the table, the first column lists fractions. The second column lists the decimal equivalents in inches, and the third column lists decimal equivalents in millimeters. If any of these values are known, the others can be found quickly and easily.

Table 2 Equivalent Linear Measures

12 Inches	1 Foot
3 Feet	1 Yard
1,760 Yards	1 Mile

Table 3 Decimal Equivalents of Common Fractions

Fraction	Decimal Equivalent	
	Standard (In.)	Metric (Mm)
1/64	.015625	0.3969
1/32	.03125	0.7938
3/64	.046875	1.1906
1/16	.0625	1.5875
5/64	.078125	1.9844
3/32	.09375	2.3813
7/64	.109375	2.7781
1/8	.1250	3.1750
9/64	.140625	3.5719
5/32	.15625	3.9688
11/64	.171875	4.3656
3/16	.1875	4.7625
13/64	.203125	5.1594
7/32	.21875	5.5563
15/64	.234375	5.9531
1/4	.250	6.3500

2.2.0 Mathematical Tables

Mathematical tables can simplify or eliminate the long calculations that are often necessary in mathematical problems. If mathematical tables are used, solutions for larger units can easily be calculated. For example, *Table 4* shows the conversion of standard and metric cubic measures, using the basic units and converting them into equivalent units of measure.

As an example, convert 7 cubic feet into gallons:

Step 1 Find the quantity in the first column.

7 units

Step 2 Find the column heading under which the correct conversion unit is listed.

Cubic Feet to Gallons

Step 3 Find the number in this column that is in the same row with the units from Step 1.

Answer: 52.36

If the number of units to be converted is not on this type of table, it can be calculated using simple addition or multiplication. For example, to convert 40 cubic feet to cubic meters, (one cubic foot equals 0.0283 cubic meters), multiply 0.0283 by 40 cubic feet which equals 1.13 cubic meters. (See the *Appendix* for conversion factors.)

Table 4 Conversion of Standard and Metric Cubic Measures

Unit(s)	Cubic Inches to Cubic Centimeters	Cubic Centimeters to Cubic Inches	Cubic Feet to Cubic Meters	Cubic Meters to Cubic Feet	Cubic Yards to Cubic Meters	Cubic Meters to Cubic Yards	Gallons to Cubic Feet	Cubic Feet to Gallons
1	16.39	0.06102	0.0283	35.31	0.7646	1.308	0.1337	7.481
2	32.77	0.1220	0.05663	70.63	1.529	2.616	0.2674	14.96
3	49.16	0.1831	0.08495	105.9	2.294	3.924	0.4010	22.44
4	65.55	0.2441	0.1133	141.3	3.058	5.232	0.5347	29.92
5	81.94	0.3051	0.1416	176.6	3.823	6.540	0.6684	37.40
6	98.32	0.3661	0.1699	211.9	4.587	7.848	0.8021	44.88
7	114.7	0.4272	0.1982	247.2	5.352	9.156	0.9358	52.36
8	131.1	0.4882	0.2265	282.5	6.116	10.46	1.069	59.84
9	147.5	0.5492	0.2549	371.8	6.881	11.77	1.203	67.32

2.0.0 Section Review

1. The simplest types of tables are _____.

   a. comparative tables
   b. logarithmic tables
   c. mathematical tables
   d. conversion tables

2. Tables can simplify long mathematical calculations.

   a. True
   b. False

SECTION THREE

3.0.0 LIVE AND DEAD LOADS ON SCAFFOLDS

Objectives

Identify types of live and dead loads on scaffolds and explain how to calculate these loads.

    a. Identify common types of live and dead loads.
    b. Explain how to calculate equipment loads.
    c. Explain how to calculate human loads.
    d. Explain how to calculate material loads.

Trade Term

Load: The weight or applied force; expressed in pounds or kilograms.

To prevent equipment failure or collapse, a scaffold cannot be loaded with more of a load, or weight, than it is designed to carry. The live load is the combined weight of workers, and the dead load is anything else contributing to the total weight imposed on the scaffold.

3.1.0 Common Scaffold Loads

A load on a scaffold can be a worker, the worker's tools, the materials the worker must have to do the job, the scaffold components, and environmental conditions such as snow, ice, or wind. In short, a load is anything that will exert a force on the scaffold. All of these factors must enter into the scaffold builder's calculations for a total scaffold load.

3.2.0 Calculating Equipment Loads

The equipment load is the weight of the actual scaffold components including the weight of the platforms. These are the first forces that bear on the scaffold legs. Each scaffold manufacturer lists the weight of each component with the shipping documentation. *Figure 4* shows a sketch of a four-level scaffold built with system components.

31104-14_F04.EPS

Figure 4 Four-level scaffold built with system components.

3.3.0 Calculating Human Loads

Personnel loads are considered live loads because they are loads that are likely to move around on the scaffold. In order to simplify the personnel load calculations, OSHA has established a standard weight of 250 pounds for one worker with the tools likely to be carried.

Scaffold designs fall into three categories: light-, medium-, and heavy-duty. Some examples of tasks that might be performed on each category of scaffold are as follows:

- *Light-duty* –25 pounds per square foot (equivalent to one 250-pound person per plank)
  - Painting
  - Cleaning
  - Window washing
  - Light maintenance

- *Medium-duty* – 50 pounds per square foot (equivalent to two 250-pound persons per plank)
  - Bricklaying
  - Drywalling
  - Equipment storage
- *Heavy-duty* – 75 pounds per square foot (equivalent to three 250-pound persons per plank)
  - Stone setting
  - Heavy-machinery installation
  - Equipment storage

The scaffold builder should determine the use of the scaffold and calculate the personnel loads that will be present on the scaffold during use.

3.4.0 Calculating Material Loads

Material loads are those items placed on a scaffold that the worker needs to perform the task such as brick, block, tools, and equipment. It is often difficult to accurately determine the total load at any particular time, so the scaffold builder must consider the maximum load that will be on the scaffold during its use.

Calculating material loads requires knowledge of the weights of some construction materials. *Table 5* gives the weights of some common materials that may be placed on a scaffold. The weights are shown in pounds per cubic foot. This means that a block of a material measuring one-foot square and one-foot high will weigh a given amount.

For example, the weight of common brick is given as 120 pounds per cubic foot. The pallet may measure 4 feet by 4 feet and be stacked 3 feet high. To calculate the total cubic feet, multiply length by width by height, or $4 \times 4 \times 3 = 48$ cubic feet. With a weight of 120 pounds per cubic foot, the total weight of the pallet of bricks is 5,760 pounds ($48 \times 120 = 5,760$).

The scaffold builder should be able to use the information in *Table 4* and basic math calculations to estimate the load placed on a scaffold.

To practice the calculations using the information in *Table 5*, calculate the weight involved in the following situation: A worker is removing cement, stone, and sand from the side of a large tank. The material is placed into lightweight plastic bags that hold one cubic yard of material. Calculate the weight of each filled bag.

The following information is known or given in *Table 5*:

- One yard equals 3 feet.
- A cubic yard is 3 feet wide by 3 feet long by 3 feet high.
- The volume of the bag is given as 1 cubic yard.
- The weight of cement, stone, and sand is 144 pounds per cubic foot.
- The weight of the bag is negligible.

The following information must be calculated:

- Number of cubic feet in a cubic yard.
- Total weight of cement, stone, and sand per bag.

Perform the following steps to calculate the weight of each bag:

Step 1 Calculate the number of cubic feet per cubic yard by cubing the number of feet in a yard.

3 feet × 3 feet × 3 feet = 27 cubic feet

Step 2 Calculate the weight of each bag by multiplying the number of cubic feet by the weight of cement, stone, and sand per cubic foot.

27 cubic feet × 144 pounds per cubic foot = 3,888 pounds

Table 5A Weights of Common Materials (1 of 2)

Substance	Weight (Lb Per Cu Ft)	Substance	Weight (Lb Per Cu Ft)
Ashlar Masonry		**Minerals**	
Granite, Syenite, Gneiss	165	Barytes	281
Limestone, Marble	160	Basalt	184
Sandstone, Bluestone	140	Bauxite	159
		Borax	109
Mortar Rubble Masonry		Chalk	137
Granite, Syenite, Gneiss	155	Clay, Marl	137
Limestone, Marble	150	Dolomite	181
Sandstone, Bluestone	130	Feldspar, Orthoclase	159
		Gneiss, Serpentine	159
Dry Rubble Masonry		Granite, Syenite	175
Granite, Syenite, Gneiss	130	Greenstone, Trap	187
Limestone, Marble	125	Gypsum, Alabaster	159
Sandstone, Bluestone	110	Hornblende	187
		Limestone, Marble	165
Brick Masonry		Magnesite	187
Pressed Brick	140	Phosphate Rock, Apatite	200
Common Brick	120	Porphyry	172
Soft Brick	100	Pumice, Natural	40
		Quartz, Flint	165
Concrete Masonry		Sandstone, Bluestone	147
Cement, Stone, Sand	144	Shale, Slate	175
Cement, Slag, Etc.	130	Soapstone, Talc	169
Cement, Cinder, Etc.	100		
		Stone, Quarried, Piled	
Various Building Materials		Basalt, Granite, Gneiss	96
Ashes, Cinders	40–45	Limestone, Marble, Quartz	95
Cement, Portland, Loose	90	Sandstone	82
Cement, Portland, Set	183	Shale	92
Lime, Gypsum, Loose	53–64	Greenstone, Hornblende	107
Mortar, Set	103		
Slags, Bank Slag	67–72	**Bituminous Substance**	
Slags, Bank Screenings	98–117	Asphaltum	81
Slags, Machine Slag	96	Coal, Anthracite	97
Slags, Slag Sand	49–55	Coal, Lignite	84
		Coal, Peat, Turf, Dry	78
		Coal, Charcoal, Pine	47
Earth, Etc., Excavated		Coal, Charcoal, Oak	23
Clay, Dry	63	Coal, Coke	33
Clay, Damp, Plastic	10	Graphite	75
Clay And Gravel, Dry	100	Paraffine	131
Earth, Dry, Loose	76	Petroleum	56
Earth, Dry, Packed	95	Petroleum, Refined	54
Earth, Moist, Loose	78	Petroleum, Benzine	50
Earth, Moist, Packed	96	Petroleum, Gasoline	46
Earth, Mud, Flowing	108	Pitch	69
Earth, Mud, Packed	115	Tar, Bituminous	75
Riprap, Limestone	80–85		
Riprap, Sandstone	90	**Coal And Coke, Piled**	
Riprap, Shale	105	Coal, Anthracite	47–58
Sand, Gravel, Dry, Loose	90–105	Coal, Bituminous, Lignite	40–54
Sand, Gravel, Dry, Packed	100–120	Coal, Peat, Turf	20–26
Sand, Gravel, Wet	118–120	Coal, Charcoal	10–14
		Coal, Coke	23–32

The specific gravities of solids and liquids refers to water at 4°C, those of gases to air at 0°C and 760 mm pressure. The weights per cubic foot are derived from average specific gravities, except where stated that weights are for bulk, heaped or loose material, etc.

Table 5B Weights of Common Materials (2 of 2)

Substance	Weight (Lb Per Cu Ft)	Substance	Weight (Lb Per Cu Ft)
Metals, Alloys, Ores		**Timber, U.S. Seasoned**	
Aluminum, Cast, Hammered	165	Moisture Content By Weight:	
Brass, Cast, Rolled	534	Seasoned Timber 15 To 20%	
Bronze, 7.9 To 14%	509	Green Timber Up To 30%	
Bronze, Aluminum	481	Ash, White, Red	40
Copper, Cast, Rolled	556	Cedar, White, Red	22
Copper Ore, Pyrites	262	Chestnut	41
Gold, Cast, Hammered	1205	Cypress	30
Iron, Cast, Pig	450	Fir, Douglas Spruce	32
Iron, Wrought	485	Fir, Eastern	32
Iron, Spiegel Eisen	468	Elm, White	46
Iron, Ferro Silicone	437	Pine, Oregon	32
Iron Ore, Hematite	325	Pine, Red	30
Iron Ore, Hematite In Bank	160–180	Pine, White	26
Iron Ore, Hematite Loose	130–160	Pine, Yellow, Long-Leaf	44
Iron Ore, Limonite	237	Pine, Yellow, Short-Leaf	38
Iron Ore, Magnatite	172	Poplar	30
Lead	710	Redwood, California	26
Lead Ore, Galena	465	Spruce, White, Black	27
Magnesium, Alloys	112	Walnut, Black	38
Manganese	475	Walnut, White	26
Manganese Ore, Pyrolusite	259		
Mercury	849	**Various Liquids**	
Monel Metal	556	Alcohol, 100%	49
Nickel	565	Acids, Muriatic 40%	75
Platinum, Cast, Hammered	1330	Acids, Nitric 91%	94
Silver, Cast, Hammered	459	Acids, Sulfuric 87%	112
Steel, Rolled	490	Lye, Soda 86%	106
Tin, Cast, Hammered	459	Oils, Mineral, Lubricants	57
Tin Ore, Cassiterite	418	Water, 4°C Max. Density	62.428
Zinc, Cast, Rolled	440	Water, 100°C	59.830
Zinc Ore, Blended	253	Water, Ice	56
		Water, Snow, Fresh Fallen	8
Various Solids		Water, Sea Water	64
Glass, Common	156		
Glass, Plate Or Crown	161	**Gases**	
Glass, Crystal	184	Ammonia	.0478
Leather	59	Carbon Dioxide	.1234
Paper	58	Carbon Monoxide	.0781
Rubber Goods	94	Gas, Illuminating	.028–.036
Salt, Granulated, Piled	48	Gas, Natural	.038–.039
Sulphur	125	Hydrogen	.00559
		Nitrogen	.0784
		Oxygen	.0892

The specific gravities of solids and liquids refers to water at 4°C, those of gases to air at 0°C and 760 mm pressure. The weights per cubic foot are derived from average specific gravities, except where stated that weights are for bulk, heaped or loose material, etc.

3.0.0 Section Review

1. The weight of materials placed on a scaffold is considered a _____.

   a. proportional load
   b. dead load
   c. transient load
   d. live load

2. A load is anything, other than weight of the workers, that exerts a force on the scaffold.

   a. True
   b. False

3. A typical use for a medium-duty scaffold would be _____.

   a. window washing
   b. stone setting
   c. drywalling
   d. painting

4. A scaffold builder must consider the maximum load that will be on a scaffold during its use.

   a. True
   b. False

4.0.0 PROPER LOAD PLACEMENT

Objectives

Explain how to calculate loads as to their placement on scaffold platforms.

   a. Explain how to calculate concentrated loads.
   b. Explain how to calculate distributed loads.
   c. Explain how to calculate cantilevered loads.

Trade Terms

Allowable load: The maximum load a component can carry without exceeding the safety factor required by OSHA.

Concentrated load: A load that extends over such a small part of the scaffold that it may be considered to act on a single point.

Distributed load: A load that is spread over a substantial portion of the scaffold surface.

Safety factor: The difference between the allowable load and the ultimate load that will cause an actual collapse of the scaffold.

When a scaffold has been installed, but has not been loaded in any way, materials and equipment can be placed at any point along the length of the platform. This section will address the safe placement of objects on the scaffold.

In addition to the actual amount of load on a scaffold, the location of the load is important. A load can be over a single point, distributed over a large area of the scaffold platform, concentrated near one end of a platform, or even suspended on a cantilevered platform such as an outrigger. Each of these situations presents the scaffold builder with a unique set of calculation requirements. Allowable loads for scaffold equipment are published by scaffold manufacturers. Allowable loads are typically listed for the leg loading and the bearer or horizontal member loading. Platform loads are transferred by the scaffold planks or platform to the bearers and then to the legs.

4.1.0 Calculating Concentrated Loads

Once the scaffold load is known or can be reliably estimated, the next step is to determine what portion of the load is shared by the bearers at each end of the scaffold platform.

If the load is in the middle of the platform or plank, the load will be shared equally between both bearers. Each of the two bearers will support one-half of the weight and one-half the weight of the scaffold platform. *Figure 5* shows a 100-pound load (weight) placed in the center of the span.

> **NOTE**
>
> Some installations require tarping. When calculating loads, the weight of the tarp must be included in the overall weight being carried by a scaffold.

If the 100-pound concentrated load is directly over one of the bearers, the entire weight of the load is supported by that bearer, and the weight of the scaffold plank is shared equally between both bearers. *Figure 6* shows an example of a concentrated load over one bearer.

To practice calculating bearer loads, calculate the load on each bearer when a 250-pound load is placed three feet from one bearer of a 10-foot scaffold platform. *Figure 7* shows the load and bearer locations.

From the previous examples we know that the weight is not evenly supported, because it is not in the center of the span; and it will not be supported entirely by one bearer because it is not directly over one bearer. The amount of weight supported by each bearer relates to the distance from the load to the bearer. Most of the weight will be supported by the closest bearer, and the remainder will be supported by the other bearer.

From the information given the following information is known:

- The weight of the load is 250 pounds.
- The load is placed on a scaffold platform with 10 feet between bearers.
- The load is placed three feet from one end (bearer) of the platform.

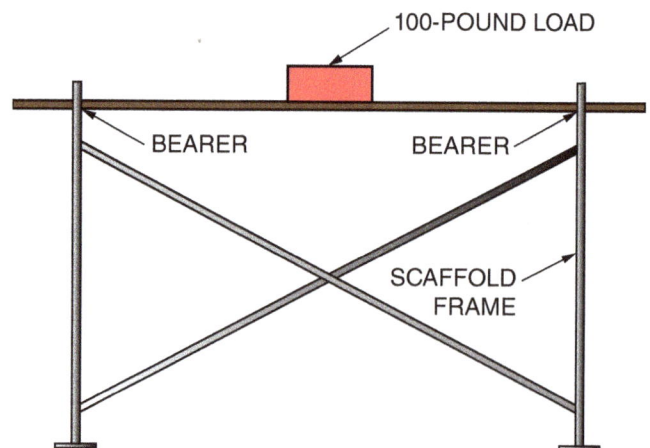

Figure 5 Load in center of span.

31104-14_F05.EPS

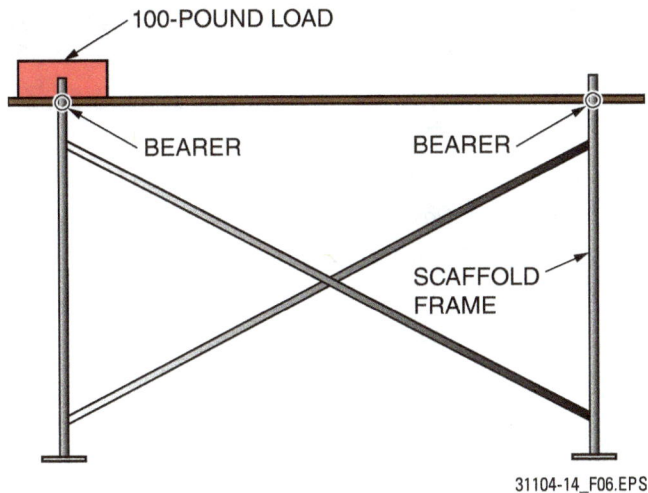

Figure 6 Load over bearer.

The following information must be calculated:

- The distance from the load to the far bearer.
- The amount of the load supported by each bearer.

Perform the following steps to calculate the weight supported by each bearer:

Step 1 Calculate the distance to the far bearer by subtracting the distance to the near bearer from the total length of the platform.

10 feet − 3 feet = 7 feet

Step 2 Calculate the proportional weight by dividing the total weight by the distance between the scaffold-platform bearers.

250 pounds ÷ 10 = 25 pounds

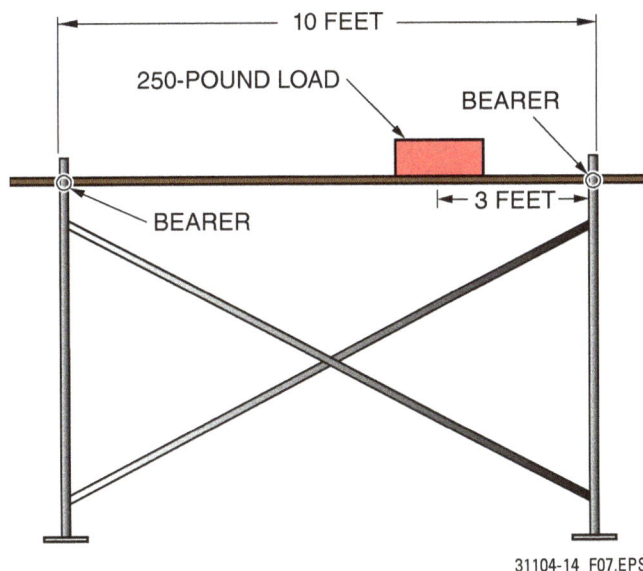

Figure 7 Load and bearer locations.

Step 3 Calculate the weight supported by the near bearer by multiplying the proportional weight by the distance to the far bearer.

25 pounds × 7 = 175 pounds

Step 4 Calculate the weight supported by the far bearer by multiplying the proportional weight by the distance to the close bearer.

25 pounds × 3 = 75 pounds

Step 5 Check your calculations by adding the two weights; the total should be the same as the original weight.

175 pounds + 75 pounds = 250 pounds

4.2.0 Calculating Distributed Loads

Distributed loads are loads that are spread over the entire scaffold platform. Materials to be used for construction, large windows or pieces of equipment, or a layer of snow or ice built up on the platform are examples of distributed loads. These loads will be supported by the bearers equally, the same as a load in the center of a span.

Use the following example and the information in *Table 5* to calculate the weights of the materials. Determine whether the scaffold is overloaded with siding or if the scaffold is still safe for the worker.

Example: A worker is installing red-cedar siding on a building. Each siding plank is ¾-inch thick, 12-inches wide, and 10-feet long. The worker's assistant has loaded 50 planks onto the center span of a light-duty scaffold using 5-foot wide by 7-foot long platforms. Use the OSHA standard worker weight of 250 pounds as the worker's weight.

From the information given above and in *Table 5*, the following information is known:

- The load is 50 planks of red-cedar siding.
- The weight of red cedar is 22 pounds per cubic foot.
- Each plank is ¾-inch thick, 12-inches wide, and 10-feet long.
- The load is placed on a 5-foot wide by 7-foot long platform.
- The scaffold is light-duty category, capable of supporting up to 25 pounds per square foot.
- The OSHA standard 250 pounds will be used as the worker's weight.

The following information must be calculated:

- The number of cubic inches in a cubic foot.
- The volume of each siding plank is in cubic inches: $T \times W \times L = V$, where T the thickness of the plank, W is the width of the plank, L is the length, and V is the volume. All dimensions are in inches.
- The safe working load of the scaffold platform: $L \times W \times P = SWL$, where L is the length of the platform, W is the width of the platform, P is the pounds per square foot allowed on the platform, and SWL is the safe working load.
- The weight of the loads.
- The weight of the loads compared to the safe working load of the scaffold platform.
- The number of boards to be removed, if any.

Perform the following steps to determine if the scaffold is overloaded or safe for the worker. In this example, all calculations have been rounded to one decimal point.

Step 1 Find the weight of red cedar from *Table 5*.

22 pounds per cubic foot

Step 2 Calculate the number of cubic inches in a cubic foot.

12 inches × 12 inches × 12 inches
= 1,728 cubic inches

Step 3 Each board is 0.75-inch (¾-inch) thick, 12-inches wide, and 120-inches (10-feet) long. To calculate the volume of each board, multiply the thickness (0.75 inch) by the width (12 inches) by the length (120 inches).

0.75 inch × 12 inches × 120 inches
= 1,080 cubic inches

Step 4 Calculate the volume of all 50 boards by multiplying the volume of each board (1,080 cubic inches) by the total number of boards.

1,080 cubic inches × 50 = 54,000 cubic inches

Step 5 Calculate the number of cubic feet of red cedar by dividing the total cubic inches (54,000) by the number of cubic inches in a cubic foot (1,728).

54,000 cubic inches
÷ 1,728 cubic inches per cubic foot
= 31.25, rounded to 31.3 cubic feet

Step 6 Calculate the weight of the siding by multiplying the number of cubic feet (31.3) by the weight per cubic foot (22).

31.3 cubic feet × 22 pounds per cubic foot
= 688.6 pounds

Step 7 Calculate the load limit for the scaffold platform.

5 feet × 7 feet × 25 pounds per square foot
= 875 square feet

Step 8 Compare the siding weight (688.6 pounds) to the load limit of the scaffold platform.

875 pounds load limit −
688.6 pounds weight of siding = 186.4

> **NOTE**
> This is still within the safe limits of the scaffold.

Step 9 Calculate the total weight on the scaffold platform with the addition of the 250-pound worker.

688.6 pounds of siding planks
+ 250-pound worker
= 938.6 pounds total load

Step 10 Compare the total weight (938.6 pounds) to the load limit of the scaffold platform.

875 pound load limit − 938.6-pound total load
= −63.6 pounds

> **NOTE**
> The total weight is over the safe load limit for the scaffold platform.

Step 11 Calculate the volume of boards to remove by dividing the weight to be removed (63.6 pounds) by the weight per cubic foot of red cedar (22 pounds per cubic foot).

63.6 pounds ÷ 22 pounds per cubic foot
= 2.89, rounded to 2.9 cubic feet

Step 12 Calculate the number of boards to remove by converting the 2.9 cubic feet to be removed into cubic inches, and dividing by the number of the cubic inches per board (1,080 cubic inches).

2.9 cubic feet
× 1,728 cubic inches per cubic foot
= 5,011.2 cubic inches
5,011.2 cubic inches
÷ 1,080 cubic inches per board
= 4.64, rounded to 5 boards minimum

4.3.0 Calculating Cantilevered Loads

Cantilevered loads are loads that are suspended by a bracket or beam extending beyond the supporting structure. These include end and side brackets mounted on built-up scaffolds as well as outrigger scaffolds. Loads placed on these cantilevered platforms can cause the scaffold to tip over. Cantilevered loads must be countered with additional bracing on built-up scaffolds and by anchoring or counterweights on outrigger scaffolds.

Cantilevered loads can be pictured like a seesaw. If there is a large weight on one end and a small weight on the other end, the position of the plank on the pivot point must be adjusted. The loads are balanced when the weight of the large load multiplied by its distance (moment arm) to the pivot point (fulcrum) equals the weight of the small load multiplied by its distance to the pivot point. The formula for this is $W \times A = w \times B$, where W is the large load, A is the distance from the large load to the fulcrum, w is the small load, and B is the distance from the small load to the fulcrum. *Figure 8* shows a balanced load.

There are two distinct problems to be considered when using a cantilevered-type scaffold. First is the stability of the scaffold. It is necessary to determine the amount of counterweight needed to prevent the scaffold from tipping and to produce an adequate safety factor. Second, the strength of the beam used to support the overhanging load must be adequate. In most cases, a professional engineer must determine the dimensions of the support beam.

Once the load on a cantilevered platform has been determined, the moment arm of the load must be determined so that the total tipping force can be calculated. Just as with the seesaw, enough weight must be provided on the other end of the beam to counter the tipping force of the loads. When calculating the counterweight or anchoring needed for a cantilevered-scaffold platform, the safety factor must be added to the equation. Generally, a safety factor of four is added to the calculations. The formula then becomes $W = (4 \times A \times L) \div B$, where W is the amount of counterweight, B is the distance between the counterweight and the fulcrum, L is the load on the platform, and A is the distance from the load to the fulcrum. *Figure 9* shows the counterweight configuration.

Scaffold builders must be extremely careful with counterweights. The calculation only considers the stability of the structure. The strength of the individual components has not been addressed. When assembling cantilevered-scaffold structures, be sure not to exceed the manufacturer's loading recommendations.

To practice the calculations for a cantilevered scaffold, consider the following example.

The gutter around the 40-foot diameter turret of a historical building with a high-pitched roof must be replaced. There are 12 windows located at regularly spaced locations around the circumference of the upper floor. The eaves can easily be reached from a platform level with the window sill. An outrigger scaffold has been selected for this task. The outrigger scaffold will be built on 20-foot beams. One beam will extend out each window. The cantilevered section of the scaffold will extend 4 feet beyond the window sill. Scaffold platform and guardrails mounted on the outrigger portion weigh 350 pounds per section. During use, there may be one or two workers with an OSHA weight of 250 pounds each, and 50 pounds of material on the platform.

From the information above the following is known:

- The entire beam length is 20 feet.
- The cantilevered section is 4 feet long.
- The weight of the platform alone is 350 pounds.

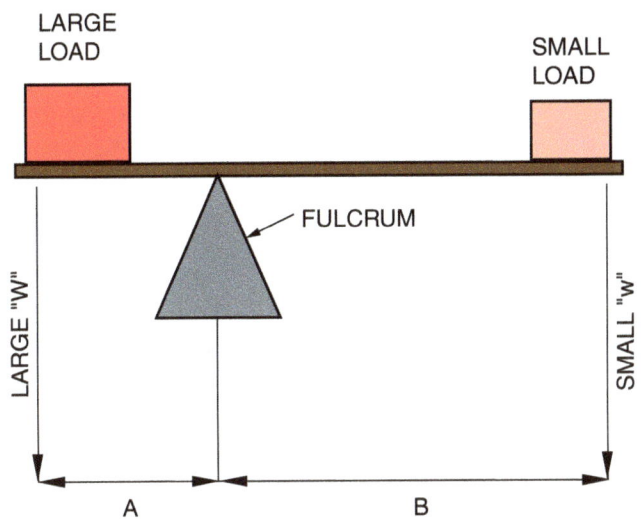

Figure 8 Balanced load.

31104-14_F08.EPS

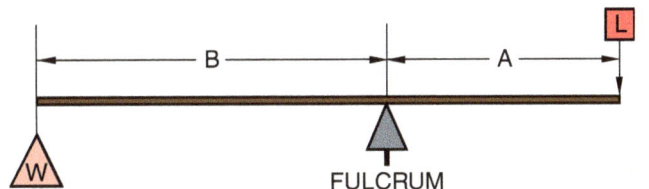

Figure 9 Counterweight configuration.

31104-14_F09.EPS

Wind Loads

Wind loads are unique. A wind load is simply another form of distributed load acting on the exposed area of the scaffold. The force of wind applied to the scaffold can change rapidly. The forces can act to push a scaffold toward the structure, away from the structure, or even lift planking or platforms off the scaffold. Wind loads and the steps taken to satisfy those loads must be determined by a professional engineer.

- The weight of workers on the platform may be as much as 500 pounds.
- The weight of materials on the platform will be 50 pounds.

The following must be calculated:

- The total weight on the cantilevered portion.
- The length of the counterweight moment arm.
- The weight of the counterweight.

Figure 10 shows the cantilevered-scaffold arrangement.

Perform the following steps to calculate the amount of counterweight required for this example:

Step 1 Calculate the total weight of the loaded cantilevered section.

> 350-pound platform
> + 500 pounds (2 workers)
> + 50 pounds of materials = 900 pounds

Step 2 Calculate the tipping force of the loaded cantilevered section.

> 900 pounds loaded cantilevered section
> × 4 feet moment arm
> = 3,600 feet per pound tipping force

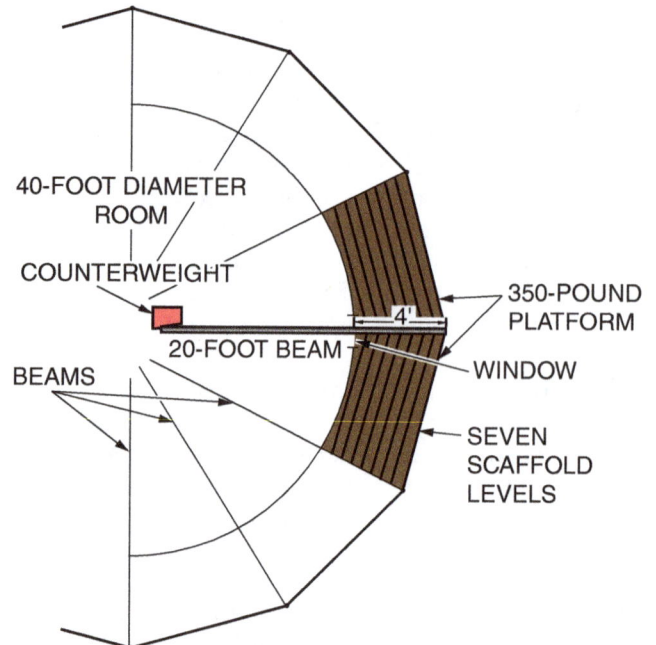

31104-14_F10.EPS

Figure 10 Cantilevered-scaffold arrangement.

Step 3 Calculate the length of the counterweight moment arm.

> 20-foot beam length −
> 4-foot cantilevered moment arm
> = 16-foot counterweight moment arm

Step 4 Calculate the amount of counterweight required using the following formula: $(4 \times A \times L) \div B = W$.

> $(4 \times 4$ feet $\times 900$ pounds$) \div 16$ feet
> = 900 pounds

> **NOTE**
>
> The 900-pound counterweight must be added to each outrigger in the scaffold system.

1. The largest load a scaffold component can support without exceeding OSHA-established limits is the _____.

   a. allowable limit
   b. stress limit
   c. maximum load
   d. safe load factor

2. An example of a distributed load is _____.

   a. a worker standing on a platform
   b. a layer of ice or snow
   c. two workers walking across a platform
   d. a stack of bricks over a bearer

3. Cantilevered loads must _____.

   a. not exceed 250 pounds
   b. be subtracted from the total load
   c. be balanced by a counterweight
   d. be within 3 feet of a support bracket

SUMMARY

It is the responsibility of the scaffold builder to erect a safe scaffold every time. In order to determine the correct type of scaffold to use and the arrangement of that scaffold, scaffold builders must be able to determine the load weight the scaffold is expected to support.

Loads include the weight of the scaffold components such as frames, braces, and platforms, the temporary loads of materials, and the live loads of the workers and their tools on the platforms. All these loads must be supported by the scaffold frames and legs. In addition to the weight of the loads, the scaffold builder must consider the location of the loads. Is the load concentrated in one location or spread over a large area? Is the load over a bearer, in the center of a span, or cantilevered outside the base of the scaffold? Each of these factors must be considered. The effects on the design and erection of the scaffold must be calculated.

Environmental conditions must also be considered. Wind, ice, and snow place loads on the scaffold frames and platforms. Additional ties may be required to ensure that the scaffold remains secure during these conditions.

Calculations required to determine the design or adjustments to a scaffold arrangement involve simple addition and multiplication skills. Calculations to determine areas and weights are also required. Tables are readily available to provide the scaffold builder with load limits and weights of common materials. OSHA regulations set limits to provide minimum safety factors for the loads placed on scaffolds. These regulations must always be followed. Providing an extra margin of safety by adding extra bracing, ties, or supports is often a good idea.

Review Questions

1. A rectangle is a figure with _____.
   a. three sides
   b. four sides
   c. five sides
   d. six sides

2. To calculate the area of a rectangle, _____.
   a. subtract width from length
   b. add together length of all four sides
   c. divide length by width, then multiply the result by 0.5
   d. multiply length by width

3. The radius of a circle is _____.
   a. one-half its diameter
   b. one-third its circumference
   c. twice its diameter
   d. two-thirds its circumference

4. A circle with a 9-foot diameter has an area of _____.
   a. 640 square feet
   b. 320 square feet
   c. 64 square feet
   d. 32 square feet

5. To find the perimeter of a rectangle, use the formula _____.
   a. $P = S \times 4$
   b. $P = (S1 + S2) \times 2$
   c. $P = (S1 - S2) + (S3 - S4)$
   d. $P = (S1 + S2)$

6. If a square building has an perimeter of 66 feet and Side 1 is 16 feet, 6 inches long, Side 2 has a length of _____.
   a. 8 feet, 9 inches
   b. 12 feet, 4 inches
   c. 16 feet, 6 inches
   d. 33 feet, 3 inches

7. The perimeter of a circle is called its _____.
   a. radius
   b. diameter
   c. tangent
   d. circumference

8. The weight of a block of material one-foot square and one-foot high is the weight per _____.
   a. cubic foot
   b. linear foot
   c. square foot
   d. standard foot

9. Tables showing relative values of measure, weight, or money are examples of _____.
   a. logarithmic tables
   b. comparative tables
   c. evaluative tables
   d. mathematical tables

10. Inches, meters, miles, and yards are all _____.
    a. linear measures
    b. area measures
    c. volume measures
    d. decimal measures

11. A measurement of 6.3500 millimeters is the decimal equivalent of _____ (refer to *Table 3*).
    a. ¹⁄₁₆ inch
    b. ³⁄₃₂ inch
    c. ⁹⁄₆₄ inch
    d. ¼ inch

12. The conversion factor from cubic inches to cubic centimeters is _____.
    a. 16.39
    b. 0.06102
    c. 1.308
    d. 35.31

13. A volume of 24 cubic meters is equivalent to _____ (refer to *Table 4*).
    a. 696 cubic feet
    b. 792 cubic feet
    c. 847 cubic feet
    d. 936 cubic feet

14. Because they are loads likely to move around on the scaffold, workers are considered _____.
    a. dynamic loads
    b. live loads
    c. variable loads
    d. allowable loads

15. Scaffolds intended for use in bricklaying would have a capacity equal to _____.

    a. one standard-weight worker
    b. two standard-weight workers
    c. three standard-weight workers
    d. four standard-weight workers

16. One-cubic foot of cast iron weighs more than one-cubic foot of _____ (refer to *Table 5*).

    a. cast bronze
    b. lead
    c. cast zinc
    d. rolled steel

17. California redwood boards in a bundle measuring 12 feet by 4 feet by 4 feet weigh _____ (refer to *Table 5*).

    a. 2,860 pounds
    b. 3,250 pounds
    c. 4,212 pounds
    d. 4,992 pounds

18. A pallet measuring 4 feet by 3 feet with sandstone ashlar masonry stacked 1 foot high, located in the middle of a platform, would subject each bearer to a load of _____ (refer to *Table 5*).

    a. 840 pounds
    b. 1,076 pounds
    c. 1,345 pounds
    d. 1,680 pounds

19. Loads spread over the entire scaffold platform and supported equally by the bearers are _____.

    a. dispersed loads
    b. balanced loads
    c. distributed loads
    d. permissible loads

20. The difference between the allowable load and the ultimate load that will cause the scaffold to collapse is the _____.

    a. safety factor
    b. structural allowance
    c. load limit
    d. stress percentage

Trade Terms Quiz

Fill in the blank with the correct term that you learned from your study of this module.

1. A(n) _____ is spread over a substantial portion of the scaffold surface.

2. A(n) _____ may be considered to act on a single point because it extends over such a small part of the scaffold.

3. The _____ is the maximum load a component can carry without exceeding the safety margin required by OSHA.

4. The difference between the allowable load and the ultimate load that will cause an actual collapse of the scaffold is the _____.

5. Weight or an applied force is known as a(n) _____.

Trade Terms

Allowable load
Concentrated load
Distributed load

Load
Safety factor

Trade Terms Introduced in This Module

Allowable load: The maximum load a component can carry without exceeding the safety factor required by OSHA.

Concentrated load: A load that extends over such a small part of the scaffold that it may be considered to act on a single point.

Distributed load: A load that is spread over a substantial portion of the scaffold surface.

Load: The weight or applied force; expressed in pounds or kilograms.

Safety factor: The difference between the allowable load and the ultimate load that will cause an actual collapse of the scaffold.

Appendix

Table A-1 SI Conversion Factors

APPROXIMATE CONVERSIONS TO SI UNITS					APPROXIMATE CONVERSIONS FROM SI UNITS				
Symbol	When You Know	Multiply By	To Find	Symbol	Symbol	When You Know	Multiply By	To Find	Symbol
LENGTH									
in	inches	25.4	millimeters	mm	mm	millimeters	0.039	inches	in
ft	feet	0.305	meters	m	m	meters	3.28	feet	ft
yd	yards	0.914	meters	m	m	meters	1.09	yards	yd
mi	miles	1.61	kilometers	km	km	kilometers	0.621	miles	mi
AREA									
in^2	square inches	645.2	square millimeters	mm^2	mm^2	square millimeters	0.0016	square inches	in^2
ft^2	square feet	0.093	square meters	m^2	m^2	square meters	10.764	square feet	ft^2
yd^2	square yards	0.836	square meters	m^2	m^2	square meters	1.195	square yards	yd^2
ac	acres	0.405	hectares	ha	ha	hectares	2.47	acres	ac
mi^2	square miles	2.59	square kilometers	km^2	km^2	square kilometers	0.386	square miles	mi^2
VOLUME									
fl oz	fluid ounces	29.57	milliliters	mL	mL	milliliters	0.034	fluid ounces	fl oz
gal	gallons	3.785	liters	L	L	liters	0.264	gallons	gal
ft^3	cubic feet	0.028	cubic meters	m^3	m^3	cubic meters	35.71	cubic feet	ft^3
yd^3	cubic yards	0.765	cubic meters	m^3	m^3	cubic meters	1.307	cubic yards	yd^3

NOTE: Volumes greater than 1,000 l should be shown in m^3.

APPROXIMATE CONVERSIONS TO SI UNITS					APPROXIMATE CONVERSIONS FROM SI UNITS				
MASS									
oz	ounces	28.35	grams	g	g	grams	0.035	ounces	oz
lb	pounds	0.454	kilograms	kg	kg	kilograms	2.202	pounds	lb
T	short tons (2000 lb)	0.907	megagrams (or "metric ton")	Mg (or "t")	Mg (or "t")	megagrams (or "metric ton")	1.103	short tons (2,000 lb)	T
TEMPERATURE (exact)									
°F	Fahrenheit temperature	5(F - 32)/9 or (F - 32)/1.8	Celsius temperature	°C	°C	Celsius temperature	1.8C + 32	Fahrenheit temperature	°F
FORCE AND PRESSURE OR STRESS									
lbf	pound-force	4.45	newtons	N	N	newtons	0.225	pound-force	lbf
lbf/in^2	pound-force per square inch	6.89	kilopascals	kPa	kPa	kilopascals	0.145	pound-force per square inch	lbf/in^2

* SI is the symbol for the International System of Units. Appropriate rounding should be made to comply with Section ASTM E380.

Additional Resources

This module presents thorough resources for task training. The following resource material is suggested for further study.

"Scaffolding." OSHA. **www.osha.gov/SLTC/scaffolding/index.html**

Fall Protection and Scaffolding Safety: An Illustrated Guide. 2000. Grace Drennan Ganget. Government Institutes.

International Building Code®, Latest Edition. Falls Church, VA: International Code Council.

International Residential Code®, Latest Edition. Falls Church, VA: International Code Council.

Figure Credits

Courtesy of PERI Formwork Systems, Inc., Module Opener

Section Review Answer Key

Answer	Section Reference	Objective
Section One		
1. d	1.1.0	1a
2. c	1.1.2	1a
Section Two		
1. a	2.1.0	2a
2. a	2.2.0	2b
Section Three		
1. b	3.0.0	3
2. b	3.1.0	3a
3. c	3.3.0	3c
4. a	3.4.0	3d
Section Four		
1. a	4.1.0	4a
2. b	4.2.0	4b
3. c	4.3.0	4c

NCCER CURRICULA — USER UPDATE

NCCER makes every effort to keep its textbooks up-to-date and free of technical errors. We appreciate your help in this process. If you find an error, a typographical mistake, or an inaccuracy in NCCER's curricula, please fill out this form (or a photocopy), or complete the online form at **www.nccer.org/olf**. Be sure to include the exact module ID number, page number, a detailed description, and your recommended correction. Your input will be brought to the attention of the Authoring Team. Thank you for your assistance.

Instructors – If you have an idea for improving this textbook, or have found that additional materials were necessary to teach this module effectively, please let us know so that we may present your suggestions to the Authoring Team.

NCCER Product Development and Revision
13614 Progress Blvd., Alachua, FL 32615

Email: curriculum@nccer.org
Online: www.nccer.org/olf

❏ Trainee Guide ❏ Lesson Plans ❏ Exam ❏ PowerPoints Other _____

Craft / Level: _____ Copyright Date: _____

Module ID Number / Title: _____

Section Number(s): _____

Description: _____

_____

_____

_____

_____

Recommended Correction: _____

_____

_____

_____

_____

Your Name: _____

Address: _____

_____

Email: _____ Phone: _____

31105-15

Supported Scaffolds

OVERVIEW

Most supported scaffolds are manufactured scaffolds made from tubular steel that can be fitted together to form complete sections. For light-duty use, scaffolds are sometimes made from aluminum or wood. Tubular welded-frame scaffolds, system, and tube-and-clamp scaffolds are the main types of supported scaffolds.

Module Five

Trainees with successful module completions may be eligible for credentialing through the NCCER Registry. To learn more, go to **www.nccer.org** or contact us at **1.888.622.3720**. Our website has information on the latest product releases and training, as well as online versions of our *Cornerstone* magazine and Pearson's product catalog.

Your feedback is welcome. You may email your comments to **curriculum@nccer.org**, send general comments and inquiries to **info@nccer.org**, or fill in the User Update form at the back of this module.

This information is general in nature and intended for training purposes only. Actual performance of activities described in this manual requires compliance with all applicable operating, service, maintenance, and safety procedures under the direction of qualified personnel. References in this manual to patented or proprietary devices do not constitute a recommendation of their use.

Objectives

When you have completed this module, you will be able to do the following:

1. Describe the safety considerations regarding supported scaffolds.
   a. Identify safety regulations for various types of supported scaffold systems.
2. Explain the basic principles of system scaffolds, and outline proper erection procedures.
   a. Explain the versatility of system scaffold components.
   b. Describe various system scaffold configurations.
   c. Outline the steps for proper erection of a system scaffold.
3. Explain the basic principles of tubular welded-frame scaffolds, and outline proper erection procedures.
   a. Identify common applications of tubular welded-frame scaffolds.
   b. Identify the components of tubular welded-frame scaffolds.
   c. Outline the steps for proper erection of a tubular welded-frame scaffold.
4. Explain the basic principles of tube-and-clamp scaffolds, and outline proper erection procedures.
   a. Identify common applications of tube-and-clamp scaffolds.
   b. Identify the components of tube-and-clamp scaffolds.
   c. Outline the steps for proper erection of a tube-and-clamp scaffold.
5. Identify other supported scaffold systems.
   a. Explain the basic principles of outrigger scaffolds, and outline proper erection procedures.
   b. Explain the basic principles of pump-jack scaffolds, and outline proper erection procedures.

Performance Task

Under the supervision of your instructor, you should be able to do the following:

1. Safely erect a section of two of the following types of scaffolds:
   - System scaffold
   - Tubular welded-frame scaffold
   - Tube-and-clamp scaffold

Trade Terms

Bearer	Cross bracing
Brace	Runner
Clamp	Screw jack

Industry-Recognized Credentials

If you're training through an NCCER-accredited sponsor, you may be eligible for credentials from NCCER's Registry. The ID number for this module is 31105-14. Note that this module may have been used in other NCCER curricula and may apply to other level completions. Contact NCCER's Registry at 888.622.3720 or go to **www.nccer.org** for more information.

Code Note

Codes vary among jurisdictions. Because of the variations in code, consult the applicable code whenever regulations are in question. Referring to an incorrect set of codes can cause as much trouble as failing to reference codes altogether. Obtain, review, and familiarize yourself with your local adopted code.

Contents

Topics to be presented in this module include:

Figures and Tables

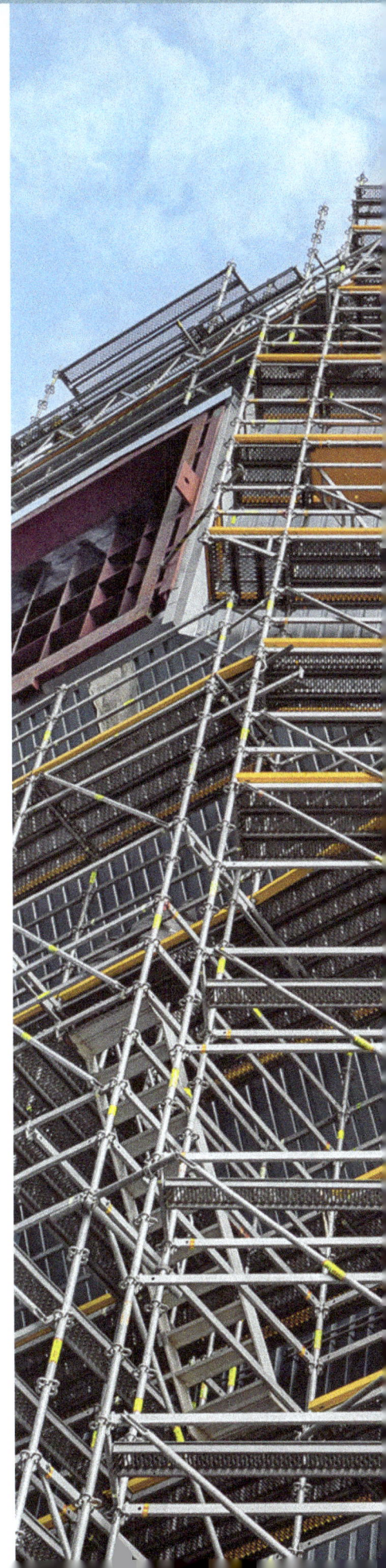

1.0.0 SAFETY CONSIDERATIONS

Objectives

Describe the safety considerations regarding supported scaffolds.

   a. Identify safety regulations for various types of supported scaffold systems.

Trade Terms

Bearer: A horizontal scaffold member (which can be supported by ledgers or runners) upon which the scaffold platform rests; joins scaffold uprights, posts, poles, and similar members.

Brace: A rigid connection that holds one scaffold member in a fixed position with respect to another member or to a building or structure.

Clamp: A device for locking together the tubes of a tube-and-clamp scaffold.

Cross bracing: Braces placed between opposite corners to keep the scaffold plumb and secure; also called transverse bracing.

Runner: The lengthwise horizontal spacing or bracing member that can support the bearers; also called a ledger or a ribbon.

As with all types of scaffolds, safety is priority number one for supported scaffolds. OSHA has a height limit of 125 feet for tube-and-clamp scaffolds. The height of scaffolds may be further limited by the number of working levels. Supported scaffolds higher than 125 feet must be designed by a professional engineer. This could vary in some locations, so always check local codes prior to planning erection.

1.1.0 Safety Regulations

While there are similarities in the safety guidelines for the various types of supported scaffolds, there are some unique considerations.

1.1.1 System Scaffolds

Guardrail systems are required on all scaffolds 10 feet or higher above a lower level as a means of safeguarding workers from fall hazards (this may vary by state). Guardrails may be required at elevations lower than 10 feet as part of a host employer rule, or contractual agreement. If a possible hazard exists, with greater than 6 feet of free fall, a guardrail or personal fall arrest system is required with a platform of any height.

A guardrail system consists of a toprail and a midrail. These rails must be placed on all open sides and ends of a scaffold platform. Regulations require toprails to be positioned at a height between 38 and 45 inches above the platform surface. A midrail must be placed halfway between the top edge of the toprail and the platform surface. If a 4-foot-high scaffold is positioned next to an open floor or elevator shaft that would allow a fall greater than 6 feet, the scaffold must have guardrails. Guardrails are also required if the scaffold is placed next to a sloping surface where a potential fall could be 10 feet or greater.

Many of the regulations regarding scaffold foundations, planking, guying, safety nets, and similar items apply to all types of supported scaffolds. For detailed safety information, review the *Trade Safety* module.

1.1.2 Tubular Welded-Frame Scaffolds

Solid sawn planks must be scaffold-grade and clearly marked. Scaffold planks must be free of saw kerfs and notches, and all nails must be removed at the time of assembly or disassembly. Other scaffold decking must be certified by the manufacturer to conform to OSHA regulations of 50 pounds per square foot or 75 pounds per square foot for medium-duty and heavy-duty scaffold ratings, respectively.

> **WARNING!**
>
> Wood planks that have been previously used as mudsills can never be used as platform decking. Since mudsills have often been in direct contact with the ground or exposed to added moisture, the strength of the planks may have been compromised.

Tubular welded-frame scaffolds must be secured or guyed to a rigid structure if the height exceeds 4 times the narrowest dimension of the base. To prevent movement, these scaffolds must be secured to the structure at vertical intervals not to exceed 20 feet for scaffolds with a base 3-feet wide or less, and 26 feet for scaffolds with a base over 3-feet wide. The horizontal interval is 30 feet regardless of the width of the base. Keep in mind that the addition of protective netting adds to the projected wind load, which will affect tieback point layout. An engineer should be consulted when netting is added to supported scaffolds.

If work is to be performed on lower levels or workers will be allowed to pass under the scaffold platform, toeboards must be provided on all open sides and edges of the platform. If materials are piled on the scaffold platform higher than the toeboards, protective netting should be placed between the toeboards and the guardrails to prevent any of the materials from falling to a lower level, if other means are not provided.

1.1.3 Tube-and-Clamp Scaffolds

Most of the general safety precautions that apply to tubular welded-frame scaffolds also apply to tube-and-clamp scaffolds: the height and locations of guardrails; bracing; guying of scaffolds more than four base-widths high or 30 feet in length; and the use of platforms and planks. The following additional safety guidelines are for the erection and use of tube-and-clamp scaffolds:

- When platforms are being moved from one level to the next level, the existing platform shall be left undisturbed until the new bearers have been set in place and braced prior to receiving the new platforms.
- Cross bracing, also called transverse bracing, forming an "X" across the width of the scaffold, must be installed at the scaffold ends and at least at every third set of legs horizontally (measured from only one end) and every fourth runner vertically. Bracing shall extend diagonally from the inner to the outer legs or runners upward to the next outer or inner legs or runners. *Figure 1* shows diagonal bracing and cross bracing. Cross bracing increases the stability and plumb of the scaffold. Building ties must be installed at the bearer levels between the transverse bracing.
- On straight-run scaffolds, longitudinal bracing across the inner and outer rows of legs must be installed diagonally in both directions, and extend from the base of the end legs upward to the top of the scaffold at approximately a 45-degree angle. On a scaffold whose length is greater than its height, such bracing must be repeated beginning at least every fifth leg. On a scaffold whose length is less than its height, such bracing must be installed from the base of the end legs upward to the opposite end post, then in alternating directions until reaching the top of the scaffold. Bracing must be installed as close as possible to the intersection of bearer and post or runner and post.

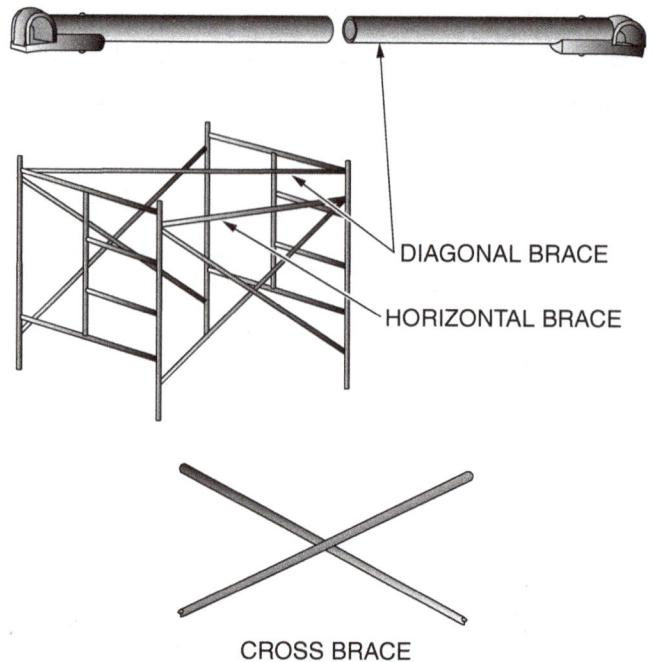

DIAGONAL BRACE

HORIZONTAL BRACE

CROSS BRACE

31105-14_F01.EPS

Figure 1 Diagonal bracing and cross bracing.

- Where conditions do not allow the braces to be attached to the legs, bracing must be attached to the runners as close to the legs as possible.
- Bearers must be installed transversely between legs, and when coupled to the legs, have the inboard clamp bear directly on the runner clamp. When the bearers are coupled to the runners, the clamps must be as close to the legs as possible.
- Bearers must extend beyond the legs and runners and provide full contact with the clamp.
- Runners must be installed along the length of the scaffold and located on both the inside and outside legs at level heights. (When tube-and-clamp toprails and midrails are used on outside legs, they may be used instead of outside runners.)
- Runners must be interlocked on straight runs to form continuous lengths, and be coupled to each post. The bottom runners and bearers must be located as close to the base as possible.
- Clamps must be made of a structural metal, such as drop-forged steel, malleable iron, or structural-grade aluminum. The use of gray cast-iron is not allowed.
- Tube-and-clamp scaffolds over 125 feet in height must be designed by a professional engineer, and be constructed and loaded in accordance with the design. Height may be limited by the number of planked and used levels.

Table 1 shows specific information covering post, runner, and brace and bearer sizes and runner spacing for light-duty, medium-duty, and heavy-duty tube-and-clamp scaffolds.

Table 2 lists the maximum number of working and planked levels for the three duty ratings of scaffold.

1.1.4 Outrigger and Pump-Jack Scaffolds

There are two significant safety considerations for workers employing outrigger scaffolds: support and fall protection. The outrigger scaffold may be used at any level of a building, from the first to the top floor, so there may be only a few feet or hundreds of feet for a worker to fall. Top rails, midrails, and toeboards are required on all outrigger scaffolds. A separate rigging point for a personal fall arrest system that meets all requirements must be provided. The interior anchor point of an outrigger beam must be rigidly secured to the structure of the building. This can be accomplished with anchor bolts inserted into the concrete slab of the floor, or attachment to a structural steel member. In any case, the beam must be secured from movement in any direction. The dangers of the beam's lifting are obvious, but

Table 2 Maximum Number of Planked Levels for Tube-and-Clamp Scaffolds

NUMBER OF WORKING LEVELS	MAXIMUM NUMBER OF ADDITIONAL PLANKED LEVELS			MAXIMUM HEIGHT OF SCAFFOLD (IN FEET)
	LIGHT-DUTY	MEDIUM-DUTY	HEAVY-DUTY	
1	16	11	6	125
2	11	1	0	125
3	6	0	0	125
4	1	0	0	125

31105-14_T02.EPS

lateral shifting or twisting could also result in the scaffold platform losing support and causing a serious fall.

Pump-jack scaffolds require the normal arrangement of toprails, midrails, and toeboards to protect the workers on or near the scaffold. Additional fall protection may be required as the platform is raised. The scaffold poles must be secured to the structure to prevent tipping. State-of-the-art scaffold equipment should be considered before using pump-jack or ladder-supported scaffolds. This is often provided by metal bracing or by wood framing.

Table 1 Minimum Size of Members for Tube-and-Clamp Scaffolds

	LIGHT-DUTY	MEDIUM-DUTY	HEAVY-DUTY
Maximum Intended Load	25 lbs/ft^2	50 lbs/ft^2	75 lbs/ft^2
Posts, Runners, And Braces	Nominal 2 In (1.90 in) OD Steel Tube or Pipe	Nominal 2 in (1.90 in) OD Steel Tube or Pipe	Nominal 2 in (1.90 in) OD Steel Tube or Pipe
Bearers	Nominal 2 in (1.90 in) OD Steel Tube or Pipe and a Maximum Post Spacing Of 4 Ft × 10 ft	NOMINAL 2 IN (1.90 IN) OD Steel Tube or Pipe and a Maximum Post Spacing of 4 ft × 7 ft Nominal 2½ in (2.375 IN) OD Steel Tube Or Pipe and a Maximum Post Spacing of 6 ft × 8 ft	Nominal 2½ in (2.375 in) OD Steel Tube or Pipe and a Maximum Post Spacing OF 6 FT × 6 ft
Maximum Runner Spacing Vertically	6 Feet, 6 Inches	6 Feet, 6 Inches	6 Feet, 6 Inches

31105-14_T01.EPS

Additional Resources

"Scaffolding." OSHA. **www.osha.gov/SLTC/scaffolding/index.html**

Fall Protection and Scaffolding Safety: An Illustrated Guide. 2000. Grace Drennan Ganget. Government Institutes.

International Building Code®, Latest Edition. Falls Church, VA: International Code Council.

International Residential Code®, Latest Edition. Falls Church, VA: International Code Council.

1.0.0 Section Review

1. Wood planks that have been used as mudsills cannot be re-used as platform decking.

   a. True
   b. False

2. OSHA limits the height of tube-and-clamp scaffolds to _____.

   a. 75
   b. 100
   c. 125
   d. 150

2.0.0 ERECTING SYSTEM SCAFFOLDS

Objectives

Explain the basic principles of system scaffolds, and outline proper erection procedures.

    a. Explain the versatility of system scaffold components.

    b. Describe various system scaffold configurations.

    c. Outline the steps for proper erection of a system scaffold.

Trade Term

Screw jack: A threaded, adjustable screw located between the legs and base plates or caster wheels of a scaffold; used to raise or lower parts of a scaffold to level it on uneven surfaces.

System scaffolds are specialized types of scaffolds that combine some of the versatility of tube-and-clamp scaffolds with the ease and speed of erection of the welded tubular-frame scaffolds. System scaffolds are sometimes called modular scaffolds. System scaffolds are well-suited for situations of limited access, obstructions, uneven surfaces, or nonrectangular shapes.

2.1.0 Versatility of System Scaffold Components

Since there are no specifications for the sizes of the system scaffold components or the connecting methods, components from different manufacturers must not be mixed, and strict attention must be paid to the manufacturer's recommendations. Some of the tubes are the same size as the tubes used in tube-and-clamp scaffolds, so these two systems may be able to be used together for additional flexibility. *Figure 2* shows a portion of a system scaffold.

System scaffolds are constructed using legs, horizontal and vertical diagonal braces, runners, bearers, bases, and screw jacks. System scaffolds can use the same platform as other supported scaffolds.

System scaffolds are similar to tube-and-clamp scaffolds because they are erected from standardized components. The legs, bearers, and runners are made from the same size tubing. The tubes can be made of steel or aluminum. The main difference is the method used to connect the components.

> **WARNING!**
>
> When using aluminum and steel tubing together, the maximum load limits for the weakest components must be used. To prevent personal injury or death, do not exceed the maximum load limit of scaffold components.

2.1.1 Vertical Legs

The legs, sometimes called standards or uprights, have special connectors welded to them at fixed, uniformly spaced intervals along the tube. Legs are available in many different lengths; the most common are 1, 2, or 3 meters (3 feet, 3 inches; 6 feet, 6 inches; and 9 feet, 10 inches). The connectors are usually spaced approximately 19½ inches apart along the entire length of the legs. These connectors allow other parts to be attached to the leg. The uniform spacing provides proper spacing for bearers for platform support, and for runners to be used as toprails and midrails for the platform. *Figure 3* shows some of the coupling methods.

There is a limited amount of flexibility with the connectors shown in *A* and *B* of *Figure 3*. These connectors provide connections only at right angles around the leg, restricting the arrangement to rectangular shapes similar to tube-and-clamp scaffolds. The connection shown in *C* uses a rosette connection that provides eight equally spaced holes around the leg.

The connections on the runners and bearers are slightly smaller than the holes. This allows even more flexibility in the angle of the connection. A connector with a ring welded to the leg is shown in *D* of *Figure 3*. This allows the runners and bearers to be connected at any point on the connector.

Legs have coupling pins permanently installed at the top for connecting to other legs. Holes are provided in the coupling pin and the bottom of legs for locking pins.

2.1.2 Horizontal Members

Runners, bearers, ledgers, and transoms are the horizontal members in a scaffold system. The horizontal members tie all the components together. They are also used as toprails and midrails on the working levels. Bearers are the horizontal members that support the platforms. Some manufacturers provide bearers with extra reinforcement to provide additional strength.

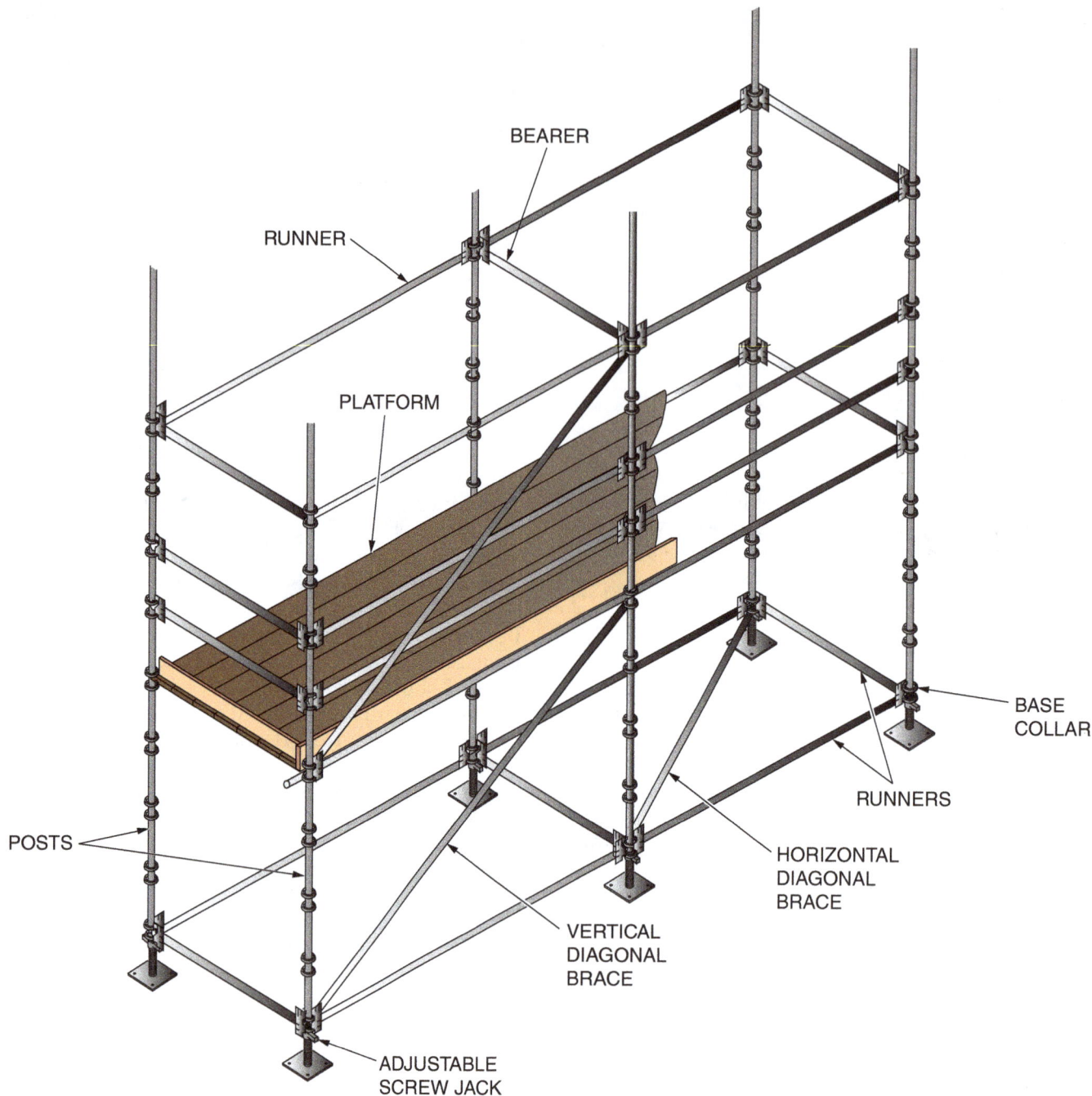

Figure 2 System scaffold.

All horizontal members are fitted with connections designed to securely attach the members to the leg connector. These fittings include some method of locking the horizontal member to the leg. Each system allows a small amount of rotation in the horizontal plane, but none in the vertical plane. This means that, once in place and locked, the runner or bearer can be moved slightly from side to side, but it will not move up or down. This arrangement allows the scaffold to be formed around a tank, yet remain rigid. Runners and bearers come in many different lengths,

usually in multiples of 1 foot. *Figure 4* shows a typical horizontal member.

2.1.3 Diagonal Braces

Diagonal braces can be placed vertically or horizontally in a system scaffold. Horizontal diagonal braces are used to ensure that the scaffold bay is square. Some manufacturers of system scaffolds provide horizontal braces with clamps to attach the brace directly to the leg. Vertical diagonal braces keep the legs vertical and prevent the scaf-

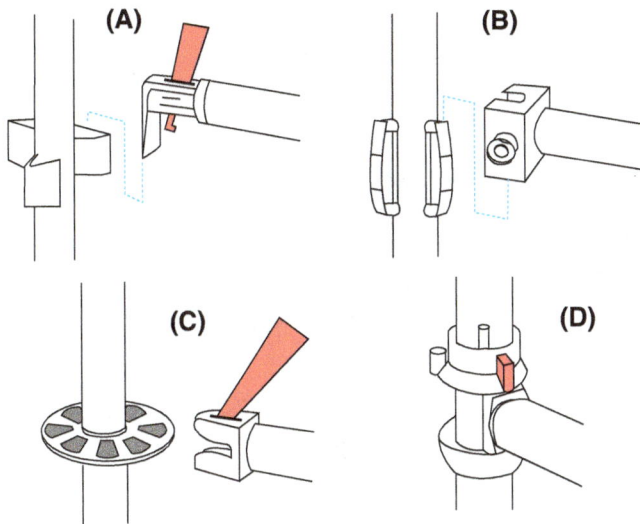

Figure 3 Coupling methods for system scaffolds.

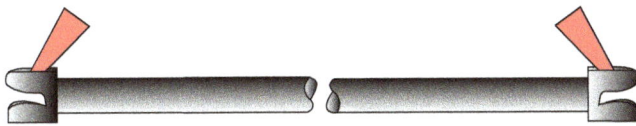

Figure 4 Horizontal member.

fold from swaying. They are attached from a low connection point on one leg to a higher point on an adjacent leg. Both types of braces are available in lengths to properly brace standard-size scaffold frames. *Figure 5* shows a vertical diagonal brace.

2.1.4 Common System Scaffold Accessories

System scaffolds have many of the same accessories as other types of supported scaffolds. Screw jacks are available to level a scaffold just as with other supported scaffolds. Fixed base plates are available for use when screw jacks are not necessary. Casters are available to erect a rolling system scaffold. The casters can be used with or without screw jacks.

Access to system scaffolds can be provided by ladders or stairways. The ladders are usually attached to the legs by ladder brackets, but some manufacturers provide ladders that attach to the bearers or runners. Both types of ladders have connector pins at the tops of the legs to join additional ladder units. *Figure 6* shows ladders and a ladder bracket.

Figure 5 Vertical diagonal brace.

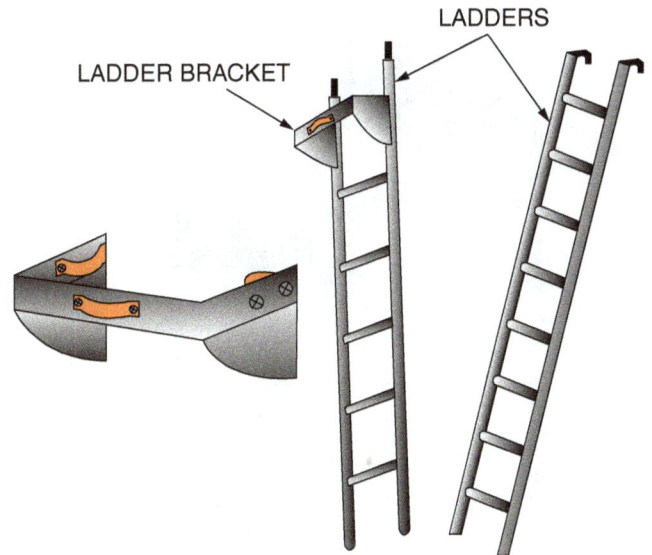

Figure 6 Ladders and ladder bracket.

Stairways for system scaffolds usually consist of separate stringers and treads. Some stringers are provided in right- and left-hand units. These stringers rest on the bearers, and the treads are attached between them. Handrails may be provided as specialized units, or vertical diagonal braces can be installed as handrails. *Figure 7* shows two configurations of system-scaffold stair units.

Figure 7 also shows the use of a side bracket to support the landings for the stair units. As with other scaffold types, the side brackets for system scaffolds must not be used for storage of material.

When long spans are needed to bridge over equipment or some other obstruction, system scaffolds use a girder, truss, or putlog. These specialized open-web components provide a strong and lightweight support for a platform. The ends are fitted with the same connections as other horizontal members. *Figure 8* shows an open-web putlog, commonly referred to as a truss.

2.2.0 System Scaffold Configurations

System scaffolds can be erected in any of the usual arrangements: tower scaffold, scaffold runs, and area scaffolds. *Figure 9* shows the system components used as a scaffold tower.

Additional components are added to the end of the basic tower to create a scaffold run, and to the sides and ends to create an area or birdcage scaffold. *Figure 10* shows the arrangement for an area scaffold, also known as multiple-bay scaffold.

System scaffold components can be arranged to provide a cantilevered platform that is wider than what can be created by side or end brackets. There is a restriction, however; the width of the platform

12-POST CONFIGURATION

6-POST CONFIGURATION

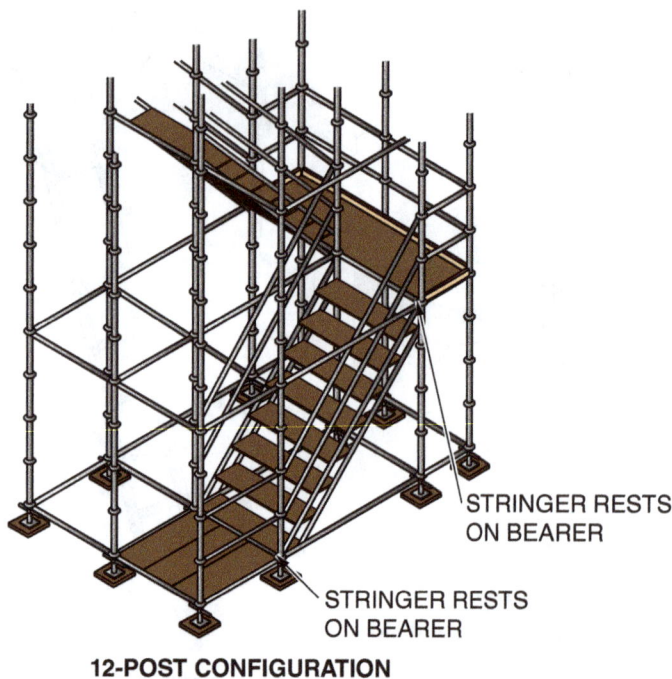

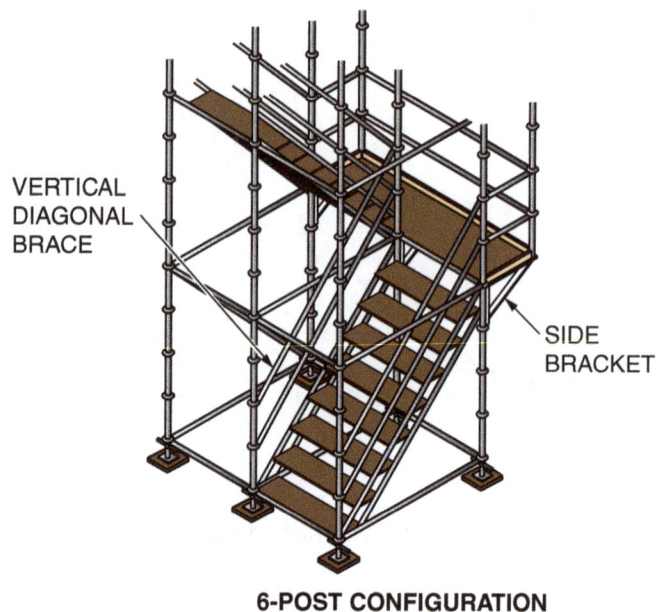

31105-14_F07.EPS

Figure 7 System scaffold stair units.

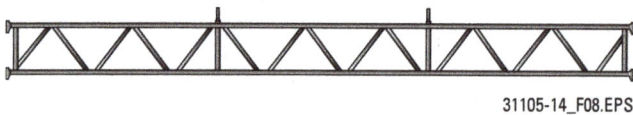

31105-14_F08.EPS

Figure 8 Open-web putlog.

must be wide enough for a vertical diagonal brace to fit between the outer edge of the cantilevered platform and one of the attachment points on the scaffold leg. Always check the manufacturer's recommendations for loading limitations. *Figure 11* shows a cantilevered platform.

> **NOTE**
>
> Check with a qualified person prior to erecting a cantilevered platform.

When using system scaffold components to erect a scaffold around a circular tank or building, certain design restrictions must be accommodated. The runners and bearers are a fixed length. Since the runners and bearers set the distance between legs, the width of the platform must match one of the bearer lengths. The main restriction will be encountered when erecting the span lengths. The legs on the inner and outer circles must each be set at a span that matches a runner length. For example, if all the runners are in even-foot lengths, a circle cannot be constructed that requires the legs to be set at 5 feet, 8 inches.

The tube-and-clamp scaffold referenced in the *Appendix* may need to be modified if it is to be built from system components.

2.3.0 System Scaffold Erection

For this exercise, an area scaffold three-bays wide, two-bays deep, and one-lift high is erected. The platform will be used as an entertainment stage on a relatively level area of grass. Mudsills are used under each of the legs, and base plates and screw jacks are used to level the platform. The scaffold frame will be covered with solid platforms. Toprails and midrails are not required for a scaffold less than 10-feet high (this may vary by state), but a guardrail will be added to the platform. Access will be by a staircase attached to the front of the platform.

Perform the following steps to erect this system area scaffold:

Step 1 Gather and inspect all scaffold equipment for the scaffold arrangement.

Step 2 Place appropriate mudsills in their approximate locations.

Step 3 Attach the screw jacks to the mudsills.

Step 4 Adjust the screw jacks to near their lowest position.

Step 5 Determine the location of the highest base.

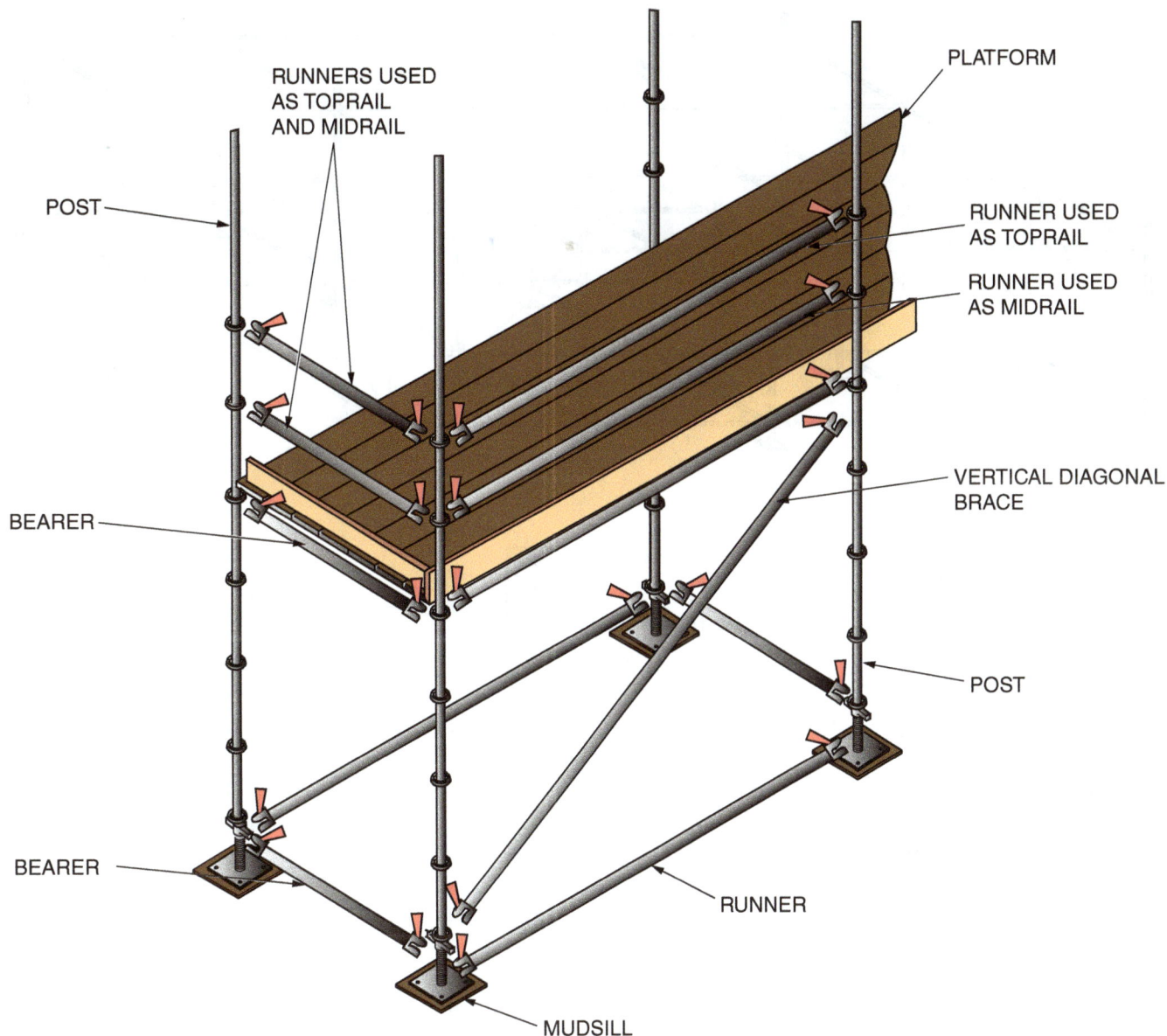

RUNNERS USED
AS TOPRAIL
AND MIDRAIL

POST

BEARER

BEARER

MUDSILL

PLATFORM

RUNNER USED
AS TOPRAIL

RUNNER USED
AS MIDRAIL

VERTICAL DIAGONAL
BRACE

POST

RUNNER

31105-14_F09.EPS

Figure 9 System components used as a scaffold tower.

Step 6 Lay the runners on the ground between the bases. *Figure 12* shows the layout of the first two components.

Step 7 Attach the runners to the bases.

> **NOTE**
>
> Start at the highest point to make leveling the scaffold easier. Do not hammer the wedges in at this time.

Step 8 Attach the horizontal diagonal brace to square the first bay. *Figure 13* shows the assembly of the first bay.

Step 9 Adjust the screw jacks so that a torpedo level placed on each runner and bearer indicates that the bay is level. *Figure 14* shows leveling the first bay.

Step 10 Attach the runners and horizontal diagonal braces to assemble the remaining bays.

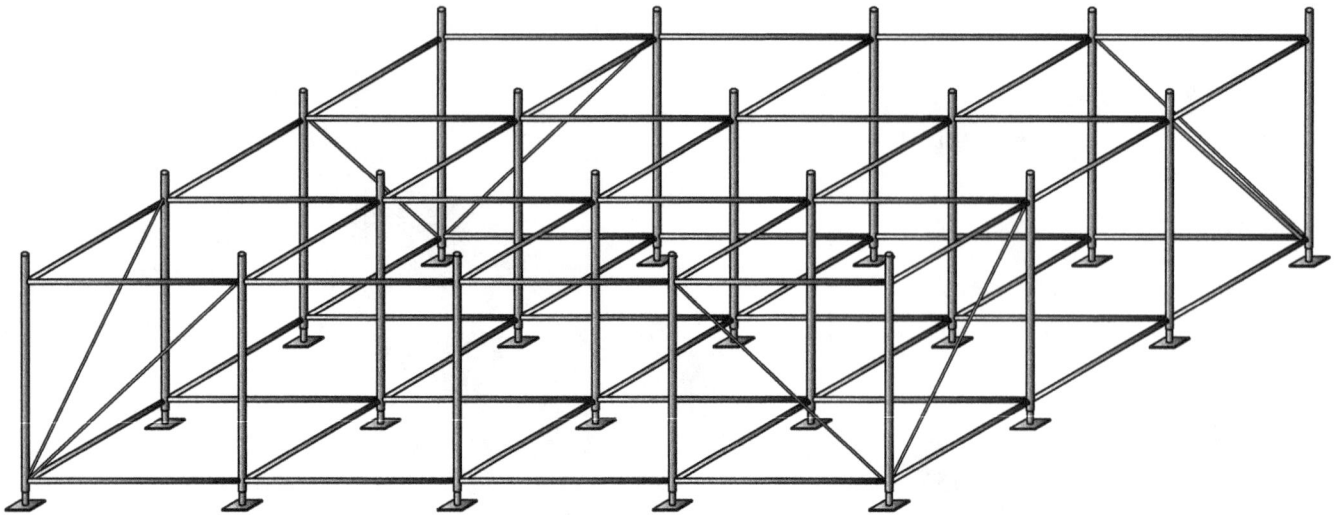

31105-14_F10.EPS

Figure 10 Area scaffold arrangement.

Step 11 Adjust the screw jacks to level the remaining bays.

Hammer each of the wedges into the connector collars to lock the runners and horizontal diagonal braces in place.

Step 13 Install a post on each base. *Figure 15* shows installing the posts.

Step 14 Attach runners and bearers at the desired level to frame the platform level. *Figure 16* shows attaching the platform-level runners and bearers.

Step 15 Connect vertical diagonal braces across the face of each bay. *Figure 17* shows installing the vertical diagonal braces.

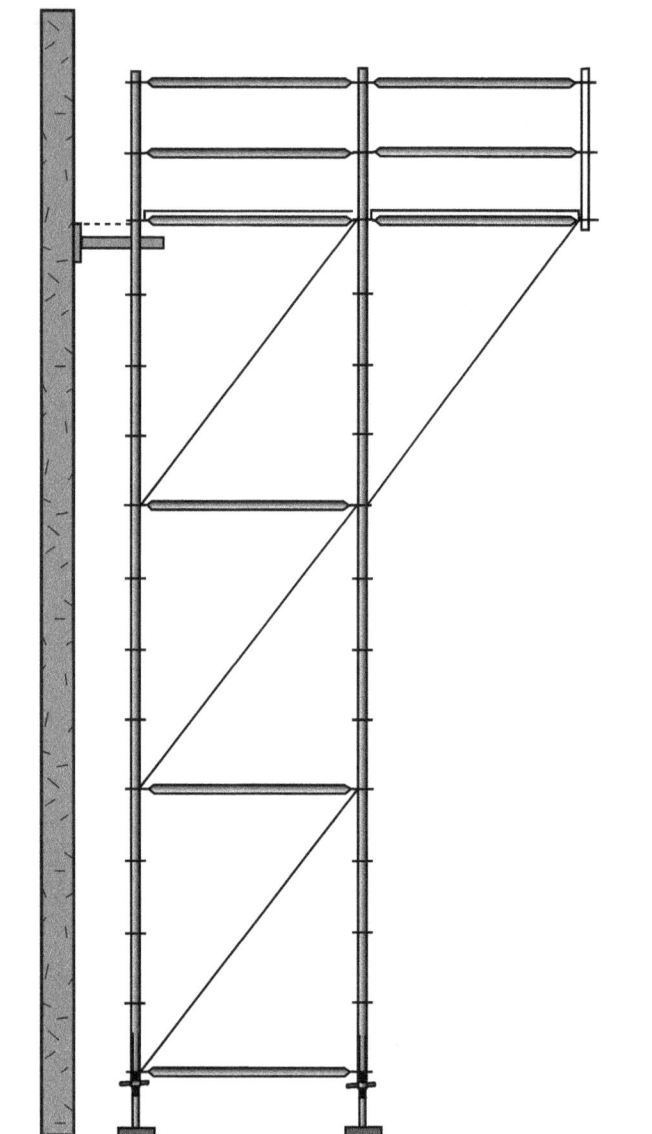

31105-14_F11.EPS

Figure 11 Cantilevered platform.

Step 16 Verify that each leg is plumb using a torpedo level.

Step 17 Install the stair stringers to the front right bay.

Step 18 Attach stair treads to the stair stringers.

Step 19 Attach handrail assemblies to the stair stringers. *Figure 18* shows the completed scaffold frame.

Step 20 Install the platforms on the platform-level runners and bearers to completely cover the platform area.

Step 21 Attach the additional legs if necessary to the legs along the back and sides of the platform.

Step 22 Attach runners to the legs as midrails and toprails, leaving a gap at the staircase for access.

Step 23 Verify that all connections are secure. *Figure 19* shows the completed area scaffold.

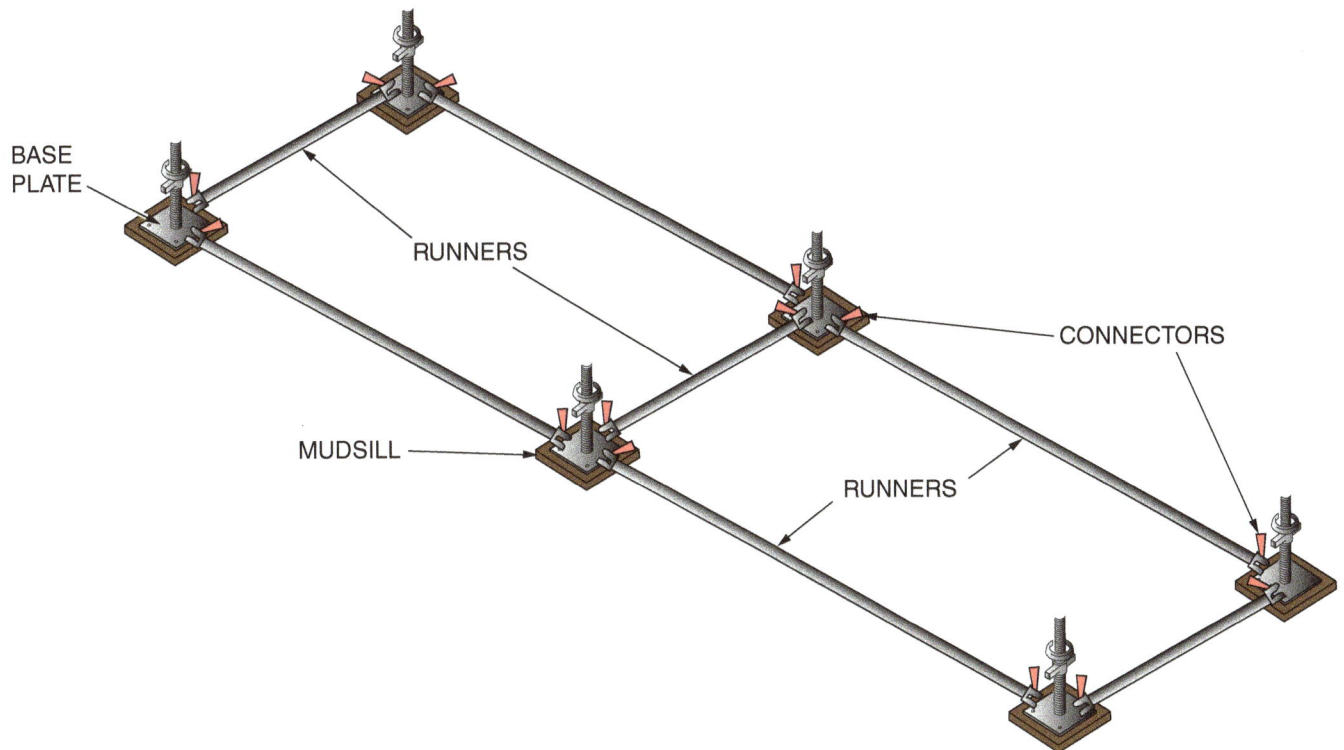

Figure 12 Layout of components.

31105-14_F12.EPS

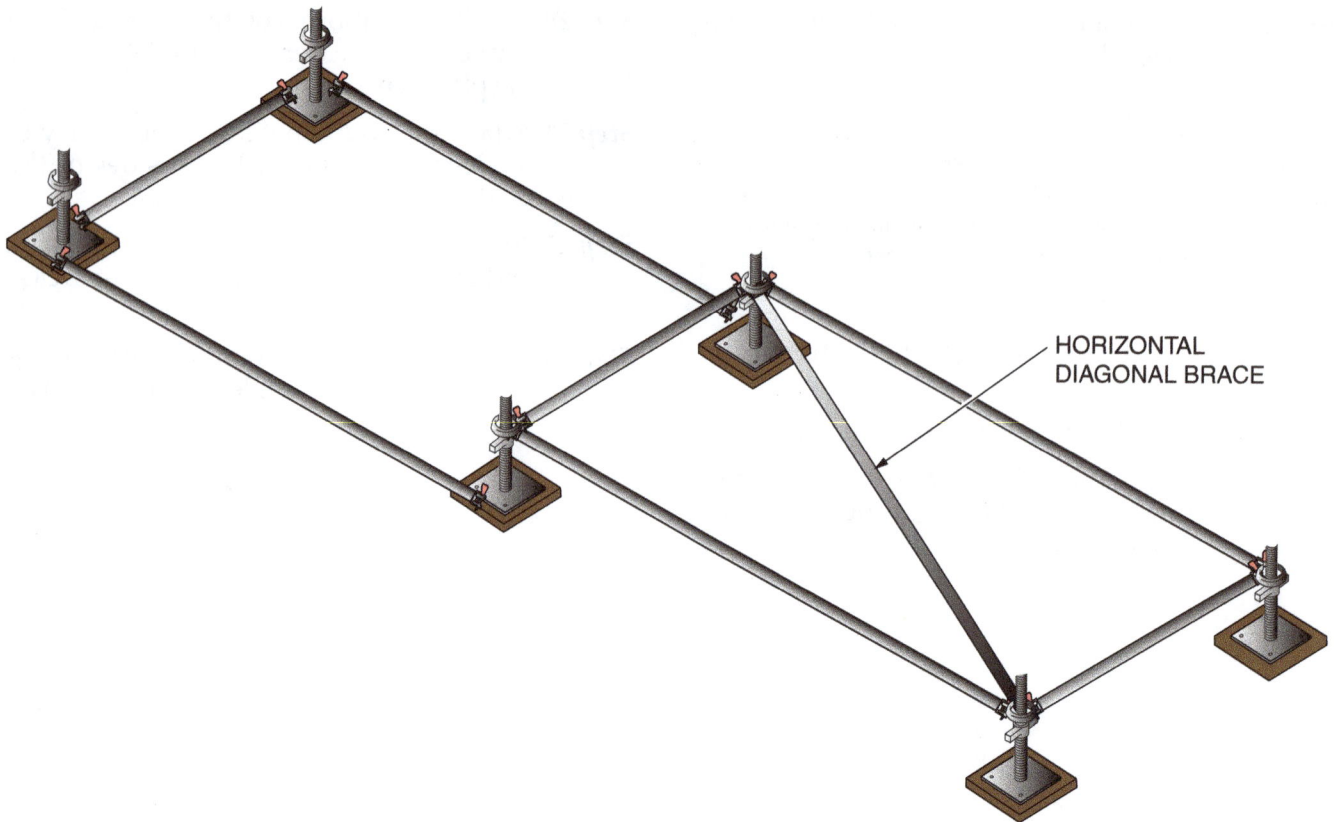

HORIZONTAL
DIAGONAL BRACE

31105-14_F13.EPS

Figure 13 Assembly of first bay.

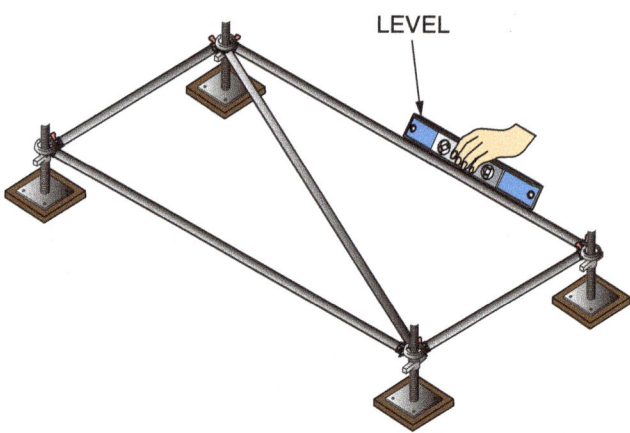

LEVEL

31105-14_F14.EPS

Figure 14 Leveling first bay.

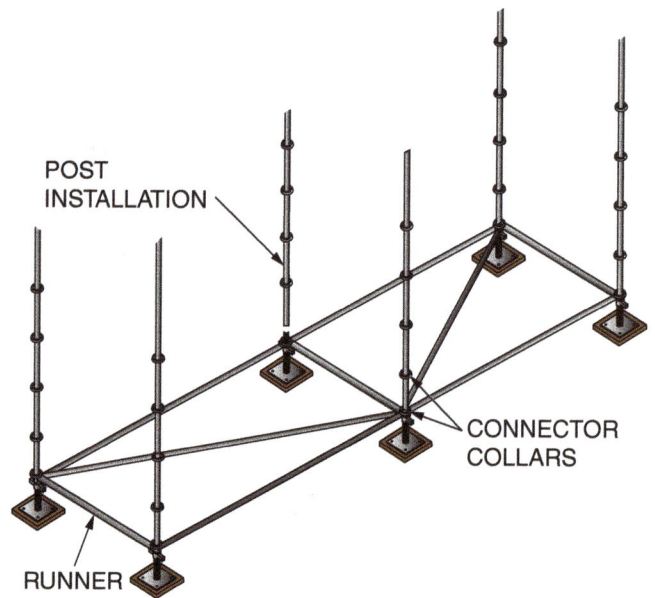

POST
INSTALLATION

CONNECTOR
COLLARS

RUNNER

31105-14_F15.EPS

Figure 15 Installing posts.

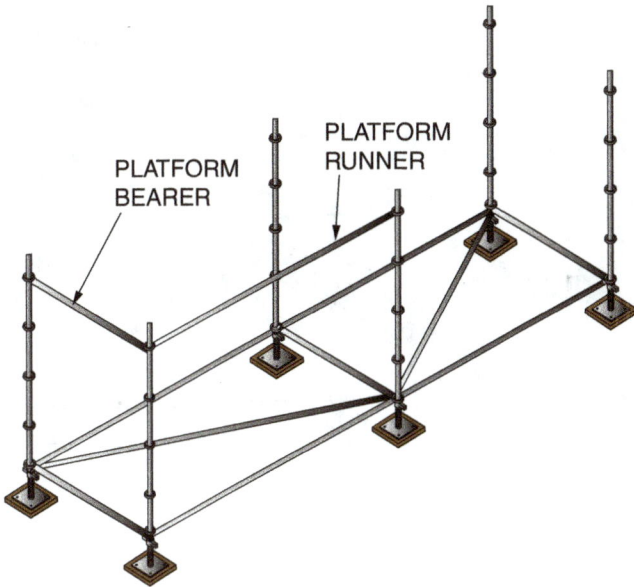

PLATFORM
BEARER

PLATFORM
RUNNER

31105-14_F16.EPS

Figure 16 Attaching platform-level runners and bearers.

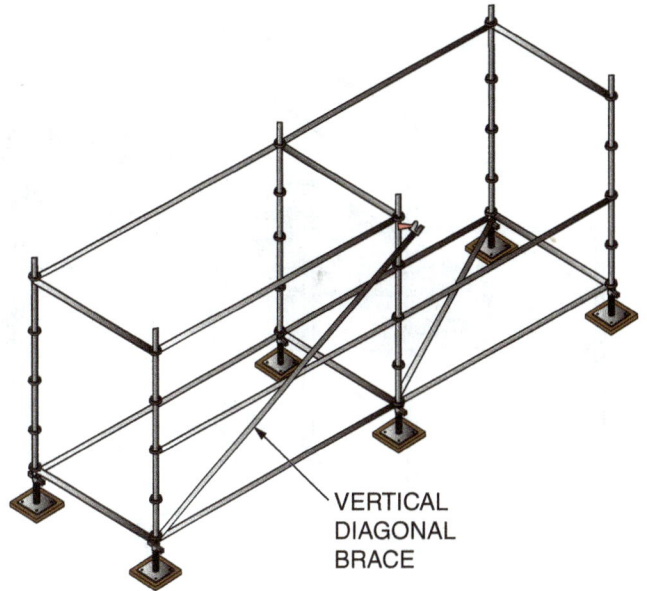

VERTICAL
DIAGONAL
BRACE

31105-14_F17.EPS

Figure 17 Installing vertical diagonal braces.

31105-14_F18.EPS

Figure 18 Completed scaffold frame.

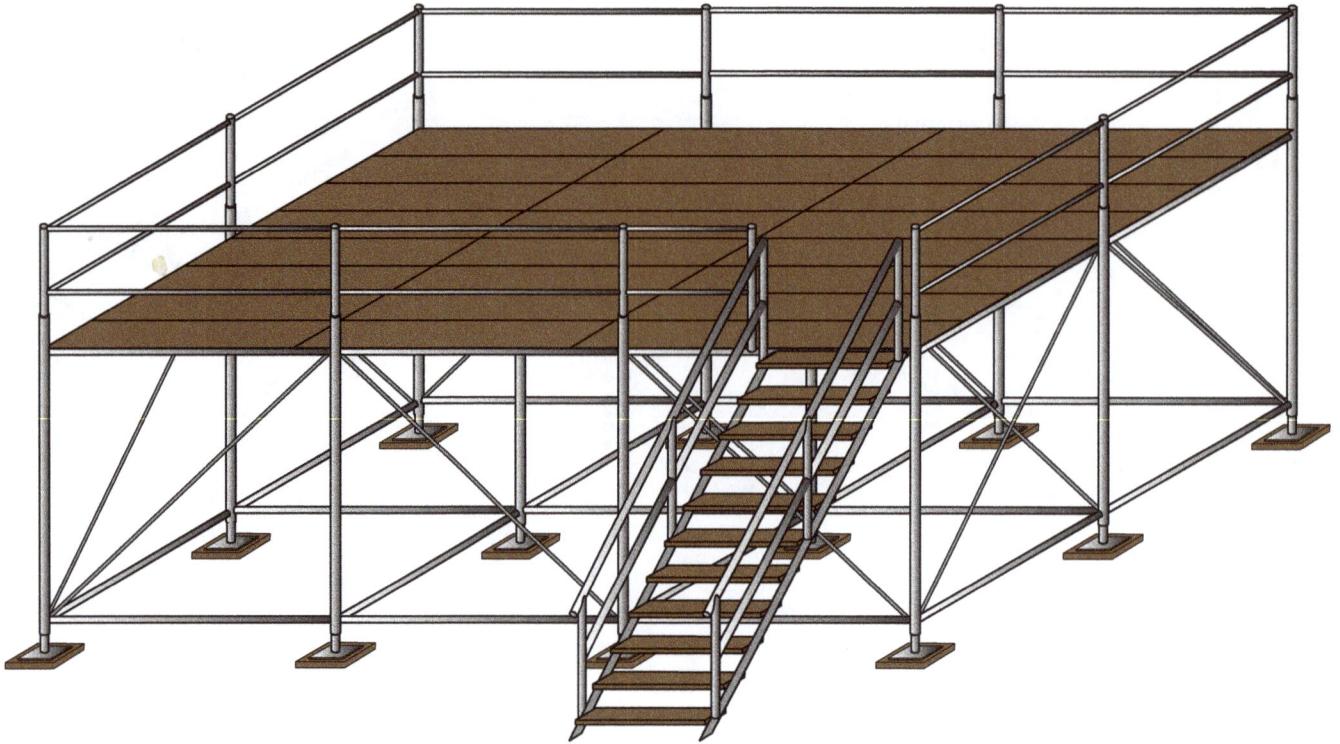

31105-14_F19.EPS

Figure 19 Completed area scaffold.

Additional Resources

"Scaffolding." OSHA. **www.osha.gov/SLTC/scaffolding/index.html**

Fall Protection and Scaffolding Safety: An Illustrated Guide. 2000. Grace Drennan Ganget. Government Institutes.

International Building Code®, Latest Edition. Falls Church, VA: International Code Council.

International Residential Code®, Latest Edition. Falls Church, VA: International Code Council.

2.0.0 Section Review

1. When using aluminum and steel tubing together, you must observe the maximum load limit of the component that is _____.

    a. strongest
    b. smallest
    c. weakest
    d. most predominant

2. In a system scaffold, all the components are tied together by the _____.

    a. horizontal members
    b. diagonal braces
    c. vertical members
    d. connectors

3. Vertical diagonal braces keep scaffold legs vertical and prevent _____.

    a. settling
    b. sagging
    c. swinging
    d. swaying

Section Three

3.0.0 Erecting Tubular Welded-Frame Scaffolds

Objectives

Explain the basic principles of tubular welded-frame scaffolds, and outline proper erection procedures.

   a. Identify common applications of tubular welded-frame scaffolds.
   b. Identify the components of tubular welded-frame scaffolds.
   c. Outline the steps for proper erection of a tubular welded-frame scaffold.

Tubular welded-frame scaffolds are used in accessible places with fairly level ground conditions. They use one or more manufactured platforms supported by welded-end frame sections, horizontal bearers, and intermediate members. Tubular welded-frame scaffolds are made in various heights and widths, and are joined with horizontal and diagonal cross braces and secured by pins. The braces have a fixed length that automatically squares and vertically aligns vertical members so the erected scaffold is always plumb, square, and rigid. The scaffold is extended by adding braces and frames until the desired length is reached. Scaffold height is increased by stacking end frames on top of each other. The bottoms of the legs of the upper end frames slide into the tops of the legs of the lower end frames and are joined together with coupling pins that are locked into position. *Figure 20* shows a tubular welded-frame scaffold system.

3.1.0 Common Applications

Tubular welded-frame scaffolds can be assembled in three basic configurations: single-bay or tower scaffold, scaffold run, or area scaffold. They are commonly used on the inside or outside of rectangular buildings where access is not restricted.

The work to be accomplished is the first factor to consider when determining the type of scaffold to use. Is the scaffold to be used to replace a light bulb, an entire ceiling, lay a brick wall, or build a tower? Each of these activities may require a different type of scaffold, or at least a different arrangement of scaffold components. The work to be performed will also determine the load requirements. Typically, the height of the tubular welded-frame scaffold and the length of the cross

THIS SIDE AGAINST BUILDING

31105-14_F20.EPS

Figure 20 Tubular welded-frame scaffold system.

braces will determine the scaffold's loadbearing capacity. A frame scaffold can be assembled in three basic configurations.

A single-bay or tower scaffold is the simplest form of the tubular welded-frame scaffold. It is made of two frames, two cross braces, and four bases. Additional sections are stacked on top of each other until the desired height is reached. Work is generally performed on the top level, but intermediate levels may be used. All scaffolds that are higher than four times the narrowest base dimension must be braced or guyed to a rigid structure. Some typical applications for a single-bay scaffolds are as follows:

- Camera stand
- Electrical work
- Window washing
- Inspection platform
- HVAC work
- Painting

A scaffold run is a series of joined tower scaffolds. The run is erected one bay wide and made as long as necessary by adding additional frames and braces to achieve the desired length. A scaffold run may be used for the following applications:

- Masons' scaffold
- Stucco work
- Wall access for painting or window installation
- Electrical work

Figure 21 shows an example of a scaffold run.

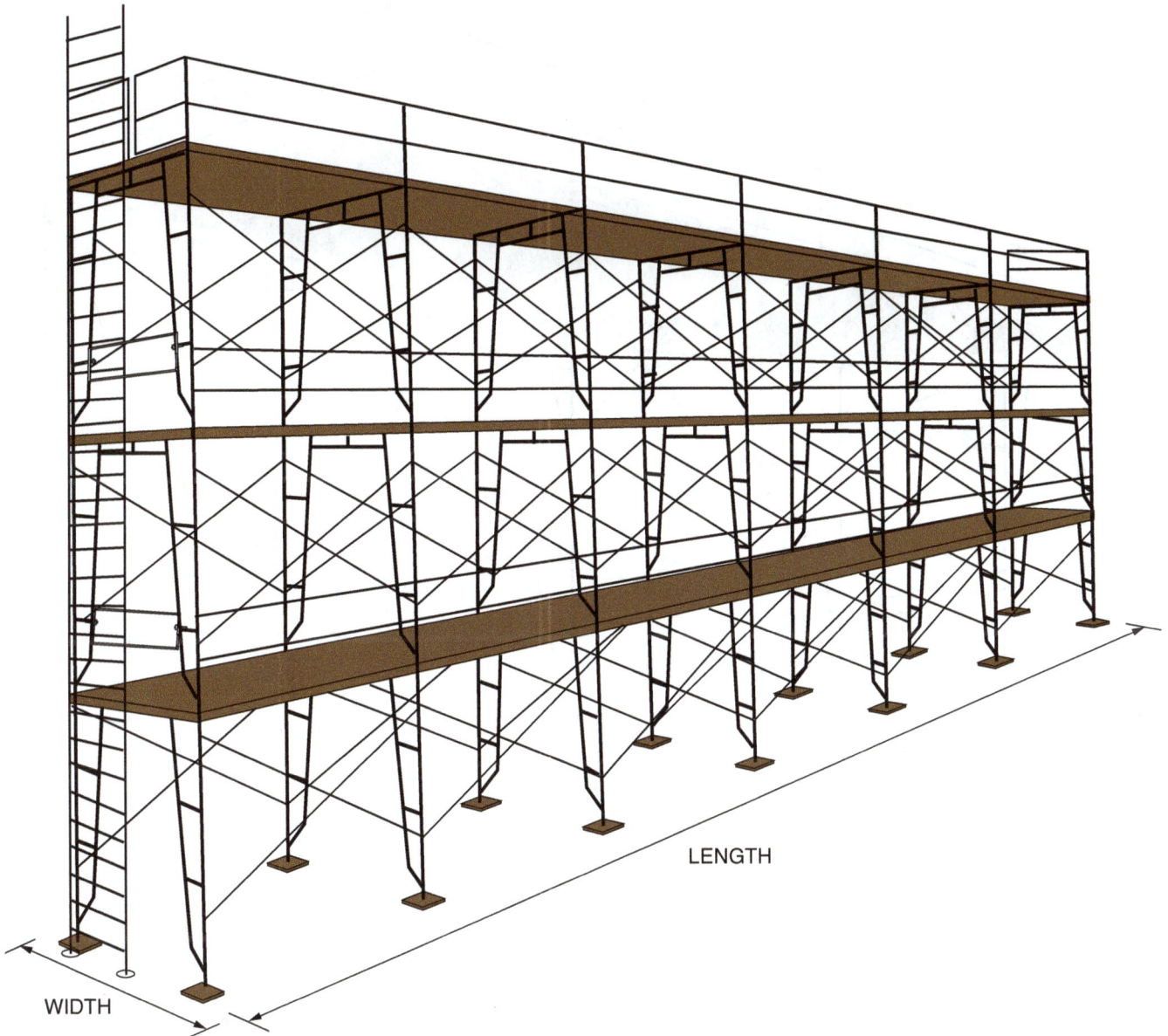

Figure 21 Scaffold run.

An area scaffold may have a length similar to a scaffold run, but it is constructed to be more than one-bay wide. An area scaffold may be as wide as it is long. The frame bays are braced in both directions to provide the greatest stability. Typical uses for an area scaffold are the following:

- Ceiling tile installation
- Dance floor
- Stage
- Auditorium floor
- Painters' scaffold

Figure 22 shows an area scaffold.

There are many different applications for tubular welded-frame scaffolds. Some additional uses are shown in Figures 23 through 26.

3.2.0 Components

Tubular welded-frame scaffold components consist of frames, locking devices, braces, base plates, putlogs, casters, platforms and planks, guardrails and gates, and ladders and stairs, as well as other miscellaneous components. These components may be assembled in a variety of arrangements to meet almost any design configuration.

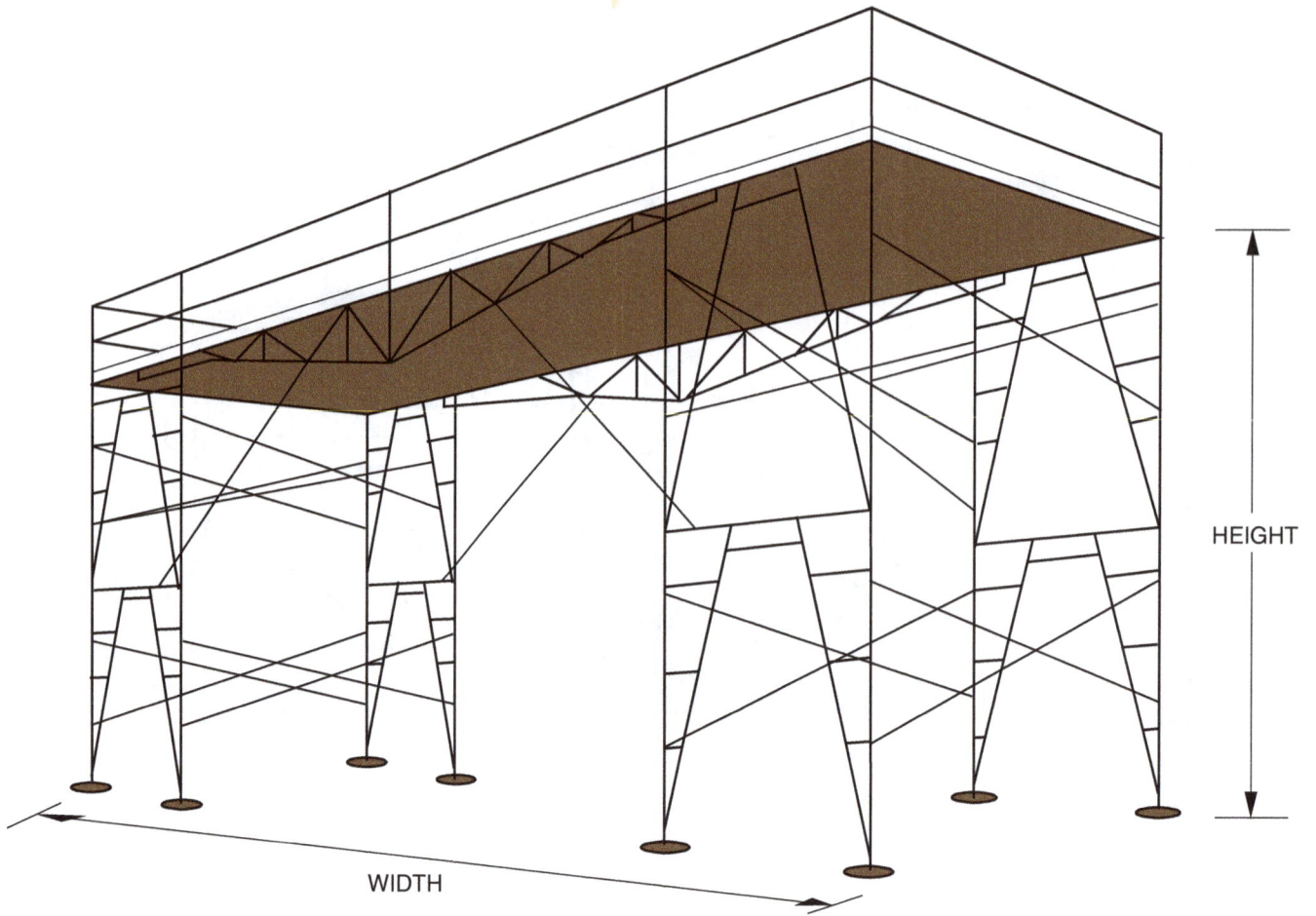

Figure 22 Area scaffold.

HEIGHT

WIDTH

31105-14_F22.EPS

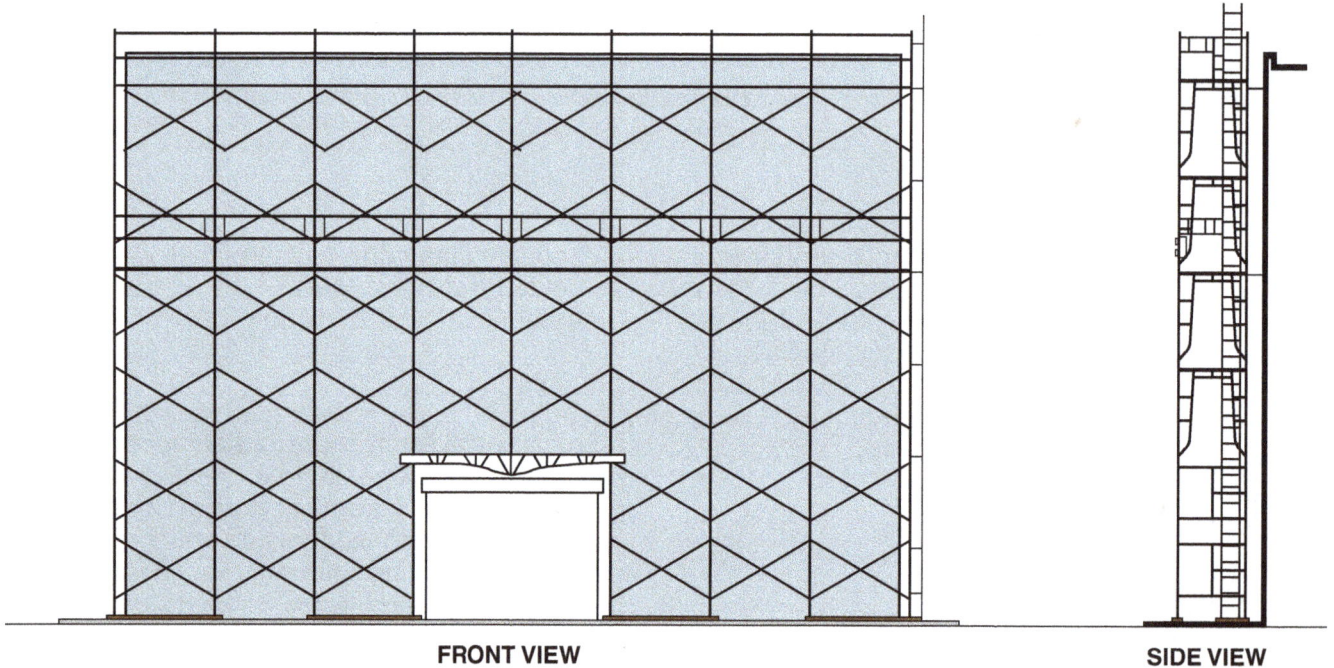

FRONT VIEW

SIDE VIEW

31105-14_F23.EPS

Figure 23 Exterior of rectangular building.

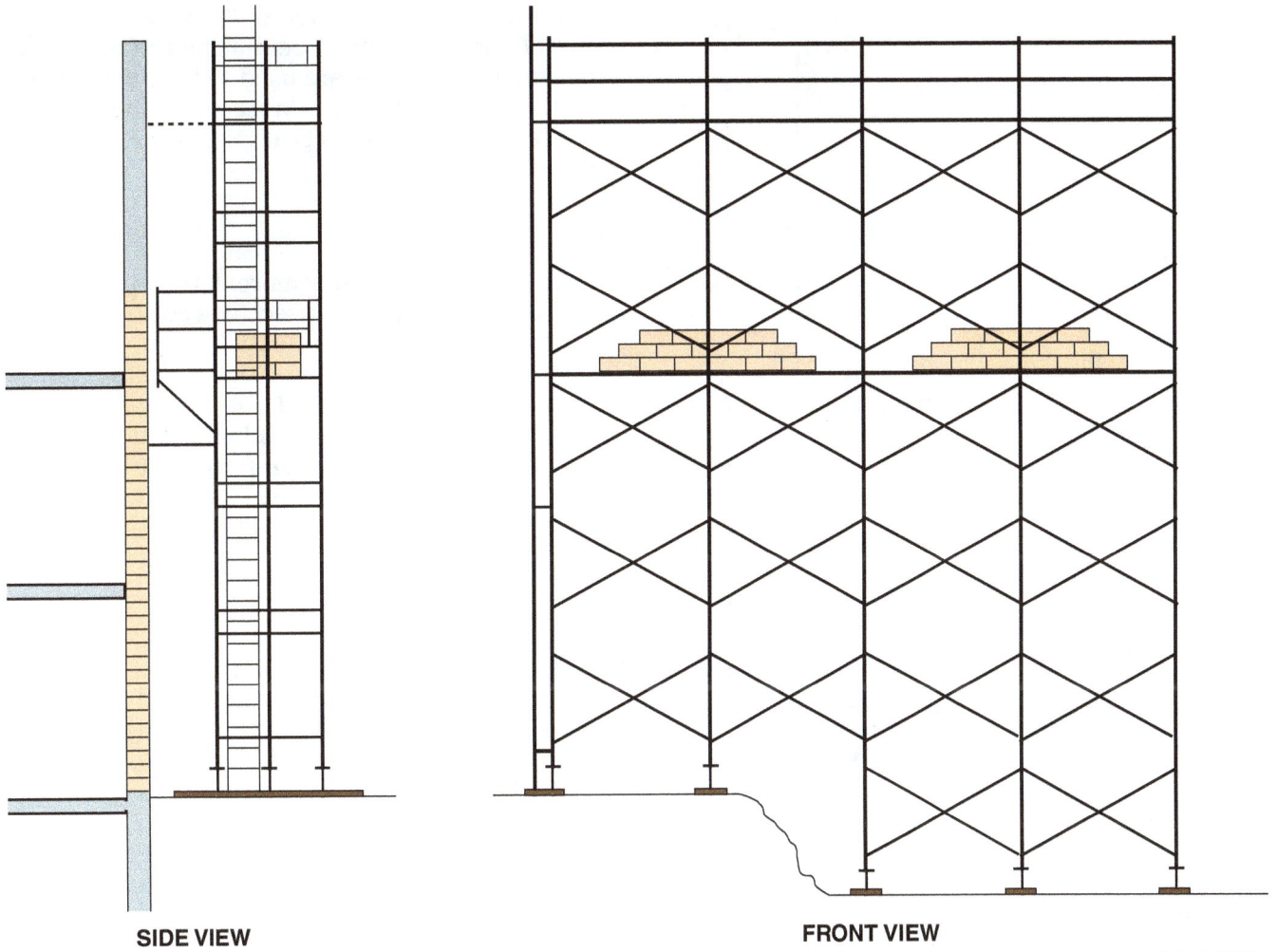

SIDE VIEW

FRONT VIEW

31105-14_F24.EPS

Figure 24 Masons' scaffold.

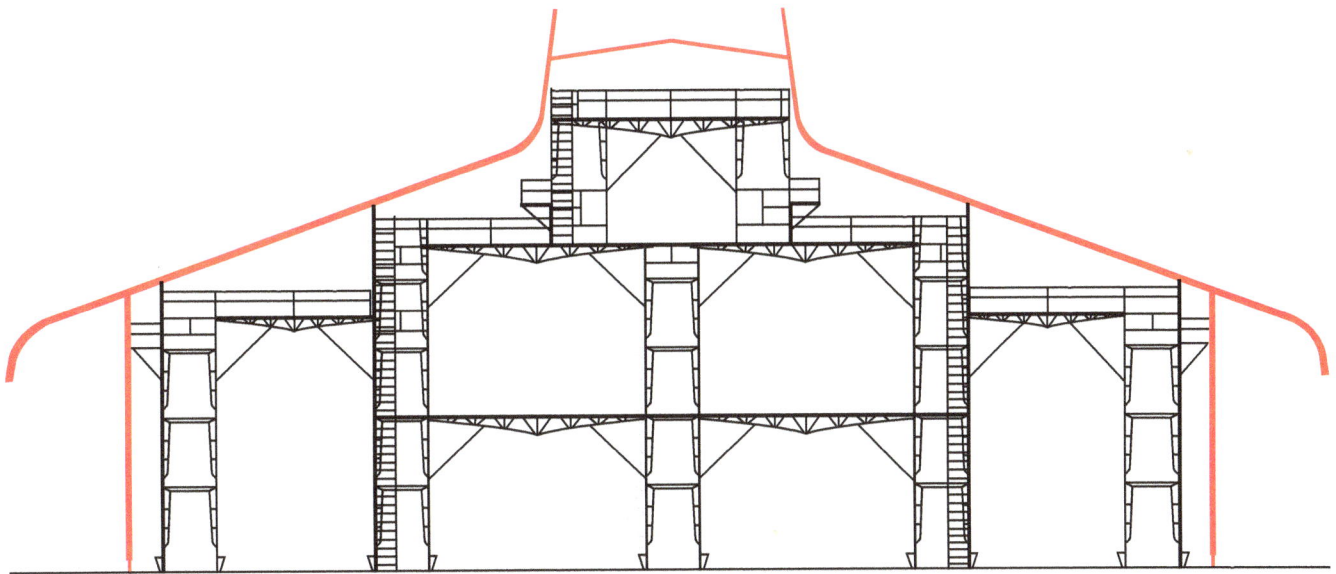

31105-14_F25.EPS

Figure 25 Interior application.

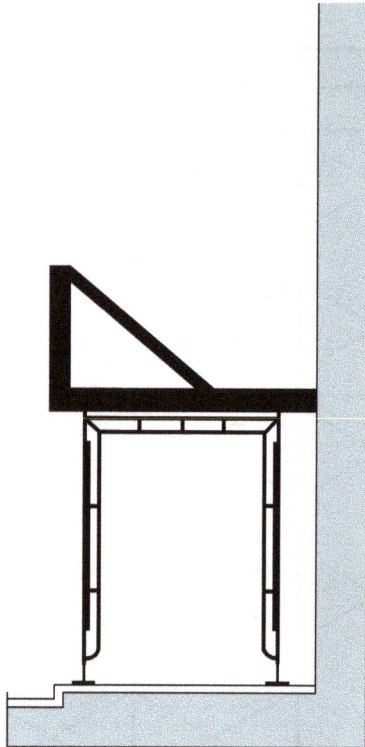

Figure 26 Canopy over a sidewalk.

The components used depend on the application. Each supplier has a complete line of components. Care should be taken not to mix components from different manufacturers, because they may not be compatible, or they may be made from dissimilar metals. When dissimilar metals come in contact with each other, they may corrode. When using system or tube-and-clamp scaffolds, always use dual-purpose clamps.

> **WARNING!**
>
> When using different components from the same manufacturer, always use the weight limits of the weakest components. Exceeding weight limits can cause scaffold failure which can result in injury or death.

3.2.1 Frames

There are a number of frame styles. Each is designed to fit a specific need while maintaining the OSHA-required safety factors. Frames of different styles can be combined to meet the needs of a particular level of a scaffold design. *Figure 27* shows a selection of tubular welded frames.

3.2.2 Braces

Strength and stability are built into the scaffold arrangement by the installation of braces. All scaffold frames are equipped with some sort of

attachment points for braces. Braces come in different lengths and must be matched to the frame spacing. Cross braces are used to keep the frames plumb, and the diagonal braces hold the scaffold square. *Figure 28* shows examples of a cross brace and diagonal brace.

3.2.3 Locks

When a scaffold requires more than a single level, a coupling pin must be used to secure the upper and lower frames together. The coupling pins must be secured to each of the scaffold frames. A rivet and hairpin, pigtail pin, or a toggle pin is inserted through the aligned holes in the scaffold legs and coupling pins. The frames also have attachment points for the cross braces. These braces must also be locked in place. *Figure 29* shows a variety of brace and coupling pin locking methods.

3.2.4 Base Plates and Screw Jacks

The base plates and screw jacks provide a foundation for the scaffold legs. They distribute the load coming down the tubing onto a larger area and prevent the legs from sinking into the bearing surface. Base plates should be used even if the scaffold is to be erected on concrete. The base plates are attached to the scaffold frames with the same locking pins as coupling pins. Screw jacks are designed to accommodate small changes (up to 9 inches) in the level of the ground. The handle on the screw jack must be securely braced against the scaffold leg to make it secure. Improper use of base plates could lead to scaffold collapse. *Figure 30* shows examples of base plates and screw jacks.

3.2.5 Putlogs

A putlog is a truss used in scaffold applications where bridging over an obstruction or opening is required. Putlogs provide excellent support for platforms or other scaffold frames. Scaffold builders must be careful not to exceed the load capacities for the putlog. Putlogs over 10 feet in length must be supported with knee braces. *Figure 31* shows examples of a putlog application.

3.2.6 Planks and Decks

Planks and decks provide a working surface for the users to stand on, support materials and tools, and provide a means to get from place to place. Planks may be made of metal or wood. Metal planks are usually 9-inches wide. Wood planks are available in a variety of lengths, and may be long enough to span more than one bay. Metal planks vary in length up to 10 feet. Planks de-

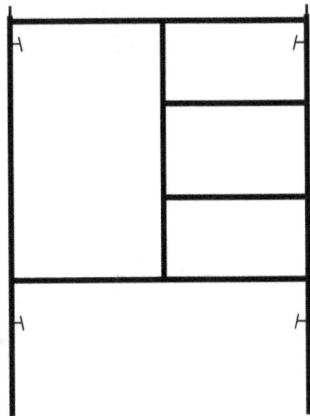

5' × 6'-7"
SINGLE-LADDER BOX

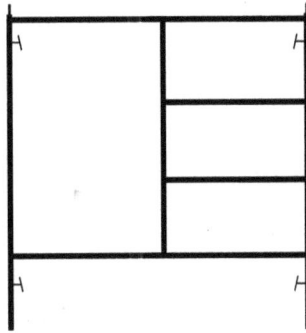

5' × 5'-1"
SINGLE-LADDER BOX

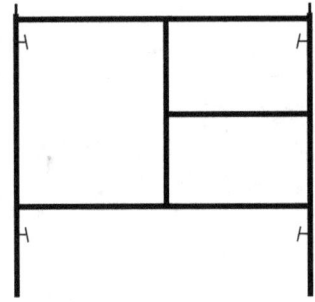

5' × 4'-1"
SINGLE-LADDER BOX

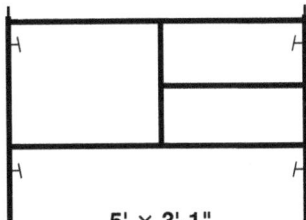

5' × 3'-1"
SINGLE-LADDER BOX

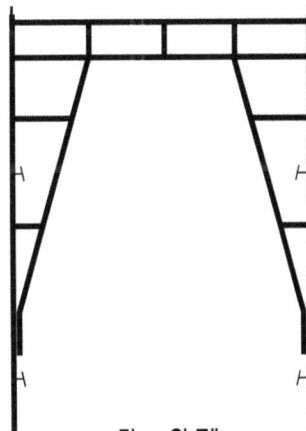

5' × 6'-7"
WALK-THROUGH

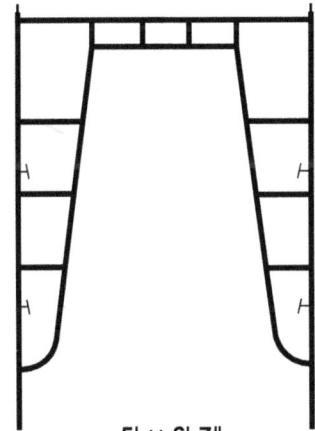

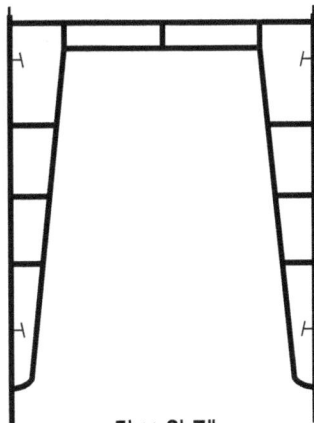

5' × 6'-7"
WALK-THROUGH

5' × 8'-7"
WALK-THROUGH

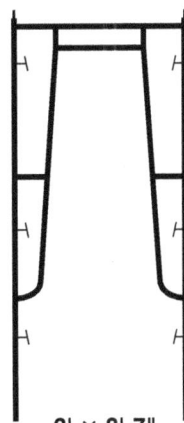

3' × 8'-7"
WALK-THROUGH

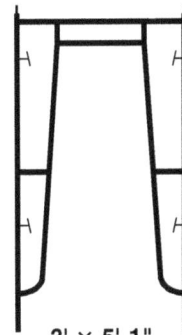

3' × 5'-1"
WALK-THROUGH

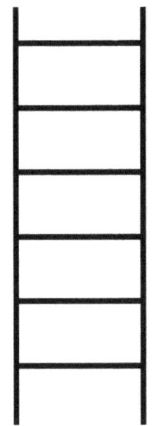

2' × 6'-7"
LADDER FRAME

31105-14_F27.EPS

Figure 27 Tubular welded frames.

DIAGONAL BRACE

HORIZONTAL BRACE

CROSS BRACE

CROSS BRACE

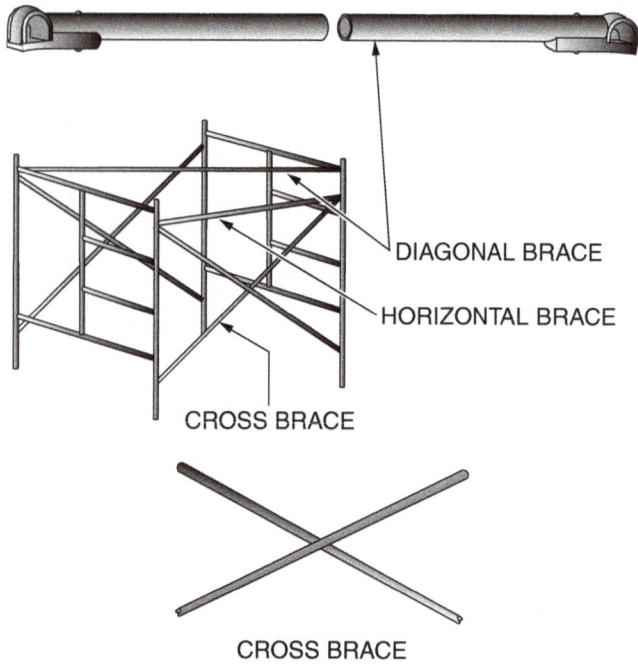

31105-14_F28.EPS

Figure 28 Cross brace and diagonal brace.

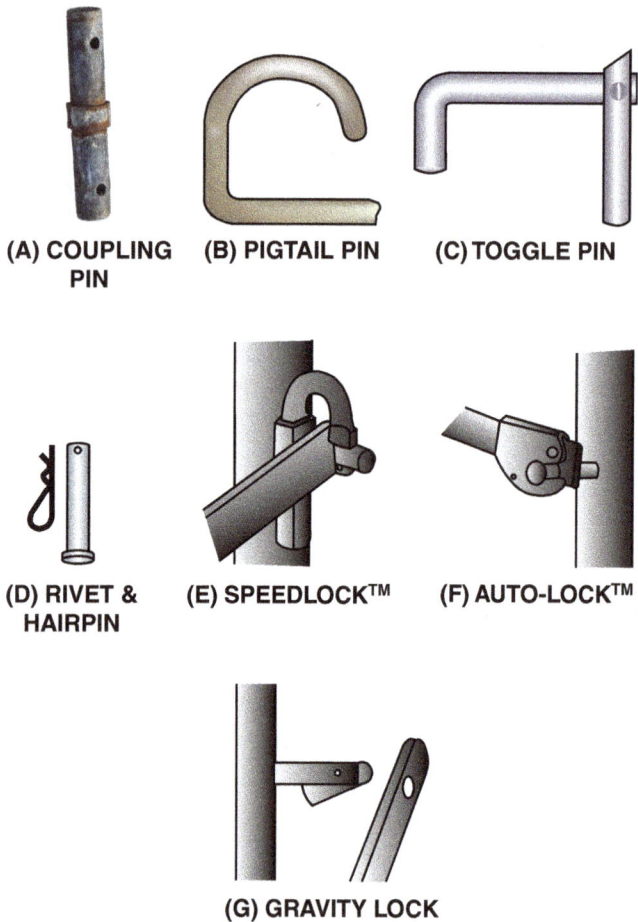

(A) COUPLING PIN

(B) PIGTAIL PIN

(C) TOGGLE PIN

(D) RIVET & HAIRPIN

(E) SPEEDLOCK™

(F) AUTO-LOCK™

(G) GRAVITY LOCK

31105-14_F29.EPS

Figure 29 Locking methods.

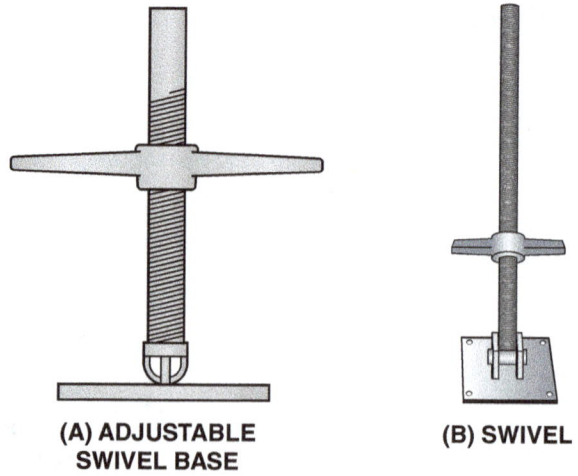

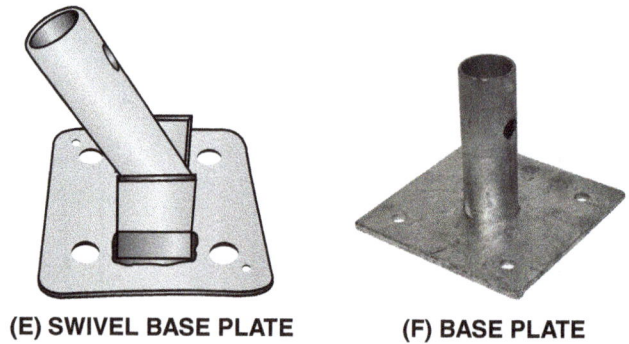

(A) ADJUSTABLE SWIVEL BASE

(B) SWIVEL

(C) SCREW JACK WITH BASE PLATE

(D) SCREW JACK WITH SOCKET

(E) SWIVEL BASE PLATE

(F) BASE PLATE

31105-14_F30.EPS

Figure 30 Base plates and screw jacks.

signed for use on tubular welded-frame scaffolds are fitted with hooks to rest on the round bearers. *Figure 32* shows examples of metal planking.

Wood planks may be sawn from solid wood, and are sometimes called solid sawn planks. The nominal size is 2 × 10, which, when planed smooth, actually measures 1½ inches × 9¼ inches. Two types of wood are specifically graded for

PUTLOGS

PUTLOG APPLICATION

31105-14_F31.EPS

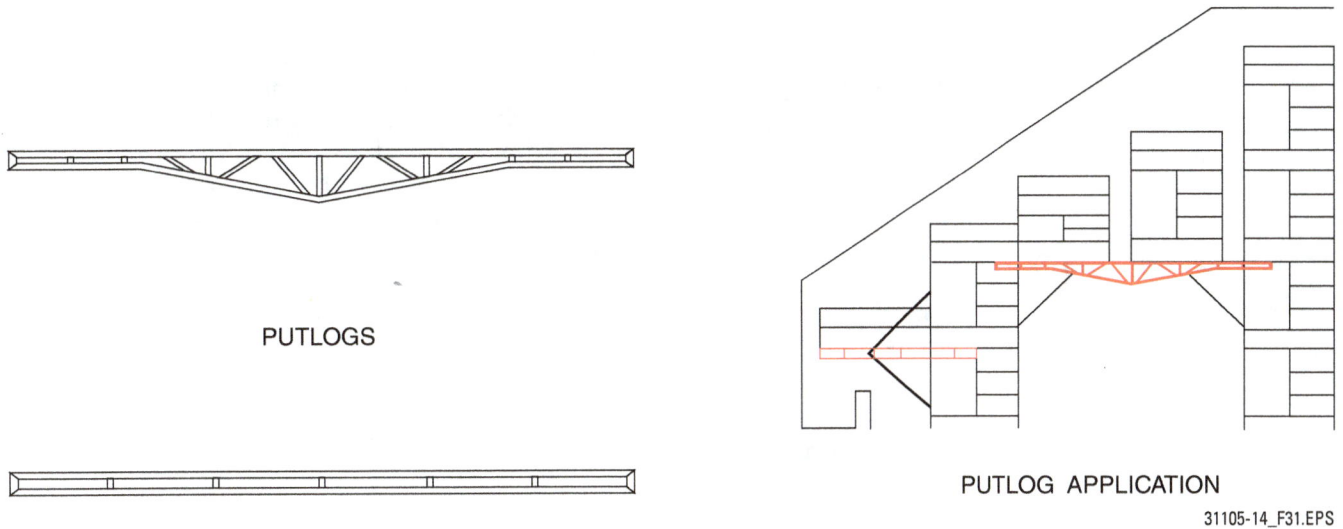

Figure 31 Putlog application.

31105-14_F32.EPS

Figure 32 Examples of metal planks.

31105-14_F33.EPS

Figure 33 Plank grade stamps.

use as scaffold decking: 171(b) Douglas Fir, and Dense Industrial Southern Pine. A scaffold plank must be strong enough to support a 250-pound load in the center of a 10-foot span without deflecting more than $\frac{1}{60}$ of its length. Planks must be free of saw kerfs, notches, and nails. Scaffold planks are stamped by an inspection agency as Scaffold Grade. The two most common inspection agencies are the Southern Pine Inspection Bureau (SPIB) and the West Coast Lumber Bureau (WCLB). If the planks are not stamped, do not use them for scaffold planks. *Figure 33* shows the two most common grade stamps for scaffold planks.

Laminated-wood planks are made from smaller pieces of wood glued together. The pieces may be edge-glued or laminated. The edge-glued planks are made up of layers, like plywood. These planks are tested and rated by the manufacturer and are not marked like solid sawn planks. *Figure 34* shows examples of laminated-wood planks.

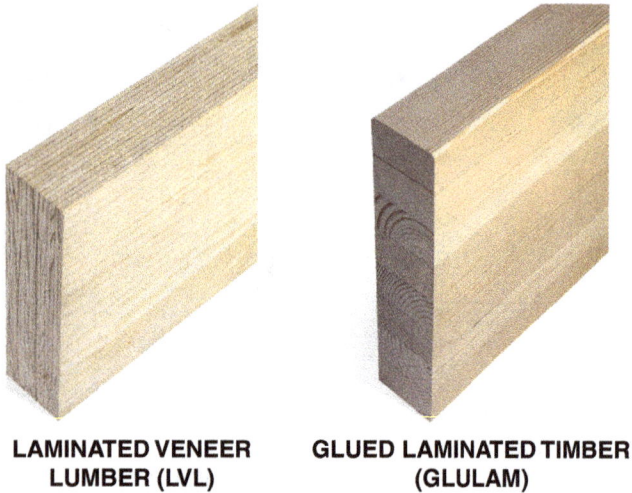

LAMINATED VENEER LUMBER (LVL)   **GLUED LAMINATED TIMBER (GLULAM)**

31105-14_F34.EPS

Figure 34 Laminated-wood planks.

> **NOTE**
>
> Consult a qualified person to determine if laminated planks can be used.

Scaffold decks are manufactured platforms that are usually 19-inches wide and long enough to span a single bay. Scaffold decks have hooks to attach to the bearers or putlogs. Decks can be made from a variety of materials. The most common decks are made entirely of aluminum or have aluminum siderails and plywood working surfaces. Scaffold decks are rated by the weight that they can safely support, usually 50 pounds per square foot or 75 pounds per square foot. *Figure 35* shows an example of a scaffold deck.

3.2.7 Guardrails

Unless a personal fall arrest system is provided, guardrail systems are required on all scaffolds 10 feet in height or higher (this may vary by state), as a means of safeguarding workers from fall hazards. A guardrail system consists of both a toprail and a midrail. These rails must be placed on all open sides and ends of a working platform. A midrail must be placed halfway between the top edge of the toprail and the platform surface. *Figure 36* shows the locations for a toprail and a midrail.

A means of access to the scaffold platform should be provided to avoid having to climb over the guardrail. Access through a guardrail system can be achieved in many ways. On narrow scaffolds, a simple chain may suffice. On other scaffolds, a swing gate or duck-through with grab rails may be used.

Swing gates must be installed so that they swing toward the platform, not away from the platform. The gate must also be positioned high enough to allow direct foot access to the platform. *Figure 37* shows the proper access to a platform through a swing gate.

Duck-through access is provided when the user must climb from below and enter a platform headfirst. This may occur when climbing a frame with built-in ladder rungs. The duck-through access system also includes a grab rail to allow the user to have a firm handhold while entering or leaving the platform. The duck-through access must be placed directly above the built-in ladder rungs. *Figure 38* shows two duck-through access panels.

Guardrail systems should be placed so that the guardrail attachment devices face toward the platform. Many of the systems will only install in this manner. Installing the guardrail system properly reduces the possibility that a toprail or midrail will come off if grabbed from the outside or if someone leans or falls against it. *Figure 39* shows the proper installation of a guardrail support post.

31105-14_F35.EPS

Figure 35 Example of a scaffold deck.

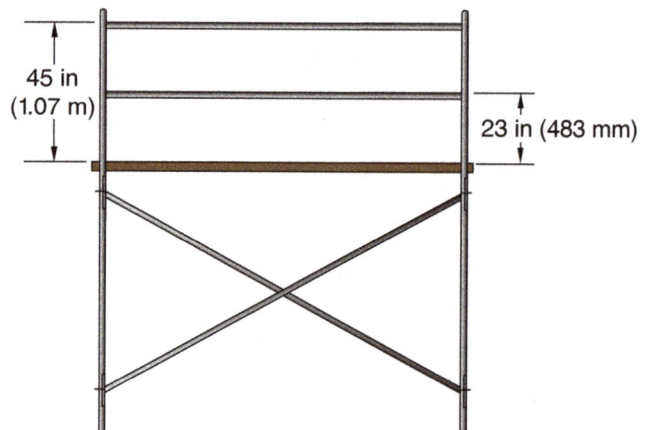

45 in (1.07 m)

23 in (483 mm)

31105-14_F36.EPS

Figure 36 Toprail and midrail locations.

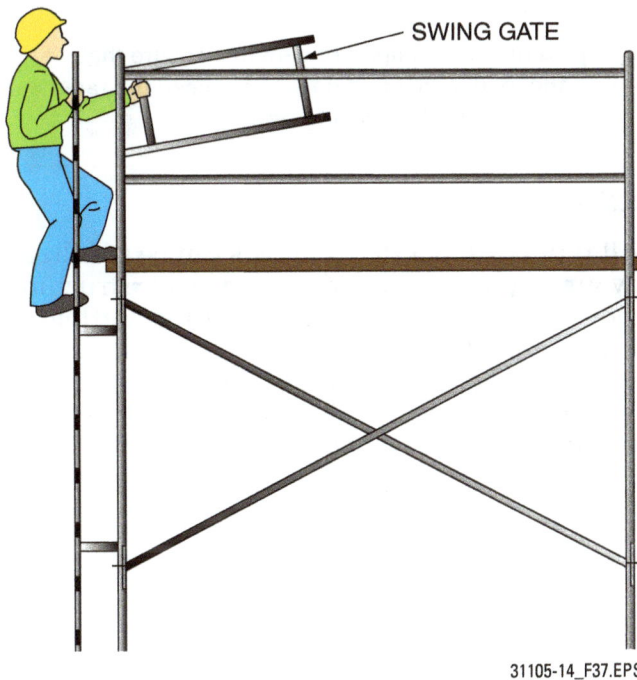

Figure 37 Access to platform through swing gate.

Figure 38 Duck-through access panels.

Placing guardrails on intermediate levels on tubular welded-frame scaffolds may present a challenge. Horizontal rails cannot be placed in the same vertical plane as a cross brace. One method of meeting the guardrail requirement is to use a cross brace plus a horizontal member as a toprail or midrail. The location of the horizontal member is determined by the height of the crossing point of the cross brace above the working platform. OSHA regulations require the addition of a horizontal toprail if the cross point falls between 20 and 30 inches above the platform, but a midrail can be omitted. If the cross point falls between 38 and 48 inches above the platform, a midrail is required, but a toprail can be omitted. Most cross brace attachment points are not long enough to accommodate both the cross brace and the guardrails, so the guardrail must be clamped to the scaffold frames above or below the cross brace. *Figure 40* shows examples of where a midrail is not required (A) and where it is required (B).

There are guardrail panels specifically designed to meet this challenge. These panels are attached to the frame vertical members directly behind the cross braces. Guardrail panels are made to fit standard cross brace spaces. *Figure 41* shows examples of these guardrail panels.

3.2.8 Ladder and Stairs

Ladders and stair units are used to allow easy access to the scaffold work platforms; they can be installed on the inside or outside of the scaffold frame. *Figure 42* shows examples of scaffold ladders and stair units.

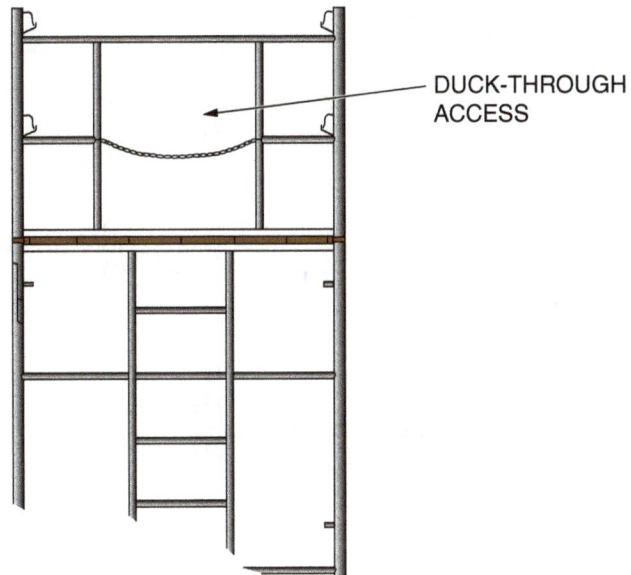

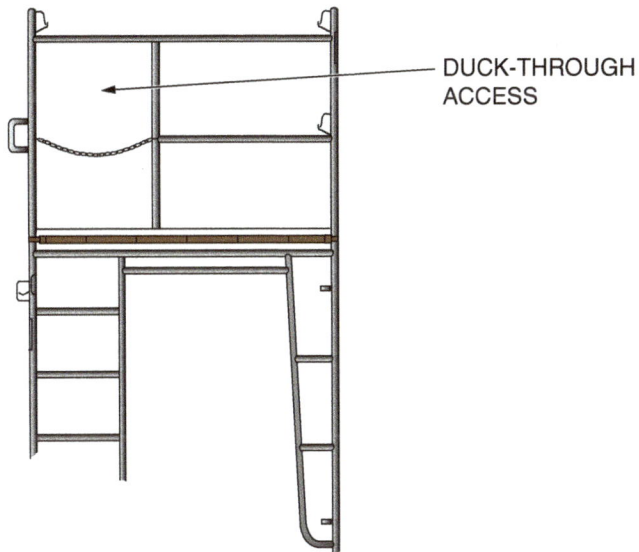

3.2.9 Accessories

Side and end brackets are used to extend the length or width of a scaffold platform. Platforms or decks supported by these brackets must never be used to store work materials or equipment. *Figure 43* shows examples of side and end brackets.

> **NOTE**
>
> If brackets are used, the maximum distance from the wall is 3 inches without handrails.

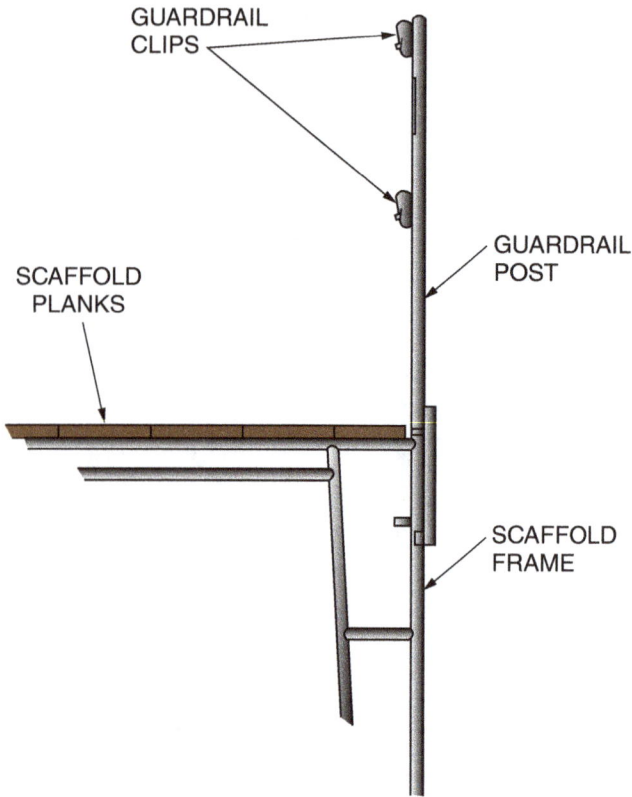

Figure 39 Installation of guardrail support post.

Hoisting arms are used to lift materials or tools to the work platform. When using hoisting arms, the manufacturer's load specifications must never be exceeded. Additional braces may be needed to counteract the overturning forces generated when making lifts. *Figure 44* shows typical hoisting arms.

3.2.10 Ties

All types of supported scaffolds can be stabilized by guying or attaching them to a stable structure. Ties are required if the height of the scaffold is greater than four times the smallest base dimension or if the length exceeds 30 feet. Guying effectively broadens the base enough to meet the requirements, but is not the most practical solution.

The most common practice of tying in a scaffold is to attach the scaffold to the structure being worked on. A tie must consist of the following three components to be effective:

• Anchor
• Tension component
• Compression component

The anchor must be a substantial point on the stable structure. The tension component keeps the scaffold from falling away from the structure. The compression component prevents the scaffold from falling toward or against the structure. It is possible to design ties where one item provides

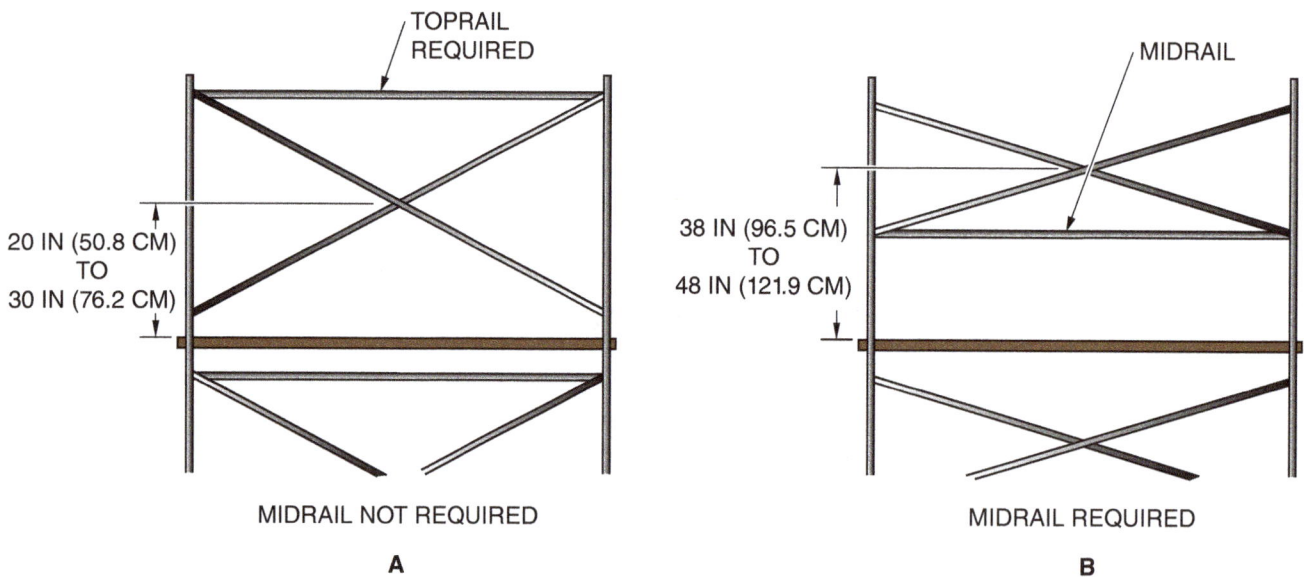

Figure 40 Guardrails and cross braces.

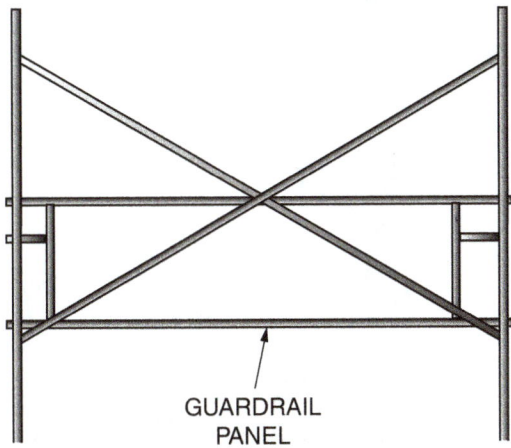

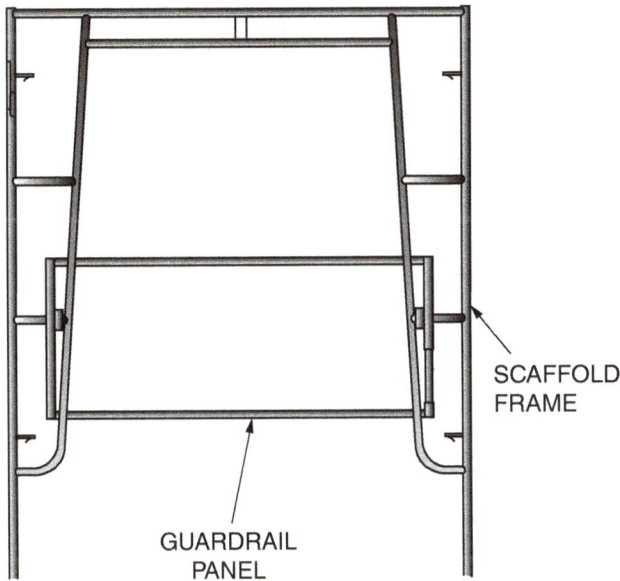

Figure 41 Guardrail panels.

31105-14_F41.EPS

both tension and compression components. *Figure 45A* shows the individual tie components, and *Figure 45B* shows an example where one item provides both tension and compression components.

The tension component has to be strong enough to withstand the forces that would pull the scaffold away from the anchor. It does not have to withstand the forces that push the scaffold toward the support; therefore, it can be made of flexible material such as wire, wire rope, or manila rope.

The compression component, however, must be strong enough to resist the force that would cause the scaffold to move toward the support. It can be made of wood blocking, steel tube, stiff steel rod, or angle iron.

The third component, the anchor, has to be able to withstand the forces that pull the scaffold away from the support. It is anchored or attached to the building in some way. It may be an insert or a lag bolt and expansion shield embedded in concrete or bolt fastened to a structural member of the building, such as a column. The method of anchoring the tie to the structure depends on the materials available in the structure and even in the face of the structure. *Figure 46* shows different methods of anchoring a tie.

NOTE

Not all of these anchoring methods are allowed in some areas. Check with local regulations before selecting an anchoring method.

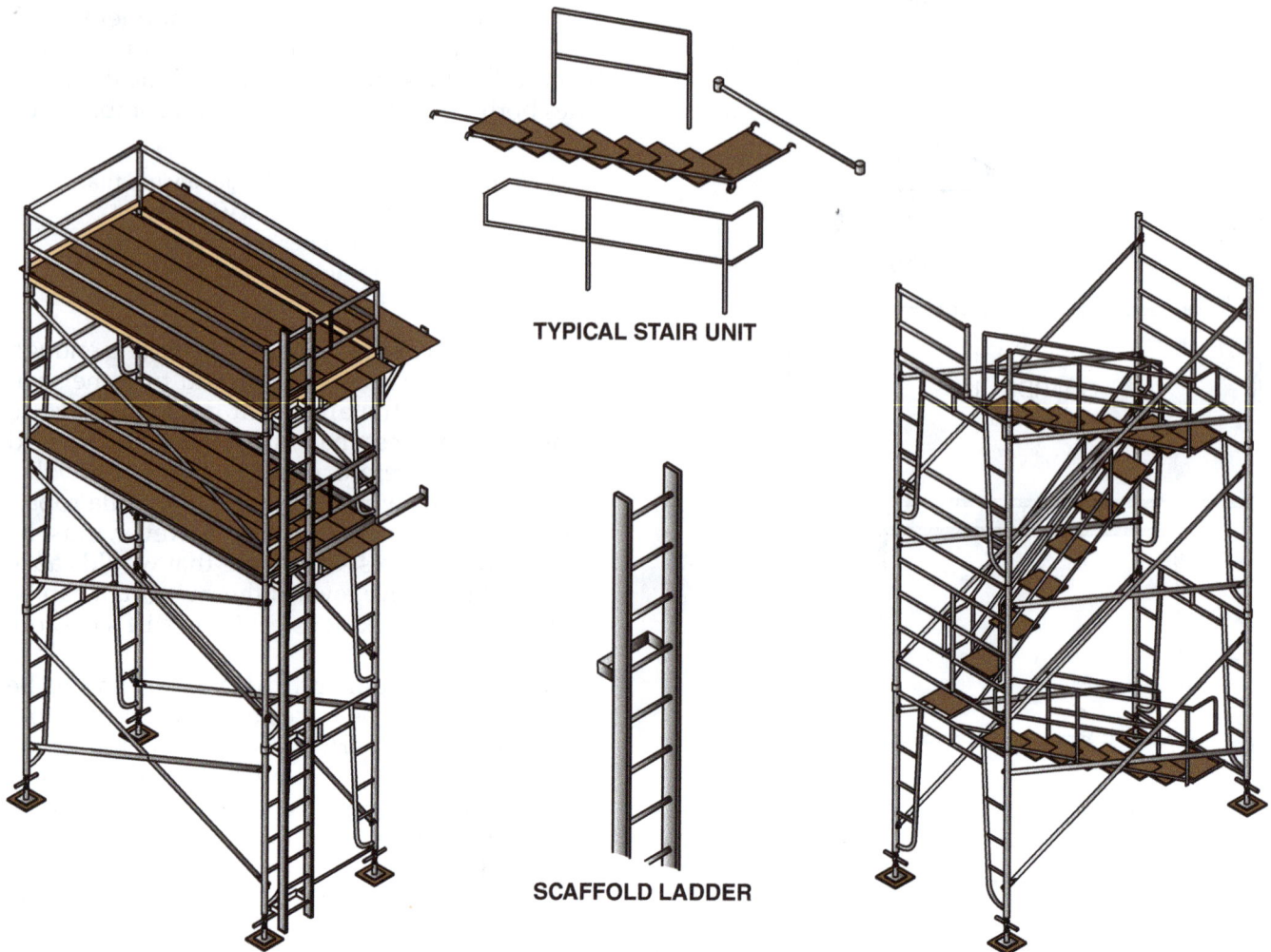

TYPICAL STAIR UNIT

SCAFFOLD LADDER

31105-14_F42.EPS

Figure 42 Scaffold ladders and stair units.

3.3.0 Proper Erection

Tubular welded-frame scaffolds are constructed from the following six essential components:

- Base plate or screw jack with base plate
- Scaffold frames
- Cross braces
- Platform
- Guardrail system (if the platform is more than 10 feet off the ground)
- Method of access, such as a ladder, stairway, or access from an adjacent structure

A tubular welded-frame scaffold system is erected using multiples of these basic components.

The following sections discuss scaffold foundations that apply to all supported scaffolds. Selection of materials for a simple scaffold system three-frames high and four-frames long is described. It will be built with a stair access on one end and a bolt-on ladder on the other end. Side brackets will be used to support a small platform to provide closer access to the adjacent building. This scaffold will be erected on a firm concrete slab. A larger multiple-bay scaffold is analyzed in the *Appendix*. *Figure 47* shows a tubular welded-frame scaffold with the components in their relative positions.

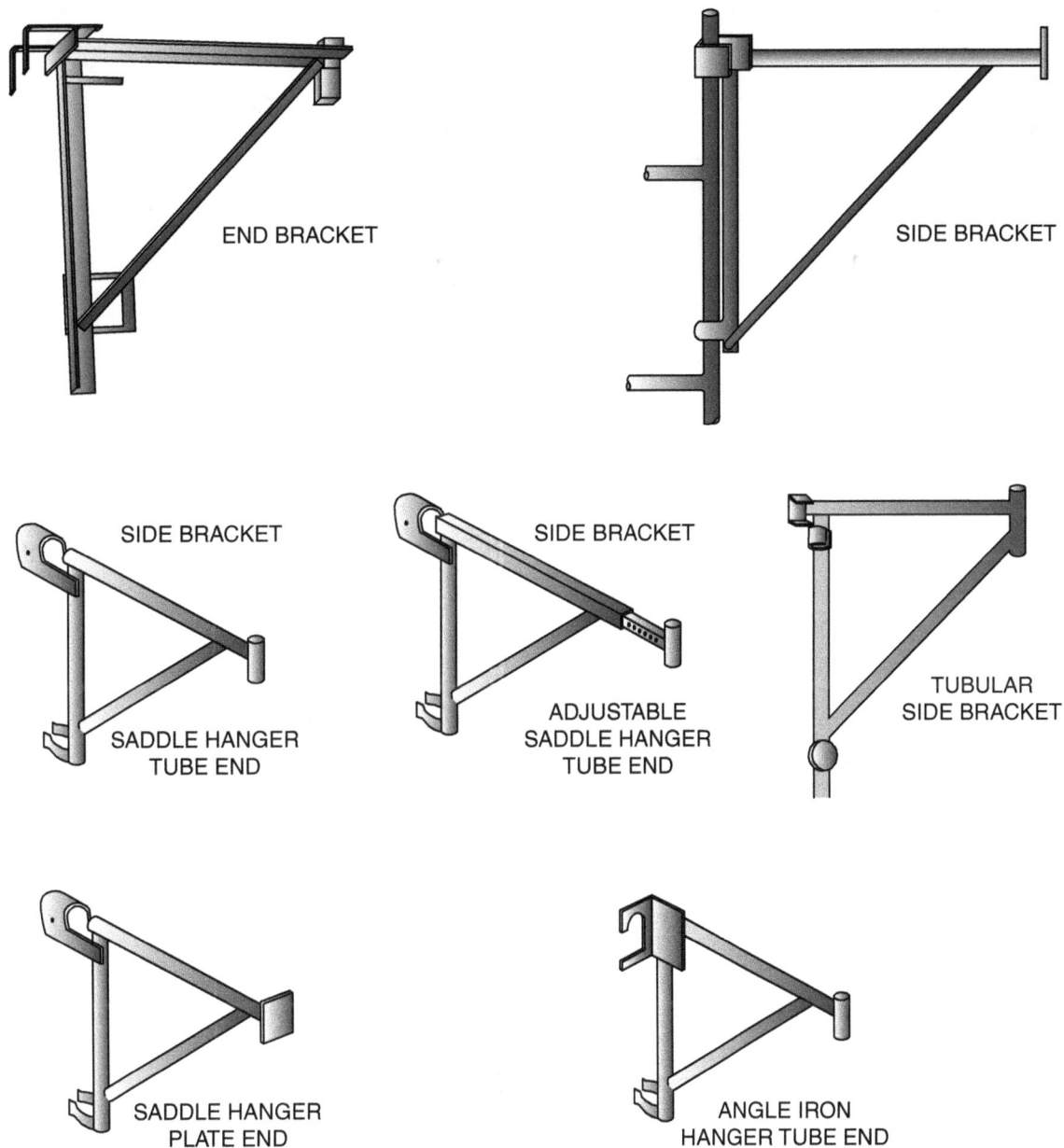

Figure 43 Side and end brackets.

END BRACKET

SIDE BRACKET

SIDE BRACKET
SADDLE HANGER
TUBE END

SIDE BRACKET
ADJUSTABLE
SADDLE HANGER
TUBE END

TUBULAR
SIDE BRACKET

SADDLE HANGER
PLATE END

ANGLE IRON
HANGER TUBE END

31105-14_F43.EPS

3.3.1 Foundation Preparation

In order to ensure the stability of a scaffold, the scaffold builder must first establish an effective foundation. An inadequate foundation can cause the scaffold to collapse. The type of support required for a given scaffold depends on two factors:

• The function of the scaffold and the loads it must carry
• The subsurface conditions

When selecting supports, the scaffold designer should do the following:

• Perform a site inspection. The bearing surface must be able to adequately support the scaffold and all the loads placed on it.
• Determine the general subsurface conditions.
• Determine which of the common types of supports could be constructed under the existing conditions, whether it would be able to carry the required loads, and whether it would settle in such a way as to cause instability. Reject any types that are obviously unsuitable.

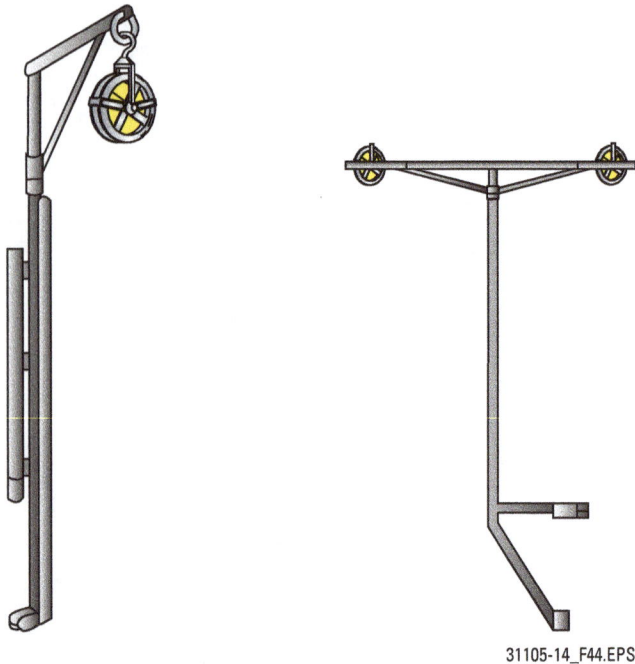

31105-14_F44.EPS

Figure 44 Typical hoisting arms.

- Make a detailed study of the remaining types of supports and tentative scaffold designs. This includes obtaining more detailed information about the scaffold loads and subsurface conditions to determine the number and size of mudsills required.
- If the foundation is an existing structure, contact a qualified person to determine whether the structure is capable of supporting the scaffold structure and loads.

The purpose of a good foundation is to spread the leg load over a wide area. The total load and the conditions of the supporting soil or floor determine the size of the foundation. When soil conditions are poor or frozen, it may be necessary to dig down to an adequate base material and fill in with good compacted material. If this is not practical, then the leg load must be spread over a much larger area using larger mudsills or a continuous deck under the scaffold legs. To support scaffolds, backfilled soils must be well-compacted and leveled. Mud or soft soil must be replaced with gravel or crushed stone that has been well-compacted. *Figure 48* shows how a bad footing is corrected by excavation and backfilling to prevent settling or system failure.

When a scaffold must be erected on sloping ground, the mudsill areas must be leveled by excavating rather than by filling and compacting. If the surface cannot be excavated, swivel jacks are used with appropriate mudsills. Bases should be nailed to mudsills using four nails on sloped

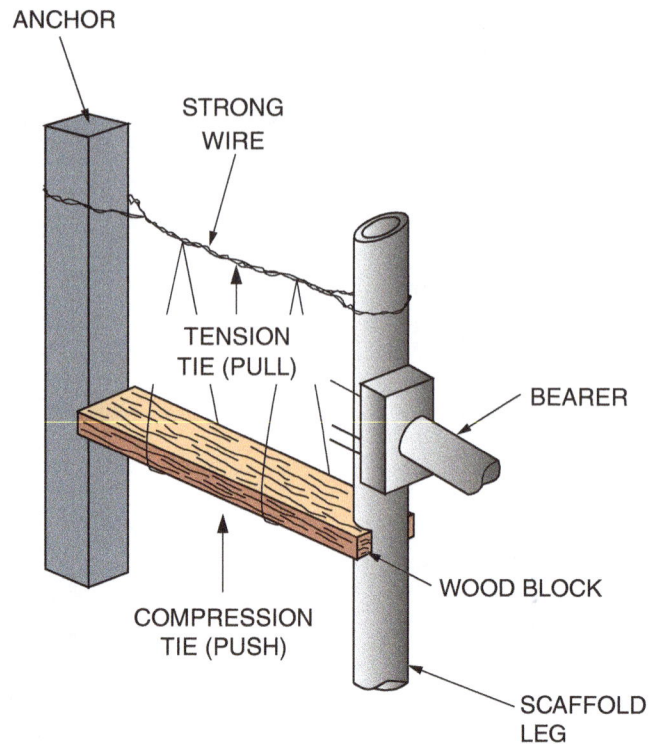

A

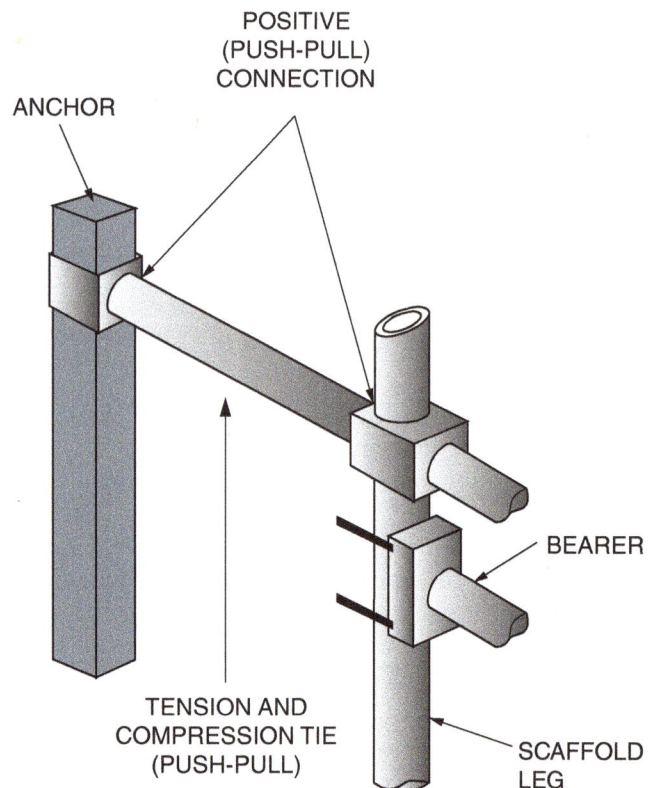

B

31105-14_F45.EPS

Figure 45 Tie components.

LAG BOLT AND
EXPANSION SHIELD

3 IN (76.2 mm)

BRACKET

GUARDRAIL

CLAMP

BLANK
WALL

SCAFFOLD
FRAME LEG

TUBE

RIGHT-ANGLE
CLAMP

SCAFFOLD
FRAME LEG

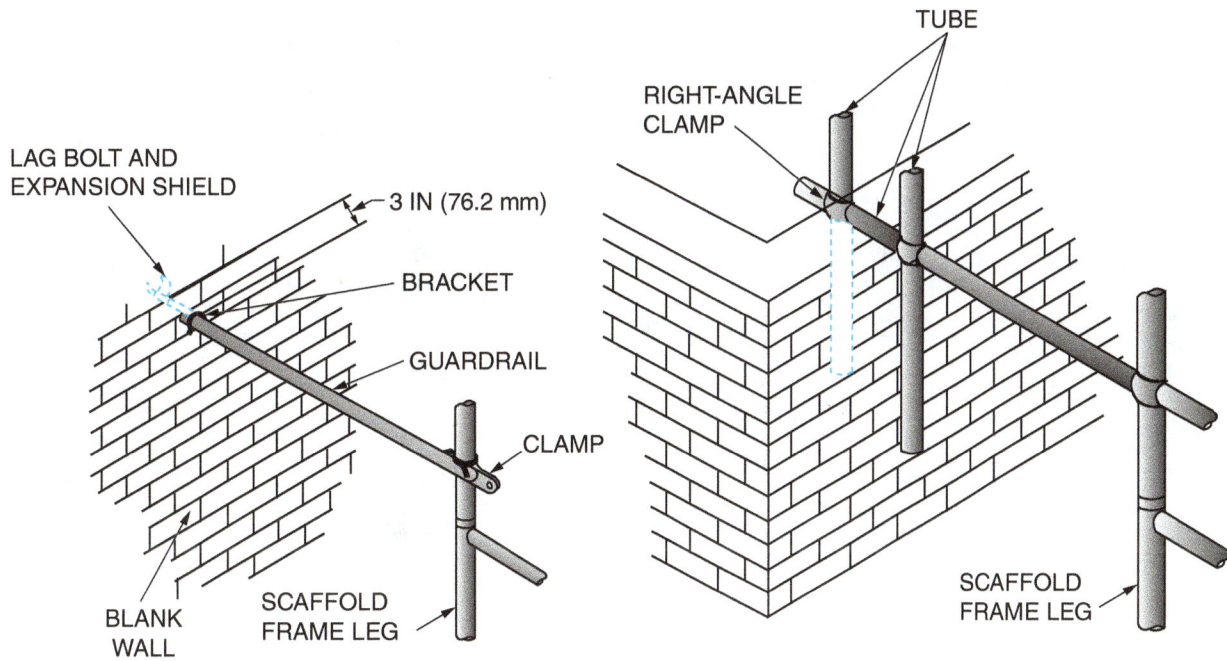

WARNING: NEVER ATTACH A RIGHT-ANGLE
CLAMP TO A BEAM. ALWAYS
USE A BEAM CLAMP.

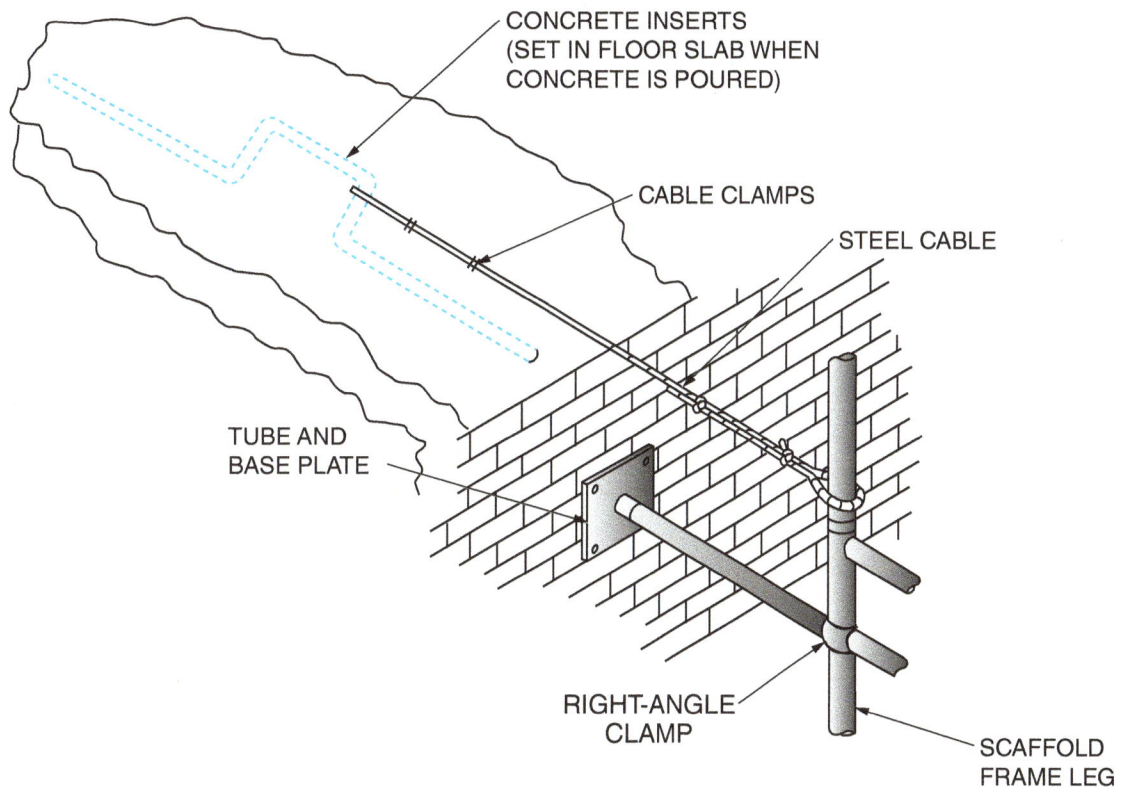

CONCRETE INSERTS
(SET IN FLOOR SLAB WHEN
CONCRETE IS POURED)

CABLE CLAMPS

STEEL CABLE

TUBE AND
BASE PLATE

RIGHT-ANGLE
CLAMP

SCAFFOLD
FRAME LEG

31105-14_F46A.EPS

Figure 46A Methods of anchoring a tie (1 of 3).

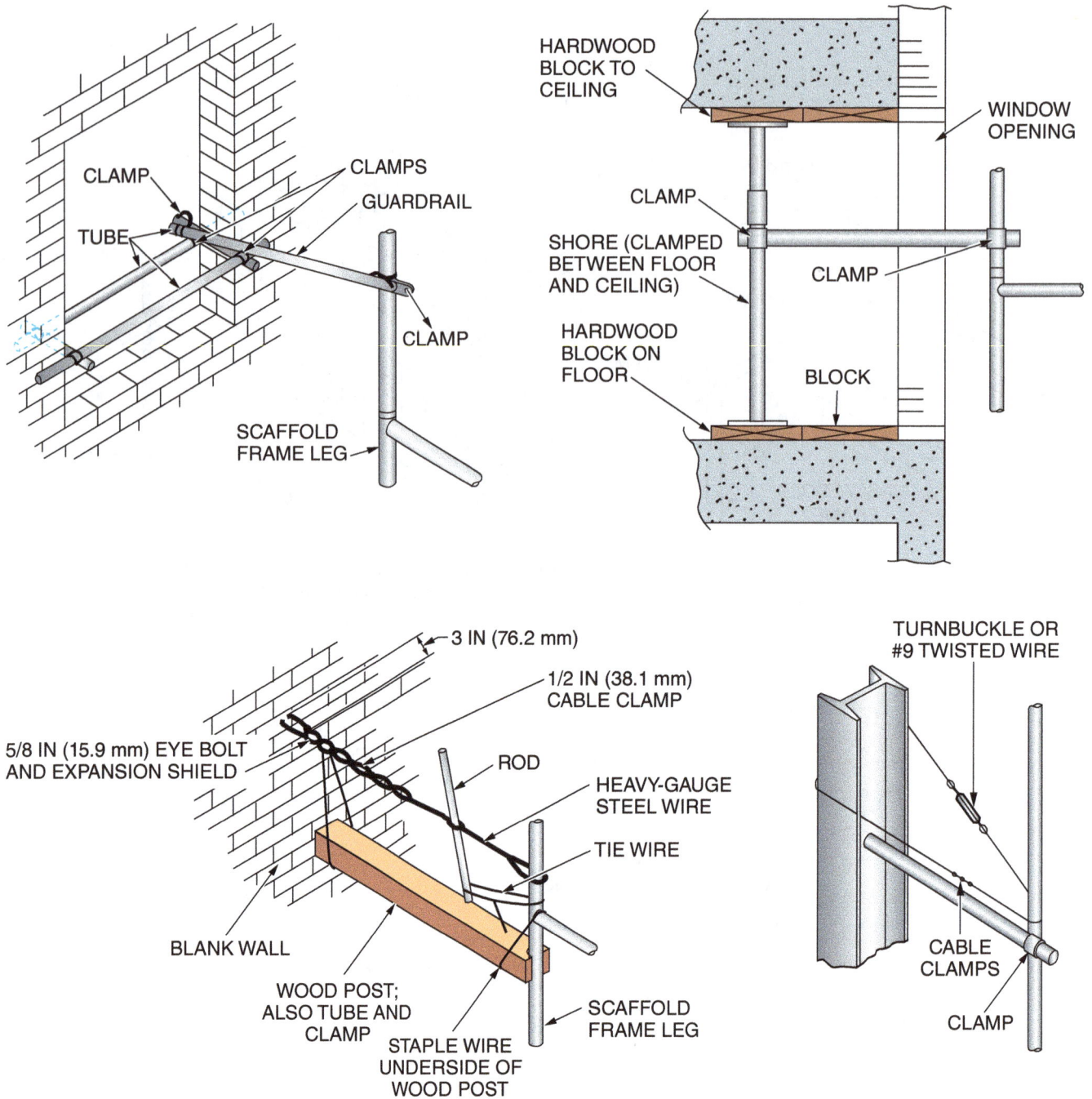

Figure 46B Methods of anchoring a tie (2 of 3).

surfaces or two nails on level surfaces. *Figure 49* shows a scaffold footing on a sloped surface. In this situation, it is more appropriate to use tube-and-clamp scaffolds.

Rainy conditions can turn an adequate foundation into a hazardous condition. Dry clay that has a bearing capacity of 8,000 pounds per square foot can drop to less than 2,000 pounds per square foot when saturated. When scaffold legs are located near ditches or other excavations, the legs should be placed at least the depth of the ditch away from the edge of the excavation. *Figure 50* shows the required distance from a ditch.

Mudsills should be used under scaffold base plates even when the scaffold is erected on an asphalt or concrete slab. This aids in the distribution of the leg load and allows better contact with the surface. The minimum recommended mudsill size is 18 inches by 18 inches.

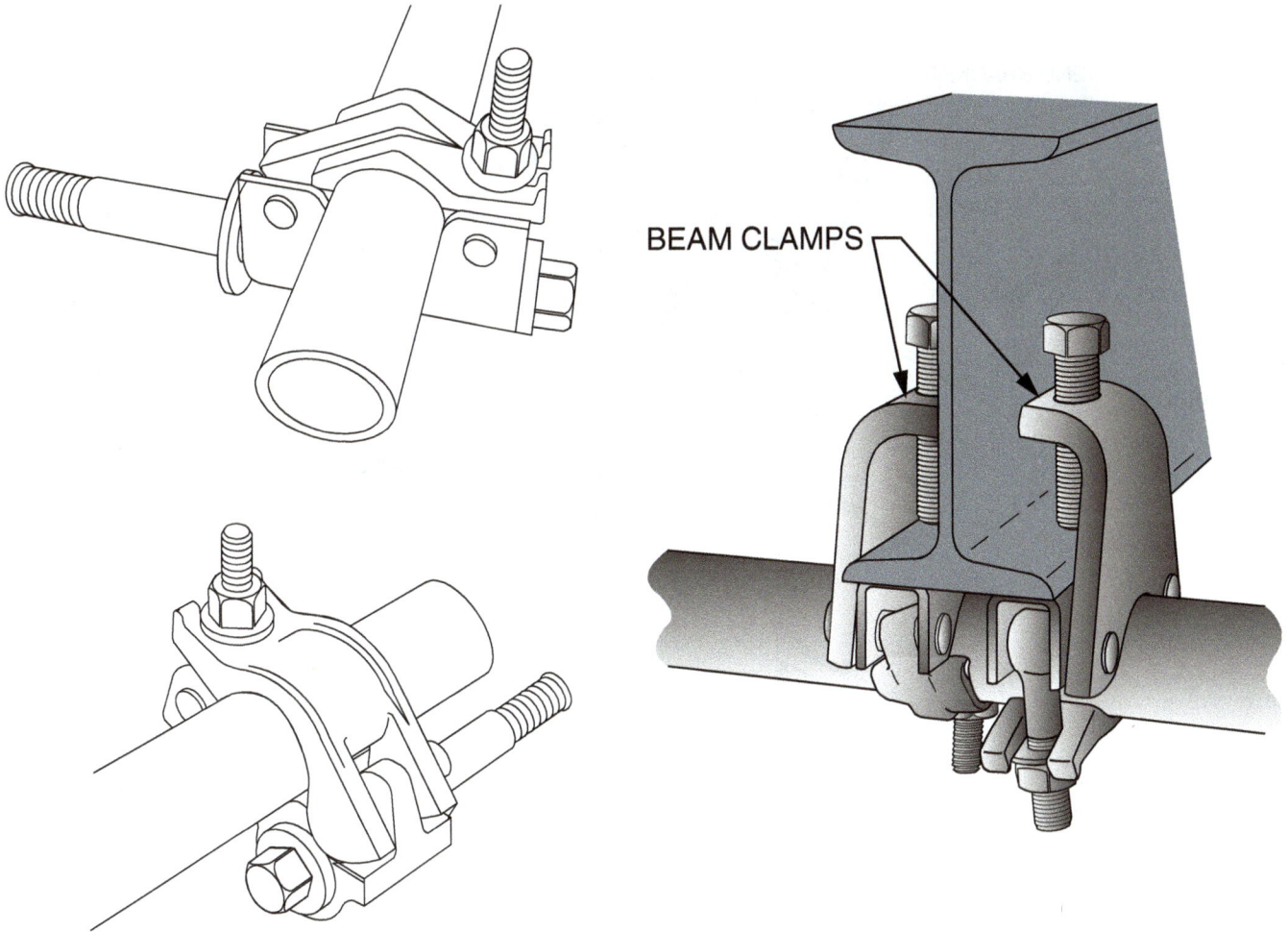

BEAM CLAMPS

31105-14_F46C.EPS

Figure 46C Methods of anchoring a tie (3 of 3). Beam clamps must always be used in pairs that are designed to be attached to both sides of the beam.

Mudsills should be made of scaffold-grade lumber. Old scaffold planks can be used as mudsills. If more than one layer of boards is necessary to build up a mudsill, each layer should be placed at a right angle to the previous layer and nailed well. This "cribbing" should never be taller than its width. The base plate should be nailed to the mudsill to prevent shifting.

3.3.2 Equipment Selection

The scaffold example presented in this section is three-frames high and four-frames long with two working levels. The left end bay will be used to support a stairway that will provide access to the top working level. There will be a bolt-on ladder attached to the right-side end frames. Screw jacks will be used on all legs to provide adjustment for small level differences in the concrete slab on which it is built. For ease of material selection and assembly, only 5-foot by 6-foot, 7-inch walk-through frames and standard solid sawn scaffold boards will be used. This assembly will be built for light-duty applications.

This scaffold system requires the following:

- 15 scaffold frames
- 10 screw jacks
- 10 base plates

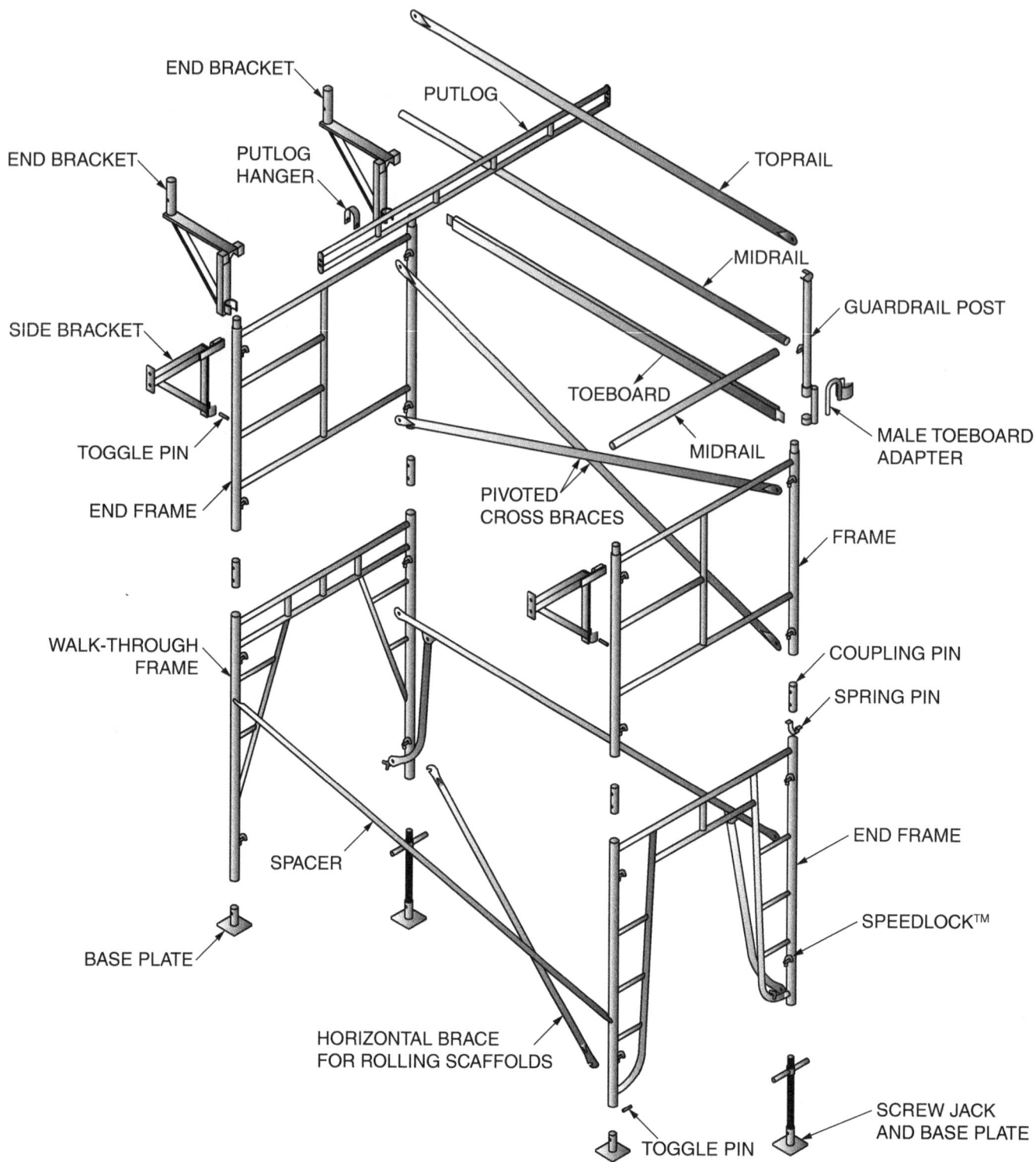

Figure 47 Tubular welded-frame scaffold components.

- 10 mudsills (18 inches × 18 inches × 1½ inches)
- 28 coupling pins
- 56 coupling pin locking devices
- 12 cross braces
- 6 guardrail legs
- 8 guardrails

- 3 guardrail panels
- 2 guardrail gate panels
- 3 stair units
- 4 side brackets with locking pins
- 42 scaffold planks
- Toeboards

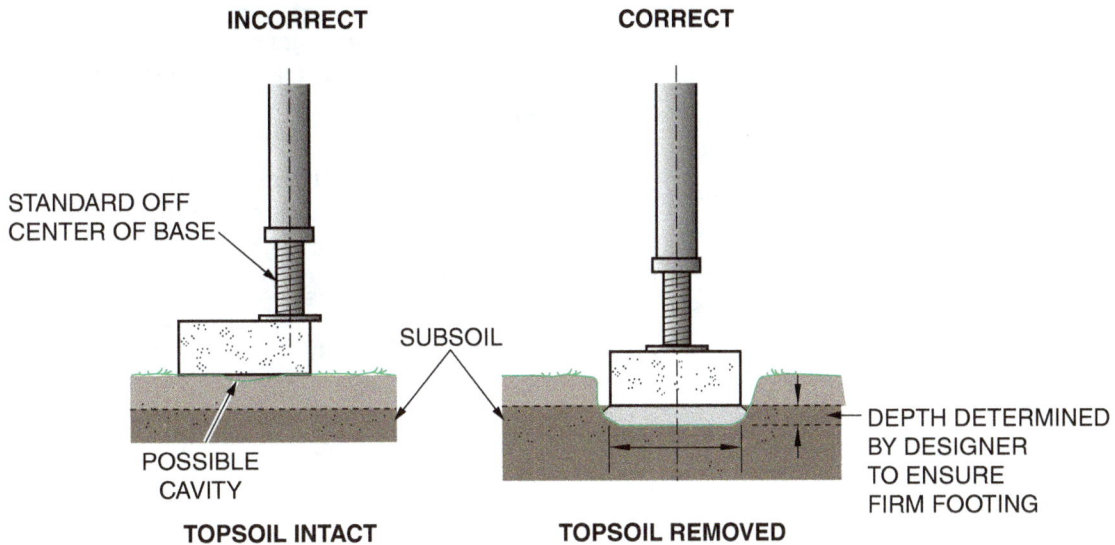

INCORRECT

CORRECT

STANDARD OFF
CENTER OF BASE

SUBSOIL

POSSIBLE
CAVITY

DEPTH DETERMINED
BY DESIGNER
TO ENSURE
FIRM FOOTING

TOPSOIL INTACT

TOPSOIL REMOVED

31105-14_F48.EPS

Figure 48 Excavation and backfilling.

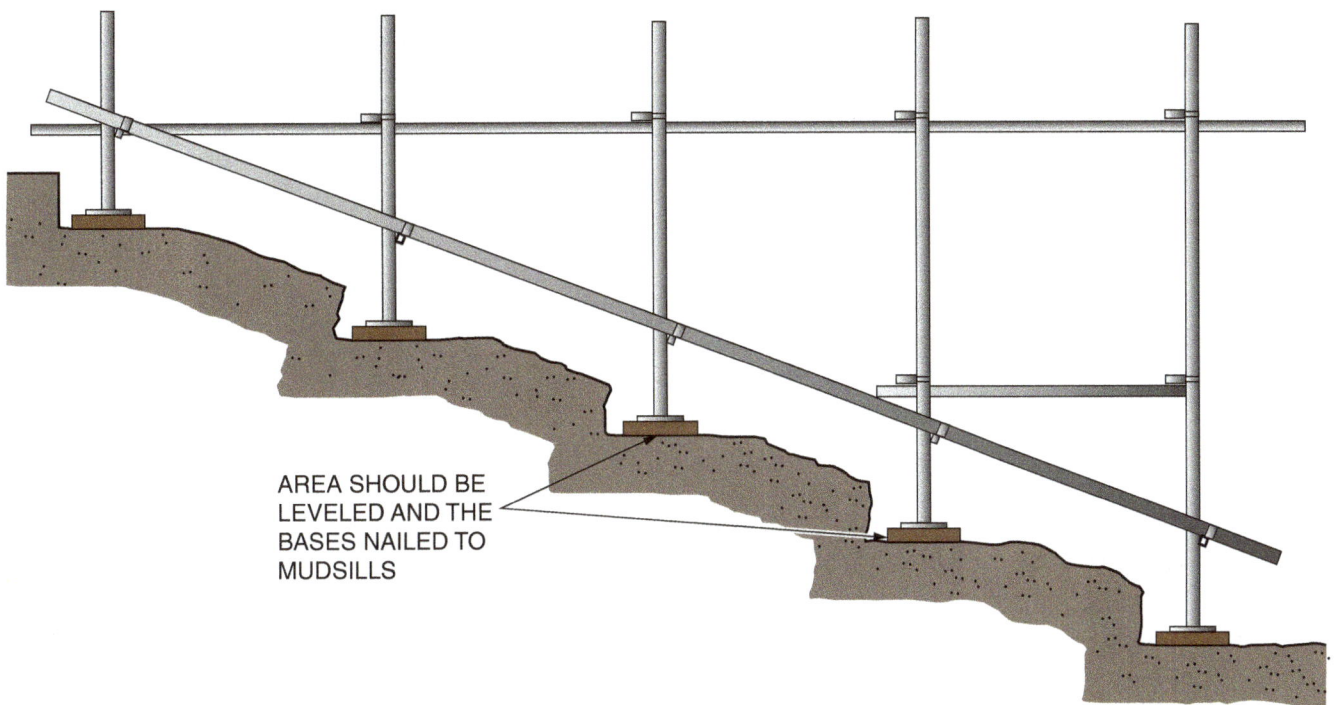

AREA SHOULD BE
LEVELED AND THE
BASES NAILED TO
MUDSILLS

31105-14_F49.EPS

Figure 49 Scaffold footing on a sloped surface.

3.3.3 Proper Assembly

Perform the following steps to assemble a four-bay scaffold:

Step 1 Barricade the area where the scaffold is to be erected.

Step 2 Gather all scaffold components as close as possible to the assembly location. Inspect each item as it is removed from storage or the delivery vehicle.

Step 3 Remove any debris from the area where the scaffold is to be erected.

Step 4 Locate the highest point that will be used to support a frame.

Step 5 Place the mudsills in their approximate locations.

Step 6 Place a base plate in the center of each mudsill and secure it with nails.

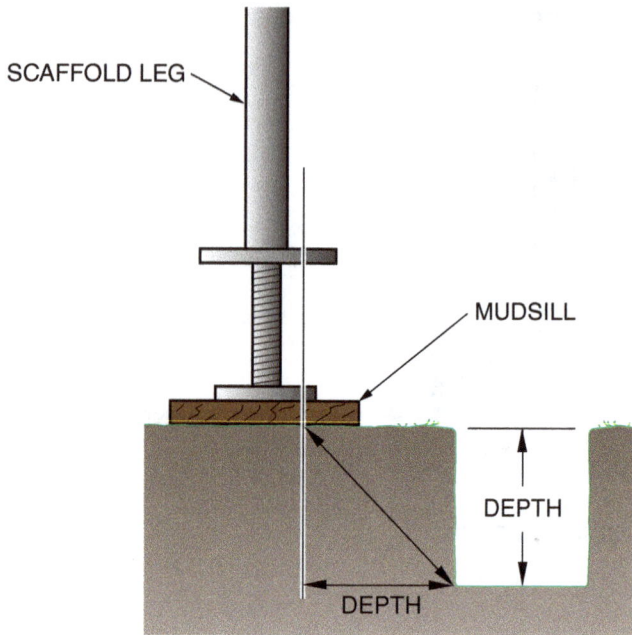

Figure 50 Required distance of scaffold legs from a ditch.

Step 7 Place a screw jack onto each of the base plates.

Step 8 Set the first scaffold frame down onto the screw jacks and base plates, starting at the highest point.

Step 9 Attach cross braces to each side of the scaffold frame, and lock in place according to the manufacturer's specifications.

Step 10 Set the second frame on the adjacent base plate/mudsill assembly.

Step 11 Attach the other ends of the cross braces to the second scaffold frame and lock into place according to the manufacturer's specifications.

Step 12 Adjust the screw jack at the highest point to approximately 1 inch from its lowest position. *Figure 51* shows the completed first frame.

Step 13 Check the plumb of each scaffold frame, using a torpedo level.

Step 14 Adjust the screw jacks as necessary to plumb and level the frames and scaffold bay. *Figure 52* shows leveling the first span.

Step 15 Adjust the location of the base plates until the diagonal distance across the bay in each direction is identical.

> **NOTE**
>
> When the diagonal distances across the bays are identical, the corner angles are at 90 degrees and the frame will be square.

Step 16 Attach the next pair of scaffold braces to the scaffold frames, according to the manufacturer's specifications.

Step 17 Set the next frame on the screw jacks.

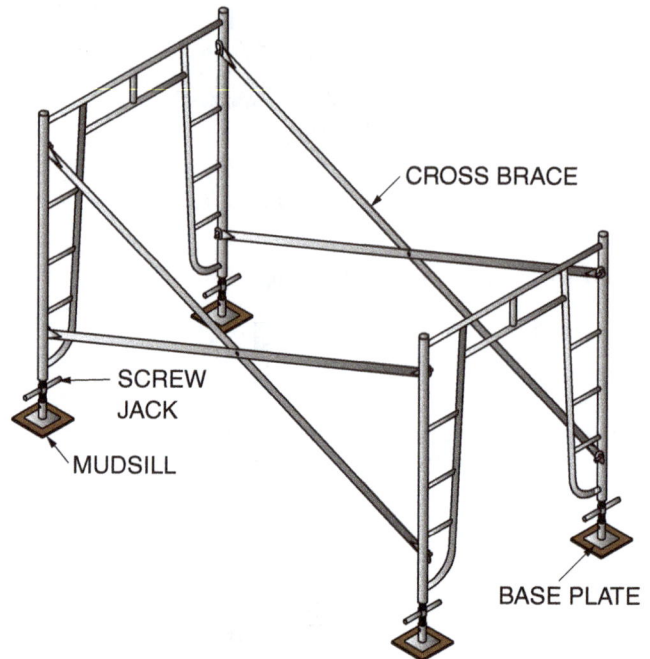

Figure 51 Completed first frame.

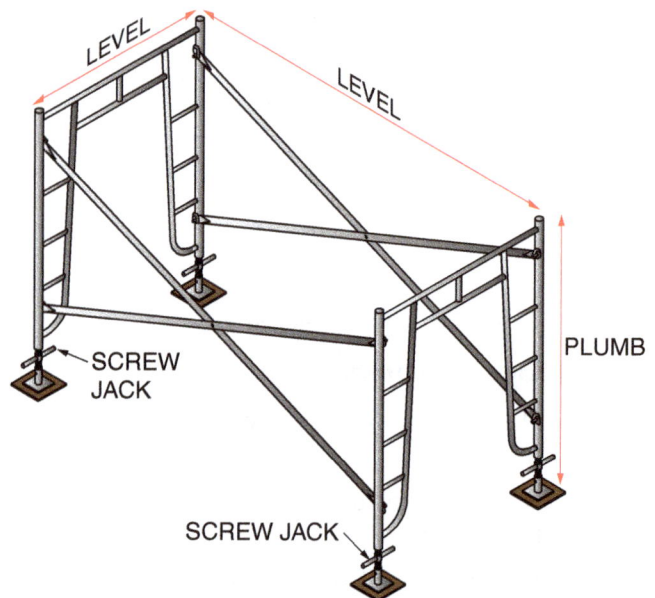

Figure 52 Leveling the first span.

Step 18 Attach the other end of the scaffold braces to the loose frame.

Step 19 Adjust the location of the base plates/mudsills as necessary to keep the scaffold run parallel with the structure.

Step 20 Level and square the bay as before, adjusting the screw jacks and base plate/mudsill locations.

Step 21 Repeat Steps 15 through 19 for the last two bays. *Figure 53* shows the completed first-level frames.

Step 22 Place coupling pins in the top of each of the scaffold frame legs and secure them in place with locking pins.

Step 23 Place a Warning tag on the scaffold, indicating that it is not complete and should not be used.

Step 24 Attach the ladder bracket and first ladder section to the right end of the scaffold, according to the manufacturer's directions.

Step 25 Place six scaffold planks on top of the scaffold frames in the first and third bays from the right. Center the planks on the span, leaving approximately 6 to 12 inches extending beyond the frame on each end.

Step 26 Place six scaffold planks on the second and fourth frames, resting the boards on top of the boards placed in Step 25. These planks should extend approximately 6 to 12 inches beyond the scaffold frames.

> **NOTE**
> This level will not be a working level. These planks are only to provide a platform to use while assembling the next level.

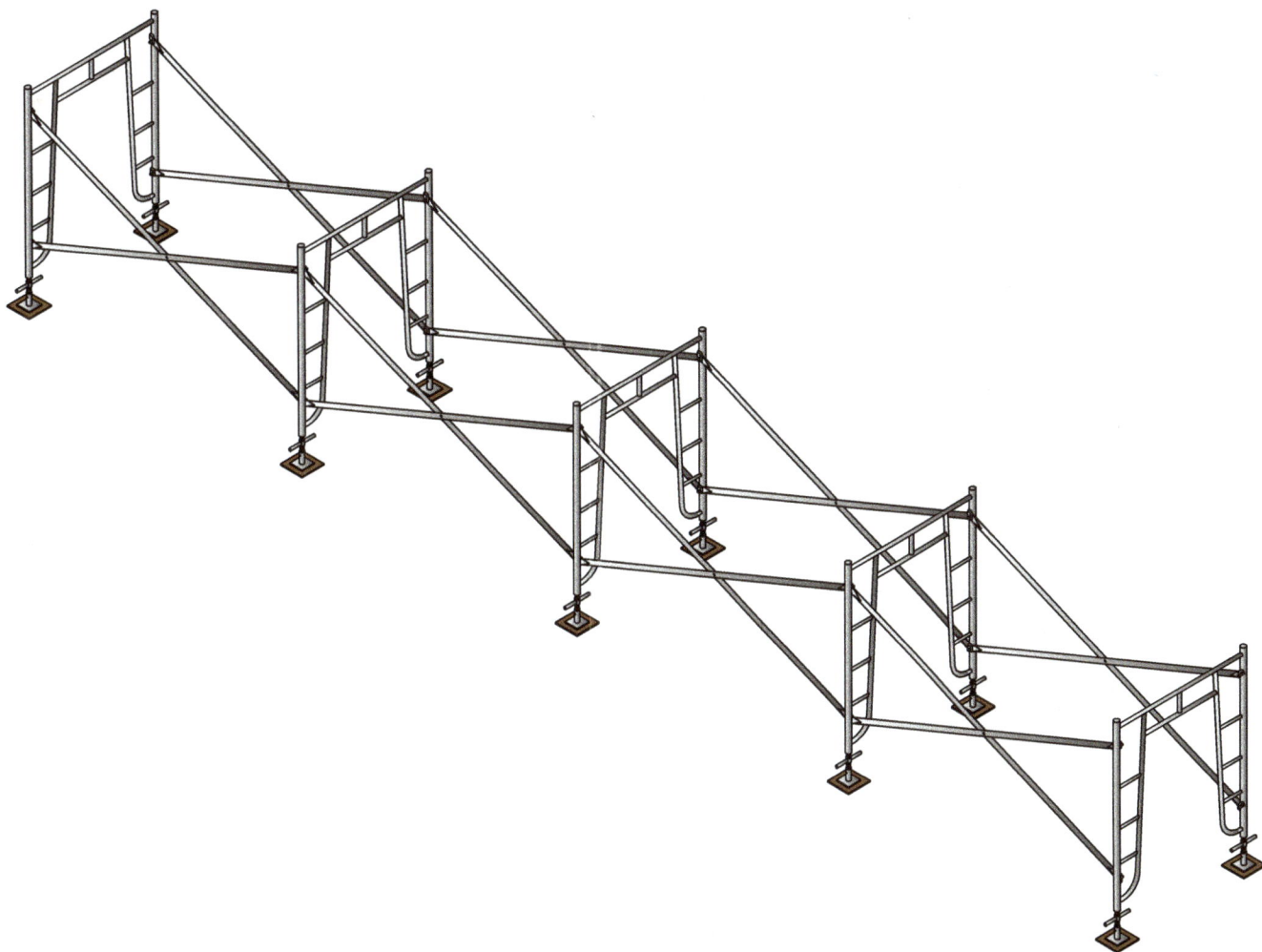

31105-14_F53.EPS

Figure 53 Completed first-level frames.

Step 27 Place coupling pins in the top of four additional scaffold frames, and lock them in place with locking pins.

Step 28 Hand a scaffold frame up and place it firmly on the coupling pins in the top of the first-level frame.

> **WARNING!**
> At this point, the scaffold builder working on the scaffold is subject to a fall hazard. To prevent personal injury or death, wear personal fall arrest equipment when assembling scaffolds.

Step 29 Secure the second-level frames to the first-level frames by inserting locking pins through the coupling pins.

Step 30 Attach a cross brace to the second-level scaffold frame and secure it in place, according to manufacturer's specifications.

Step 31 Hand up the next scaffold frame and set it firmly in place on the coupling pins of the next first-level frame.

Step 32 Secure the frame in place by placing locking pins through the coupling pins and securing the cross braces according to the manufacturer's specifications.

Step 33 Repeat Steps 28 through 32 for each of the remaining frames.

Step 34 Attach sufficient ladder sections to the existing ladder sections to reach the second level. Be sure to follow the manufacturer's specifications.

Step 35 Place six scaffold planks on top of the scaffold frames in the first and third bays from the right of the third-level frames. Center the planks on the span, with approximately 6 to 12 inches extending beyond the frame on each end.

Step 36 Remove the scaffold planks from two of the sections of the first level and place them in the second and fourth bays of the third level, overlapping the ends of the planks of the first and third bays as before. *Figure 54* shows the planked second level.

Step 37 Place coupling pins in the top of the four remaining scaffold frames, and lock them in place with locking pins.

Step 38 Hand a scaffold frame up and place it firmly on the coupling pins in the top of the second-level frame.

> **WARNING!**
> At this point the scaffold builder working on the scaffold is subject to a fall hazard. To prevent personal injury or death, wear personal fall arrest equipment when assembling scaffolds.

Step 39 Secure the third level to the lower level by inserting locking pins through the coupling pins.

Step 40 Attach a cross brace to the third level scaffold frame, and secure it in place according to the manufacturer's specifications.

Step 41 Hand up the next scaffold frame and set it firmly in place on the coupling pins of the next second-level frame.

Step 42 Secure the frame in place by placing locking pins through the coupling pins and securing the cross braces according to the manufacturer's specifications.

Step 43 Repeat Steps 38 through 42 for each of the remaining frames.

Step 44 Attach guardrail panels to the scaffold frames of the three right-side bays of the third-level scaffold frames.

Step 45 Attach guardrails across the left end of the third frame.

Step 46 Attach a gate panel to the right end of the second level to align with the attached ladder.

> **WARNING!**
> Make sure that the gate is installed to swing inward toward the platform. An improperly installed gate can result in personal injury or death.

Step 47 Attach sufficient ladder sections to the existing ladder sections to reach at least 3 feet above the third-level platform. Be sure to follow the manufacturer's specifications.

Step 48 Remove the scaffold planks from the left span of the second level and place them on the third level.

Step 49 Adjust the spacing of the scaffold planks on the second level so that there is no more than a 1-inch gap between the planks.

Step 50 Place toeboards on both sides of the second level.

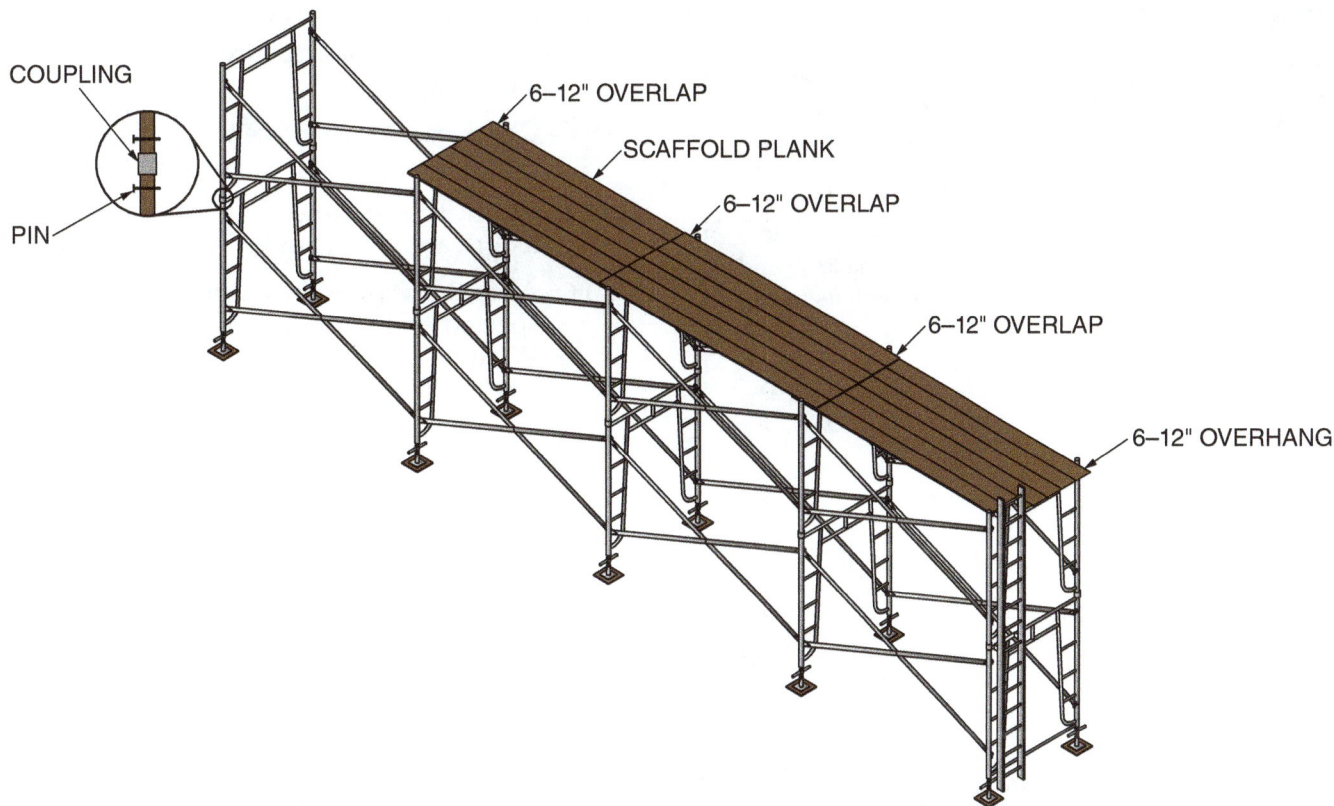

Figure 54 Planked second level.

Step 51 Place toeboards on both ends of the second level, resting them on top of the scaffold planks.

Step 52 Secure the toeboards to the scaffold frames with wire so that they are held firmly down against the scaffold planks.

Step 53 Pass the remaining scaffold planks up and place them in the first and second bays of the third level.

Step 54 Pass up and install the four guardrail legs onto the outside scaffold frame legs of the three right-side frames.

> **WARNING!**
> Guardrail legs must be installed so the guardrail brackets are on the inside toward the platform. Improperly installed guardrails can result in personal injury or death.

Step 55 Install guardrail legs on the inside legs at each end of the planked frames.

Step 56 Pass up and install the four side brackets on the inside scaffold legs of the three right-side frames, and secure them in place with locking pins.

Step 57 Place scaffold planks on the side brackets, and secure them in place with wire.

Step 58 Attach toprails and midrails to the outside guardrail legs.

Step 59 Attach a guardrail gate panel at the right end of the third level, aligning the gate with the ladder.

> **WARNING!**
> Make sure that the gate is installed to swing inward toward the platform. An improperly installed gate can result in personal injury or death.

Step 60 Install the stair sections inside the left bay in accordance with the manufacturer's specifications.

Step 61 Attach a guardrail gate panel at the left end of the planked sections, aligning the gate with the staircase.

Step 62 Attach scaffold ties from the end frame's inside scaffold leg to the building. *Figure 55* shows the completed scaffold.

Even though the scaffold ties are not required since the scaffold is not over 20-feet high (4 by 5-foot width), the side brackets provide a tipping load that must be accommodated. The scaffold ties would be necessary by California OSHA regulations because it exceeds 15-feet tall (3 by 5-foot width). Check with all local and state regulations for tying requirements at your location.

Step 63 Check each scaffold frame leg connection to ensure that it is secure and locked with a locking pin.

Step 64 Check each cross brace to ensure that it is properly secured to the scaffold frames.

Step 65 Check the scaffold for plumb along the length and across the width of the run.

Step 66 After the scaffold has been completely inspected and prepared for use, remove the Warning tag placed in Step 23 and replace it with a Ready To Use tag.

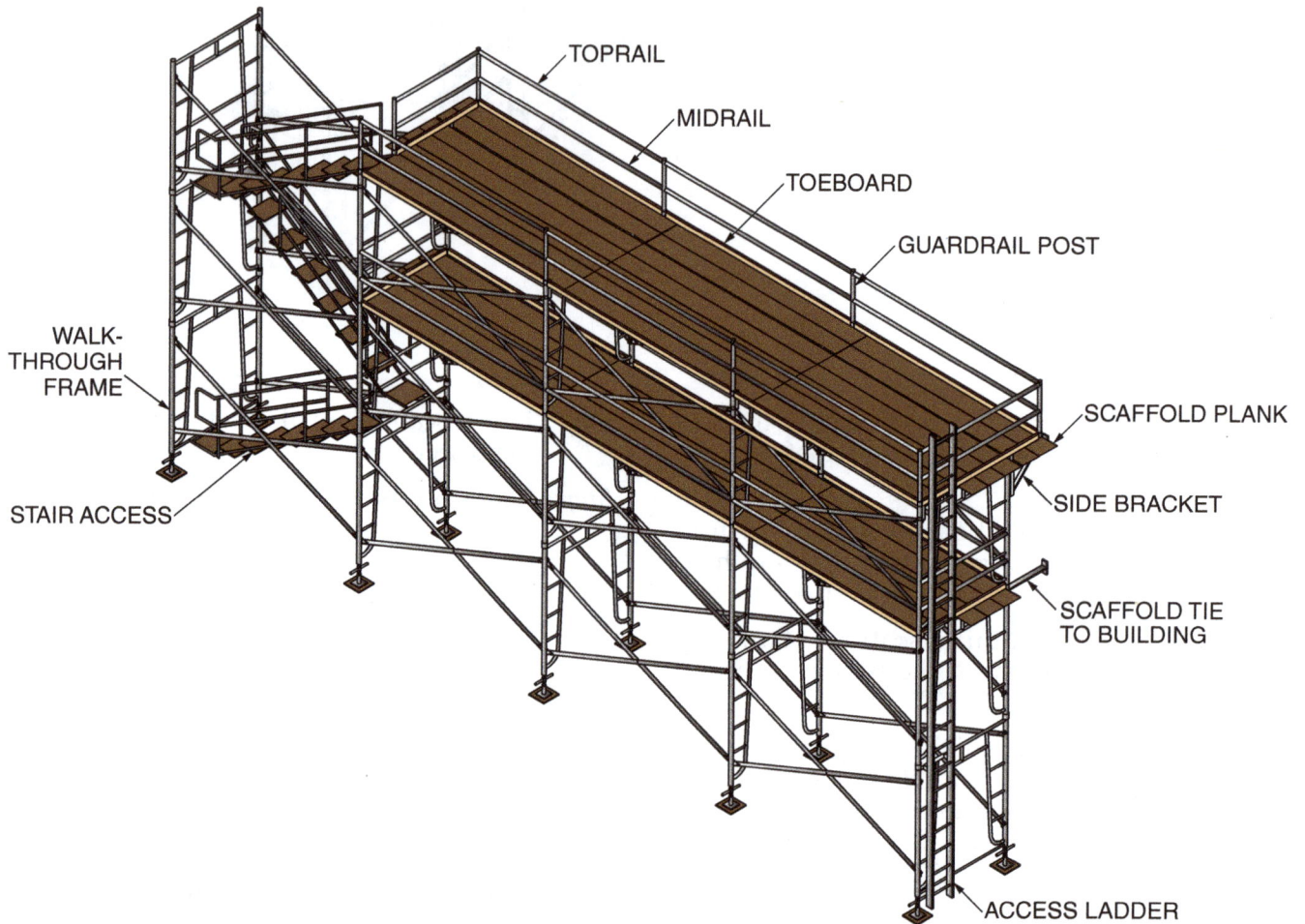

31105-14_F 55.EPS

Figure 55 Completed scaffold.

Additional Resources

"Scaffolding." OSHA. **www.osha.gov/SLTC/scaffolding/index.html**

Fall Protection and Scaffolding Safety: An Illustrated Guide. 2000. Grace Drennan Ganget. Government Institutes.

International Building Code®, Latest Edition. Falls Church, VA: International Code Council.

International Residential Code®, Latest Edition. Falls Church, VA: International Code Council.

3.0.0 Section Review

1. Tubular welded-frame scaffold end frames are secured by _____.

   a. anchor bolts
   b. pop rivets
   c. keys
   d. coupling pins

2. If side brackets are used, the maximum distance from the wall, without handrails, is _____.

   a. 2 inches
   b. 3 inches
   c. 4 inches
   d. 5 inches

3. When a scaffold must be erected on sloping ground, the mudsill area must be leveled by _____

   a. compacting
   b. spreading gravel
   c. excavating
   d. filling

4.0.0 ERECTING TUBE-AND-CLAMP SCAFFOLDS

Objectives

Explain the basic principles of tube-and-clamp scaffolds, and outline proper erection procedures.

a. Identify common applications of tube-and-clamp scaffolds.
b. Identify the components of tube-and-clamp scaffolds.
c. Outline the steps for proper erection of a tube-and-clamp scaffold.

Performance Tasks

Safely erect a section of two of the following types of scaffolds:

- System scaffold
- Tubular welded-frame scaffold
- Tube-and-clamp scaffold

Tube-and-clamp scaffolds can be either supported or suspended. In this type of scaffold, the tubes used for posts, runners, bearers, and braces are connected using clamps (couplings).

Tube-and-clamp scaffolds are typically used in difficult or inaccessible places. The various tubes that form the uprights, braces, bearers, and runners are fastened together with right-angle or swivel clamps to construct a scaffold over uneven ground and unusual shapes. End fittings are used to connect the end of one tube with another. *Figure 56* shows an assembled tube-and-clamp scaffold system and identifies some of the basic components.

> **WARNING!**
>
> Swivel clamps should not be used as right-angle clamps because they can twist and weaken the scaffold. Improper use of swivel clamps can cause equipment failure which can result in personal injury or death.
>
> Metal base plates attached to the tubular legs or posts of both tube-and-clamp and tubular welded-frame scaffolds serve as a foundation for supported scaffolds. On dirt and similar surfaces, mudsills must be placed beneath the base plates to provide a secure and level footing for the scaffold.

4.1.0 Applications

A tube-and-clamp scaffold can be erected and used any place a tubular welded-frame scaffold is used. In addition, tube-and-clamp scaffolds can be erected in areas that have space or access restrictions that prevent the use of a frame scaffold. For example, a tube-and-clamp scaffold could be erected inside a tank that has a small manhole for access. The individual parts are small enough to be passed through the access hole and then assembled on the inside. The clamps can also be configured so a scaffold could be erected around a round structure, which is practically impossible to reach with frame scaffolds. Tube-and-clamp scaffolds are more easily adapted to irregular foundations.

Due to the additional skills and time required to assemble tube-and-clamp scaffolds, they are usually reserved for locations where the scaffolds will be left in place for some time.

4.2.0 Components

Due to the design flexibility of the tube-and-clamp scaffold system, it has very few components. The system consists of base plates, tubing, and clamps.

4.2.1 Base Plates

There are two types of base plates for tube-and-clamp scaffold systems. One type of base plate has a male adapter. A tube fits onto the adapter and locks in place with the twist-lock pins. The second type of base plate is an open tube that is large enough to accommodate the outer diameter of the tube. The tube must be locked into this type of base plate by locking pins. Base plates should be used with mudsills whenever possible. *Figure 57* shows base plates.

4.2.2 Twist-Lock Fittings

The basic tube of twist-lock fittings is a nominal 2-inch outside-diameter (OD) steel. Tubes come in various lengths from 4 to 16 feet. Other structural materials may be used if they are designed to carry an equivalent load. Twist-lock fittings are provided at the ends of the tubes to secure longer runs together. Tubes without these fittings are designed for end-of-run applications. *Figure 58* shows the tube twist-lock arrangement.

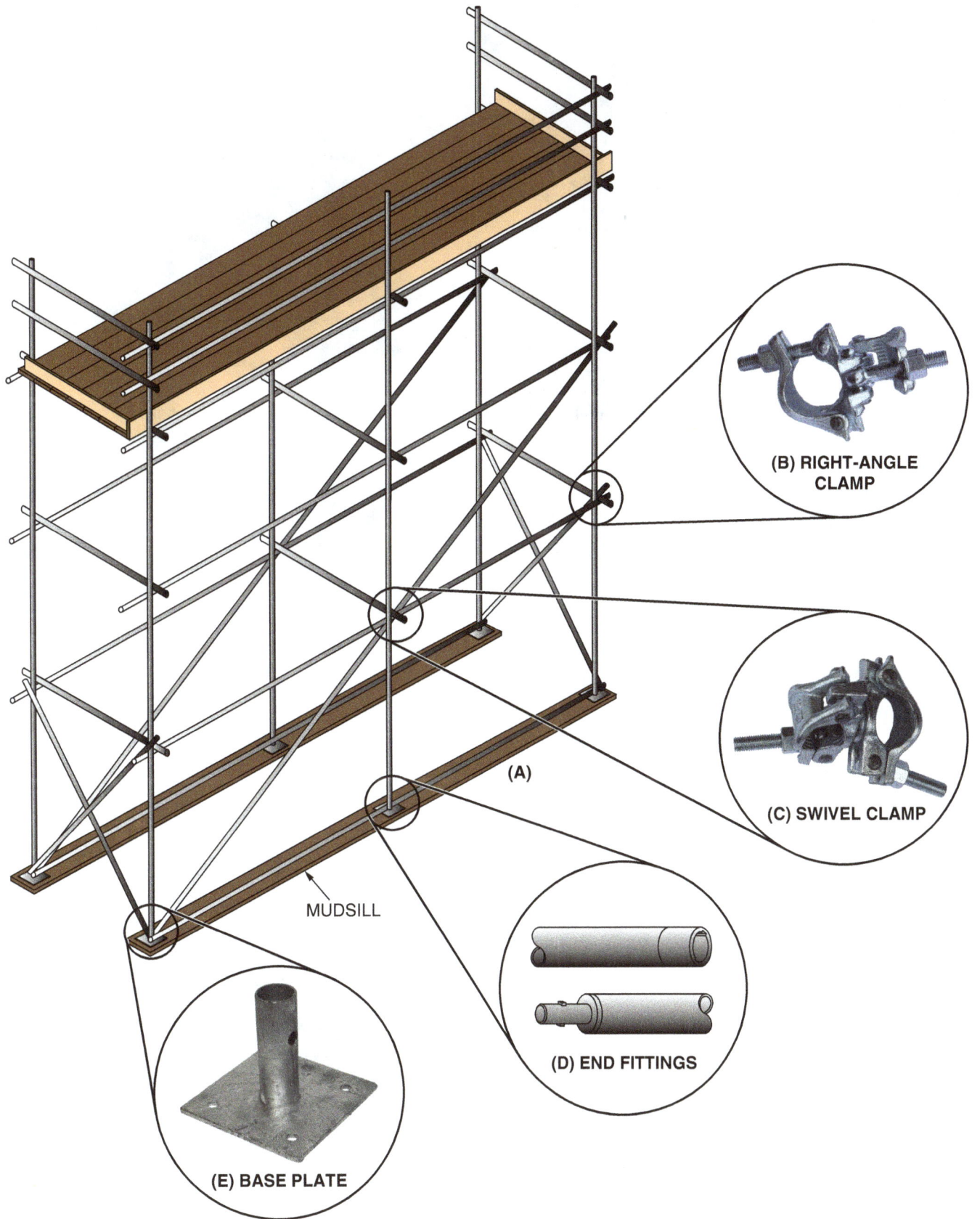

(B) RIGHT-ANGLE CLAMP

(C) SWIVEL CLAMP

(A)

MUDSILL

(D) END FITTINGS

(E) BASE PLATE

31105-14_F56.EPS

Figure 56 Tube-and-clamp scaffold system.

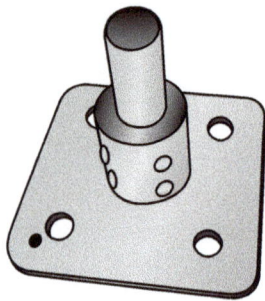

(A) TWIST-LOCK

(B) OPEN-TUBE

31105-14_F57.EPS

Figure 57 Base plates.

31105-14_F58.EPS

Figure 58 Twist-lock arrangement.

31105-14_F59.EPS

Figure 59 Right-angle clamp.

4.2.3 Right-Angle Clamps

Legs, bearers, and runners meet at right angles. Right-angle clamps are used in these applications and are designed to securely hold these joints. Generally, the maximum strength of a right-angle clamp is 2,500 to 3,500 pounds. Always check the manufacturer's specification for loading recommendations and limitations. *Figure 59* shows a right-angle clamp.

4.2.4 Swivel Clamps

Diagonal braces are attached to the supporting members using a swivel clamp. While these clamps may be positioned at any angle for a number of different applications, they are best installed when holding the component cradle-side up to minimize the possibility of failure. Always check the manufacturer's specification for loading recommendations and limitations. Generally, the maximum strength of a swivel clamp is 750 to 1,500 pounds. *Figure 60* shows a swivel clamp.

31105-14_F60.EPS

Figure 60 Swivel clamp.

4.2.5 Beam Clamps

Scaffold tubes can be secured to vertical and horizontal beams with specialized beam clamps. These clamps are secured to the beam by a bolt that is tightened to the flange of the beam. Always check the manufacturer's specification for loading recommendations and limitations. *Figure 61* shows a beam clamp.

4.2.6 Other Components

Some platform planks are fitted with hooks designed to hang on the bearers. Designed access gates, ladders, and stair units are also available to attach to the 2-inch tubes of this system. Side and end brackets are unnecessary since runners or bearers can be extended a similar distance beyond the legs.

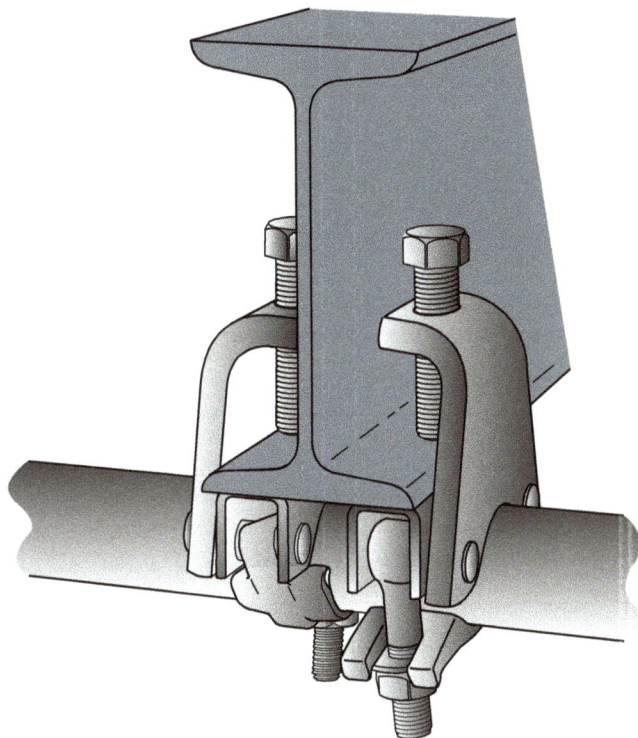

31105-14_F61.EPS

Figure 61 Beam clamp.

4.3.0 Proper Erection

The following sections discuss scaffold foundations that apply to all tube-and-clamp scaffolds. The selection of materials for a simple scaffold structure measuring 20-feet tall and 21-feet long is discussed. The scaffold will be built with a stair access on one end and a bolt-on ladder on the other end. This scaffold will be erected on a firm concrete slab. The actual assembly of the scaffold structure is also explained. A larger multiple-bay scaffold is analyzed in the *Appendix*.

4.3.1 Foundation Preparation

In order to ensure the stability of a scaffold, the scaffold builder must first establish an effective foundation. An inadequate foundation can cause the scaffold to collapse. Refer to Section 3.0.0 for instruction on foundation preparation.

4.3.2 Equipment Selection

Due to the small number of different components used in the construction of a tube-and-clamp scaffold, selection of materials merely involves ensuring the availability of the required lengths of tube, each type of clamp required, and each type of base plate required. Study the sketch of the scaffold to be built, if available, to determine the lengths of tube needed. A clamp should be installed wherever two tubes cross. This applies to legs, runners, bearers, or braces. A base plate and mudsill are required for each post. Selection of decks or planks is determined by the scaffold builder's choice and the work to be performed from the scaffold. Guardrail systems use the same tubes, and should be installed on the inside of the posts using right-angle clamps.

4.3.3 Assembly

For this exercise, a scaffold three-levels high by three 7-foot bays long will be constructed. This scaffold will have one planked level at the top and a ladder access to a cantilevered platform. It will be built on reasonably level ground, with mudsills at the base of the ladder and beneath every post. Standard solid sawn planks will be used. *Figure 62* shows the tube-and-clamp scaffold to be constructed.

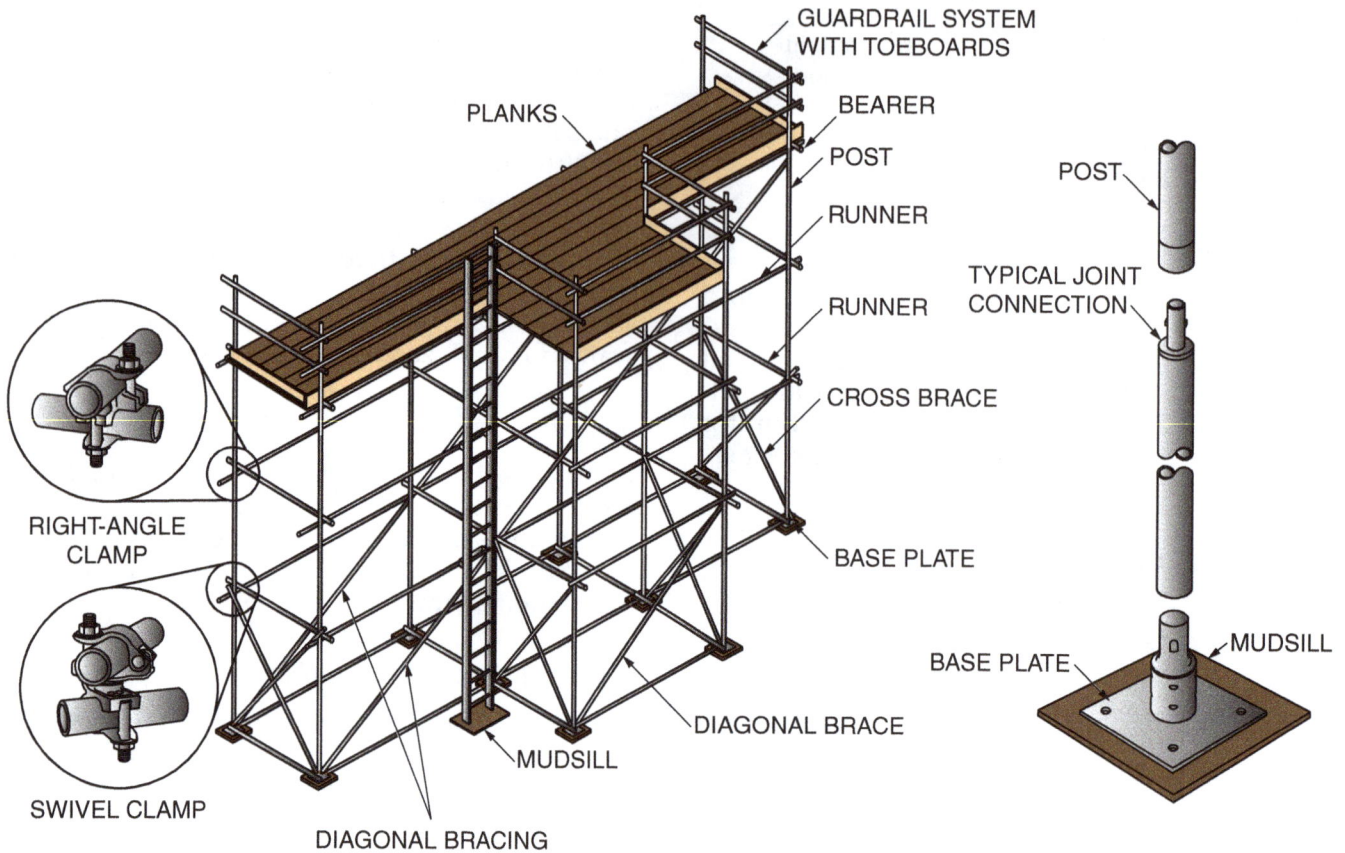

Figure 62 Tube-and-clamp scaffold.

Perform the following steps to erect the four-bay tube-and-clamp scaffold:

Step 1 Remove any debris from the base area.

Step 2 Lay out the mudsills in their approximate locations.

Step 3 Determine the scaffold width.

> **NOTE**
>
> Tube-and-clamp scaffold width should be set to allow the scaffold planks to fit without gaps. In this example, four standard 2 x 10 scaffold planks will be used. The scaffold posts should be set at 42 inches center-to-center (four 10-inch-wide planks = 40 inches, plus 2 inches for post centers).

Step 4 Secure the base plates to the mudsills on 42-inch centers.

Step 5 Set a post on the first base plate, making sure it locks in place.

Step 6 Attach a runner to the post near the base plate using a right-angle clamp.

> **NOTE**
>
> At this point, clamps should be snugged but not tightened since they will need to be adjusted to square and level the scaffold assembly. Clamps are normally installed so that the bolt is on the top and the flap is facing palm up.

> **WARNING!**
>
> To keep tubes from falling during scaffold erection, do not support the tubes with open clamps. Loose or falling tubes can cause personal injury or death.

Step 7 Attach a bearer directly on top of the runner with a right-angle clamp so that it extends along the mudsill to the other base plate.

Step 8 Set another post on the back base plate and twist it to lock it in place.

Step 9 Attach a runner to the post near the base plate using a right-angle clamp.

Step 10 Measure 7 feet along each of the runners (center-to-center) and mark the position of the next set of posts.

Step 11 Position the next mudsill and base plate assembly to fit the marks.

Step 12 Set posts on each base plate of the second mudsill and twist them to lock them in place.

Step 13 Attach the runners to the posts using right-angle clamps.

Step 14 Attach a bearer directly on top of the runner using a right-angle clamp.

Step 15 Measure the diagonal distances of the span and adjust the position of the posts to square the assembly.

Step 16 Plumb each of the posts.

Step 17 Level the runners and tighten the clamps in place.

Step 18 Level the bearers and tighten the clamps in place, resting the clamps on the clamps that secure the runners to the posts.

Step 19 Lock another tube to the runners to extend them to the next span.

Step 20 Measure 7 feet along each of the runners (center-to-center) and mark the position of the next set of posts.

Step 21 Position the next mudsill and base plate assembly to fit the marks.

Step 22 Set posts on each base plate of the second mudsill and twist them to lock them in place.

Step 23 Attach the runners to the legs using right-angle clamps.

Step 24 Attach a bearer to the vertical post directly on top of the runner using a right-angle clamp.

Step 25 Measure the diagonal distances of the span and adjust the post position to square the assembly.

Step 26 Plumb the new posts.

Step 27 Level the runners and tighten the clamps in place.

Step 28 Level the bearers and tighten the clamps in place, resting the clamps on the clamps that secure the runners to the posts.

Step 29 Repeat Steps 17 through 28 to set the remaining span.

Step 30 Install and level the second-level runners on the first span.

Step 31 Install and level second-level bearers over the first-span runners.

Step 32 Install diagonal braces across the end of the first span using swivel clamps. Attach one diagonal brace on the outside of the legs and one on the inside of the legs to form cross braces. Clamps should be placed below the runner clamps to avoid interfering with the scaffold planks and provide additional support for the bearers.

Step 33 Install diagonal braces on the outside from the first outside post up to just below the runner at the second post.

Step 34 Install and level second-level runners and bearers on the remaining spans, ensuring that the legs remain plumb.

Step 35 Install diagonal braces across the end of the third span, using swivel clamps. Attach one diagonal on the outside of the legs and one on the inside of the legs to form cross braces. Clamps should be placed below the runner clamps to avoid interfering with the scaffold planks and provide additional support for the bearers.

Step 36 Install a ladder with a mudsill to provide a safe means of access and egress. *Figure 63* shows the completed first level.

Step 37 Place scaffold planks on the first and third spans of the scaffold, making sure that 6 to 12 inches of plank extend beyond the bearer on each end.

> **NOTE**
> Planks are installed at this level only temporarily to provide a place to walk while working on the next level. The planks can be moved to the next level when it is ready.

Step 38 Place scaffold planks on the second span of the scaffold, resting the planks on top of the planks placed in Step 37, making sure that 6 to 12 inches of plank extend beyond the bearer on each end.

Step 39 Pass up and attach members to extend the posts to the second level.

> **WARNING!**
> To prevent personal injury or death, scaffold builders are required to have a personal fall arrest system if it is feasible and if it does not create a greater hazard. Most scaffold manufacturers warn against using the scaffold as an anchorage point for fall arrest systems. Check with your safety department for company policy and procedures.

Step 40 Pass up, install, and level the third-level runners.

> **NOTE**
> The maximum height for these runners is 6 feet, 6 inches above the lower-level runners.

Step 41 Pass up and install the third-level bearers, resting the bearer clamps on the runner clamps and making sure that the posts remain plumb.

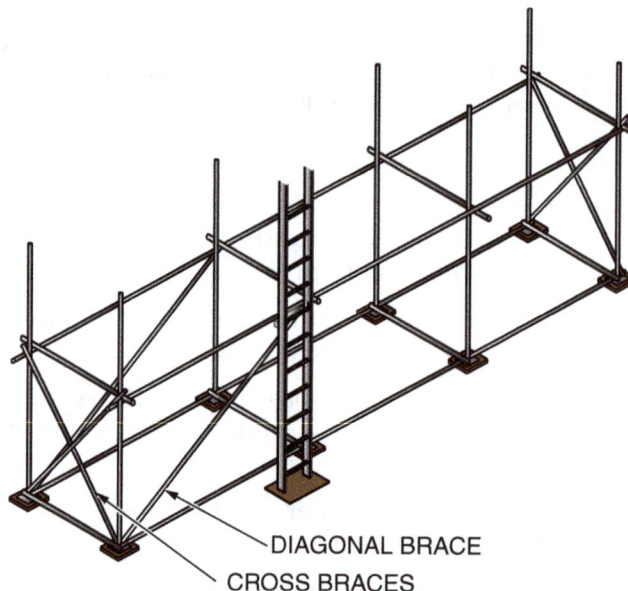

DIAGONAL BRACE
CROSS BRACES

31105-14_F63.EPS

Figure 63 Completed first level.

Step 42 Extend the front and back diagonal braces by attaching another tube to the end of the first-level brace and securing it in place with swivel clamps where it crosses a runner or post.

Step 43 Attach a ladder bracket to the third set of posts.

Step 44 Extend the ladder post up to the third level and attach it to the ladder bracket.

Step 45 Extend the ladder up to the second level, attaching it according to the manufacturer's specifications. *Figure 64* shows the completed second level.

Step 46 Place scaffold planks on the first and third spans of the scaffold, making sure that between 6 and 12 inches of plank extend beyond the bearer on each end.

> **NOTE**
> Planks are installed at this level only temporarily to provide a place to walk while working on the next level. The planks can be moved to the next level when it is ready.

Step 47 Place scaffold planks on the second span of the scaffold, resting the planks on top of the planks placed in Step 46, making sure that 6 to 12 inches of plank extend beyond the bearer on each end.

Step 48 Pass up and attach members to extend the posts to the third level.

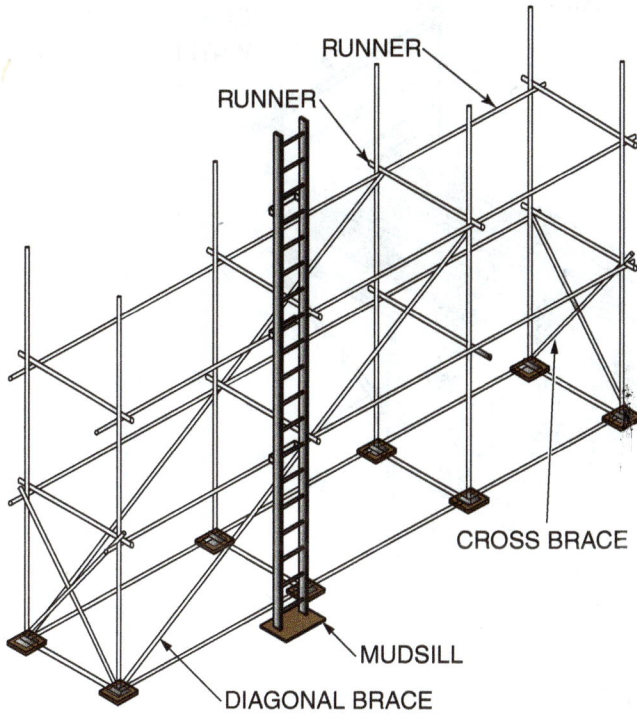

Figure 64 Completed second level.

> **NOTE**
> These posts will be used to support the guardrails for the working level, so they should extend to approximately 48 inches above the location of the third-level platform.

Step 49 Pass up and install and level the third-level runners.

> **NOTE**
> The maximum height for these runners is 6 feet, 6 inches above the lower-level runners.

Step 50 Pass up and install the third-level bearers, resting the bearer clamps on the runner clamps and making sure that the posts remain plumb.

> **NOTE**
> The bearers at each end of the second span should extend to allow for the erection of the ladder access platform.

Step 51 Attach an additional runner to the end of the ladder platform bearers.

Step 52 Extend the front and back diagonal braces by attaching another tube to the end of the second level brace and securing it in place with swivel clamps where it crosses a runner or post. The top clamps should be placed directly below the runner clamps on the fourth set of posts.

Step 53 Place scaffold planks on the first and third spans of the scaffold, making sure that 6 to 12 inches of plank extend beyond the bearer on each end.

Step 54 Place scaffold planks on the second span of the scaffold, resting the planks on top of the planks placed in Step 53, making sure that 6 to 12 inches of plank extend beyond the bearer on each end.

Step 55 Place scaffold planks to cover the ladder access platform.

Step 56 Extend the ladder up far enough to extend above the guardrail location, attaching it to the ladder platform bearers.

Step 57 Place toeboards on the sides and ends of the working platform.

Step 58 Secure the toeboards and runners in place.

Step 59 Attach and level guardrails to the inside of the legs at the required heights.

Step 60 Check each of the clamp bolts to ensure that all are properly tightened.

Step 61 After a complete inspection, place a tag on the scaffold, indicating that it is ready to use. *Figure 65* shows the completed scaffold.

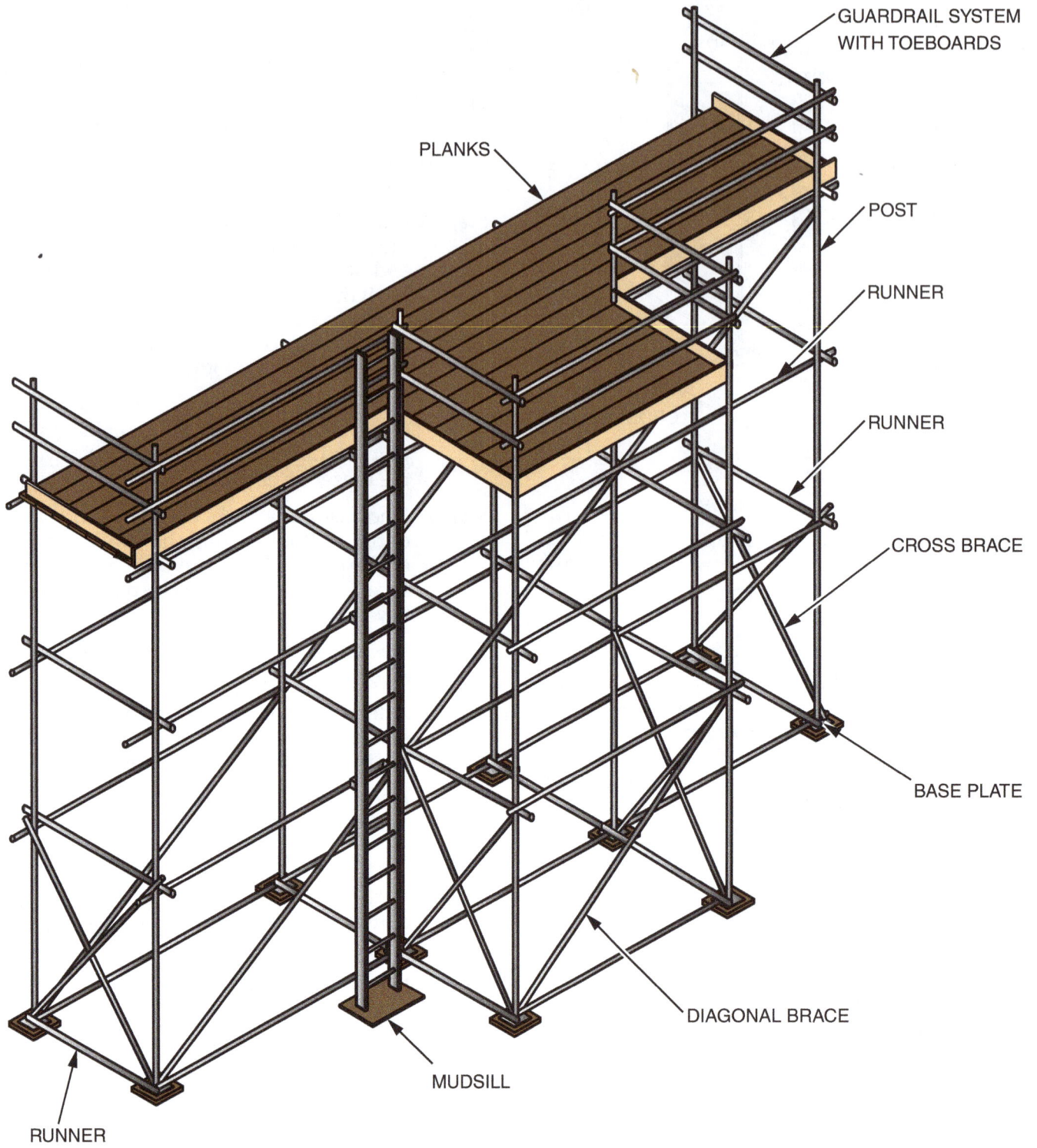

GUARDRAIL SYSTEM
WITH TOEBOARDS

PLANKS

POST

RUNNER

RUNNER

CROSS BRACE

BASE PLATE

DIAGONAL BRACE

MUDSILL

RUNNER

31105-14_F65.EPS

Figure 65 Completed scaffold.

Additional Resources

"Scaffolding." OSHA. www.osha.gov/SLTC/scaffolding/index.html

Fall Protection and Scaffolding Safety: An Illustrated Guide. 2000. Grace Drennan Ganget. Government Institutes.

International Building Code®, Latest Edition. Falls Church, VA: International Code Council.

International Residential Code®, Latest Edition. Falls Church, VA: International Code Council.

4.0.0 Section Review

1. Tube-and-clamp scaffolds can be erected in areas that have space or access restrictions.

   a. True
   b. False

2. The tube components of tube-and-clamp scaffolds are secured by _____.

   a. pins
   b. clamps
   c. rivets
   d. bolts

3. Right-angle clamps should be tightened to approximately _____.

   a. 25 foot-pounds
   b. 30 foot-pounds
   c. 45 foot-pounds
   d. 50 foot-pounds

5.0.0 OTHER SUPPORTED SCAFFOLD SYSTEMS

Objectives

Identify other supported scaffold systems.
a. Explain the basic principles of outrigger scaffolds, and outline proper erection procedures.
b. Explain the basic principles of pump-jack scaffolds, and outline proper erection procedures.

Outrigger and pump-jack scaffolds are two other types of supported scaffold systems. An outrigger scaffold shares some of the characteristics of suspended scaffolds. These scaffolds are supported completely by the structure to which they are attached. *Figure 66* shows a cutaway of an outrigger scaffold.

A pump-jack scaffold shares some of the advantages of a suspension scaffold while still being supported primarily by the ground. A pump-jack scaffold can be moved up the support pole to allow vertical access to the structure without erecting a multilevel supported scaffold. The design of the pump-jack scaffold allows the platform to be raised in increments of a few inches. This allows ready access to the entire face of the building. *Figure 67* shows a pump-jack scaffold.

5.1.0 Outrigger Scaffold Applications

Outrigger scaffolds are employed when work is necessary on the outside of a building and the erection of a traditional supported scaffold is impractical or impossible. These scaffolds are used when the work will be completed in a relatively short time and will be performed on a single level of the building.

5.1.1 Outrigger Scaffold Components

An outrigger scaffold consists of two or more beams that support a single platform. The beams can be steel or wood as long as they are strong enough to support the platform, workers, and materials. The inside portion of the beam must be secured to the structure in such a way that it will not move in any direction. Anchor bolts may be set into the floor slab, bolts can be inserted in

holes drilled completely through the floor and anchored across joists, or the beam can be bolted directly to other structural members. Bracing should be provided at the window opening to provide additional lateral support for the outrigger beam.

The work platform of an outrigger scaffold can be made of any common scaffold platform or deck materials. Since the platform is usually set in place by running it out through a window, wood or manufactured planks are generally used. Guardrails, as well as toeboards, must be erected around the back and ends of the platform. Usually the front is close enough to the face of the building so that front guardrails are not needed. Access to the scaffold is usually through the same window opening that is used for the beams.

5.1.2 Outrigger Scaffold Erection

Many different conditions are likely to be encountered when erecting an outrigger scaffold. The following are general guidelines for erection:

- The beam should be at least four times as long as the platform width. This allows a minimum of a 3-to-1 leverage advantage on the inside of the building.
- The building floor and windowsills must be inspected to ensure that they can support the weight and arrangement of the scaffold.
- Separate personal fall arrest equipment must be provided for workers during erection as well as during use of the scaffold.
- Use blocking to raise the inside of the beam so that it is level with the window frame.
- Once the beams are in place and secured, block between the beam and both sides of the window frame to provide additional lateral support.
- Secure the scaffold planks to the beam to prevent shifting. This can be done by nailing them to the beam or wiring the toeboards firmly down against the planks.

5.2.0 Pump-Jack Scaffold Applications

Pump-jack scaffolds are used when access is needed to the face of a building. Pump-jack scaffolds are used mainly in residential construction due to their height limitations (usually about 30 feet). Painters may use pump-jack scaffolds since they must reach all areas of the side of a building. These scaffolds allow the painter to raise the platform a few inches at a time while being firmly supported at all times.

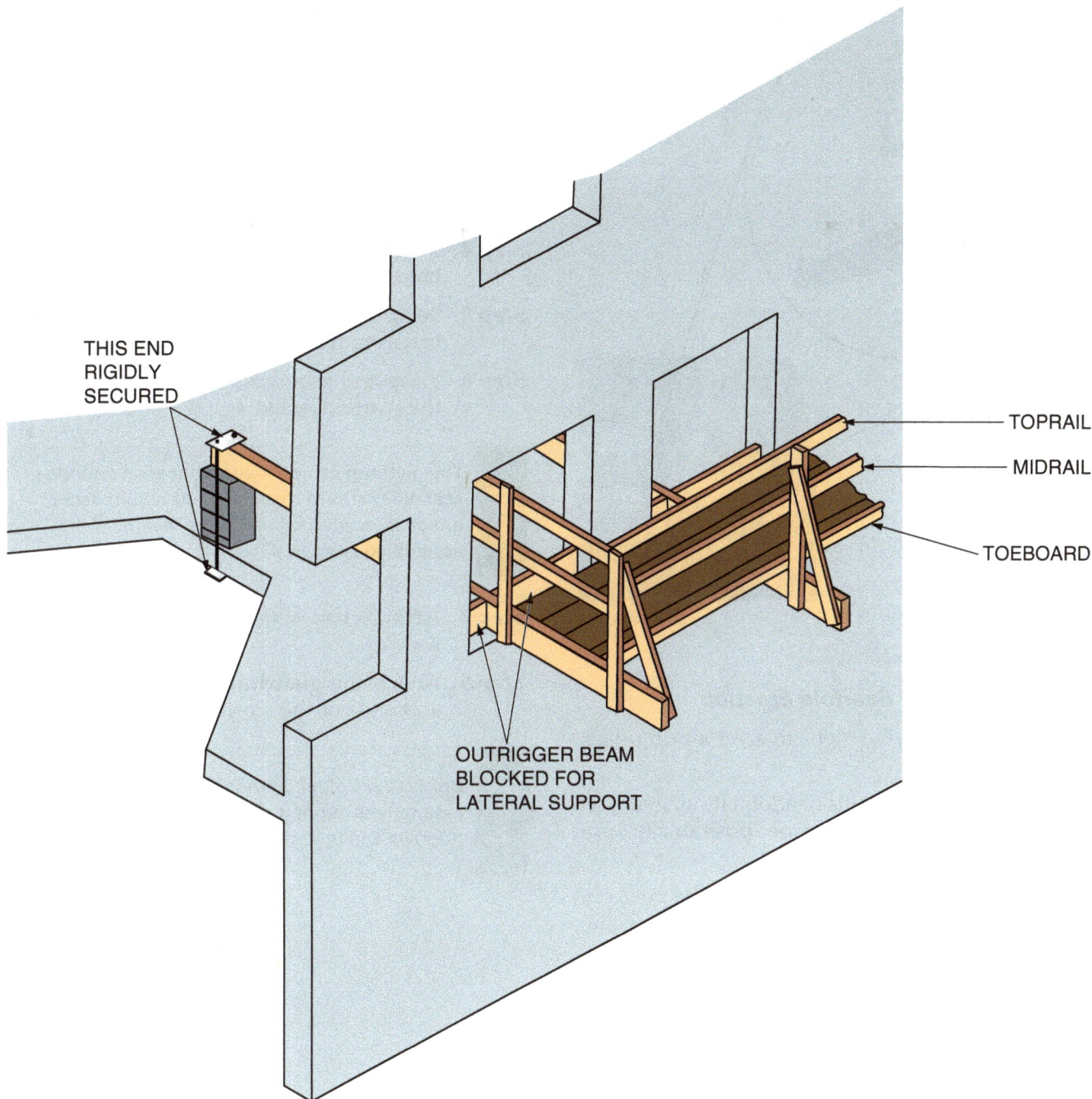

THIS END RIGIDLY SECURED

TOPRAIL

MIDRAIL

TOEBOARD

OUTRIGGER BEAM BLOCKED FOR LATERAL SUPPORT

31105-14_F66.EPS

Figure 66 Cut-away of an outrigger scaffold.

5.2.1 Pump-Jack Scaffold Components

The pump-jack scaffold is supported by a pair of vertical poles. These poles are normally wood 4 × 4 or 6 × 6 posts. The reach of a pump-jack scaffold is limited by the length of a single pole. The design does not allow splicing or piecing two or more of the posts together. The platform support can come in a variety of configurations.

The main part of a pump-jack assembly is the jack assembly. The jack assembly is clamped around the pole and inches upward when the jack is operated. A pump-jack assembly works best when workers are stationed at each end of the scaffold to operate the jacks at the same time. Workers press down on the lever with their feet to raise the platform. Both ends must be raised at the same time to prevent wedging or jamming. Additional brackets may be provided to support platforms for working materials on the outside of the poles. If provided, these areas must not be used to support workers; they are intended to be used as workbenches only.

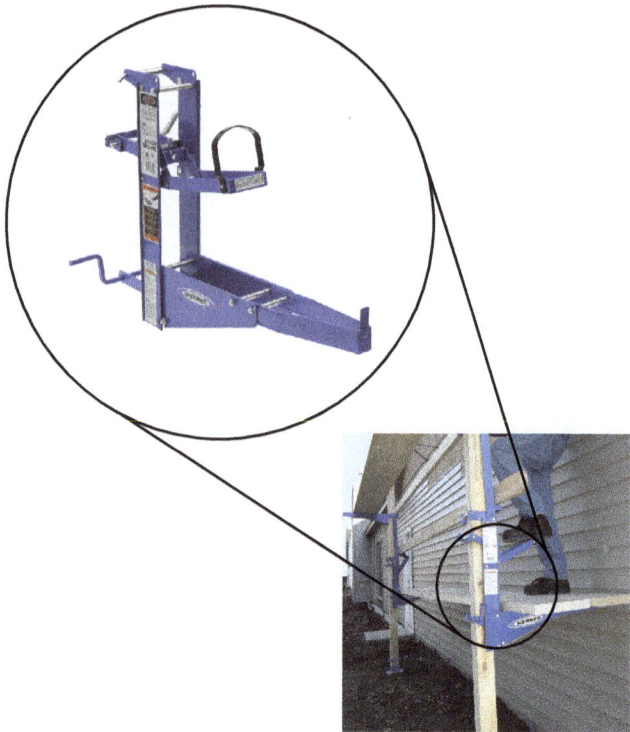

Figure 67 Pump-jack scaffold.

5.2.2 Pump-Jack Scaffold Erection

Perform the following steps to erect a pump-jack scaffold:

Step 1 Position mudsills of adequate size for soil conditions at the proper positions to support the poles.

Step 2 Slide the jack assembly on to the pole, ensuring that all rollers and jacking mechanisms are operating properly.

Step 3 Attach braces to the bottom and top of the scaffold pole.

> **NOTE**
> The specific locations of the braces will be determined by the location of adequate tie points on the building face.

Step 4 Stand the pole on the mudsill, resting the braces against the building.

Step 5 Secure the pole to the mudsill and the braces to the building.

Step 6 Place and secure the scaffold platform to the scaffold jacks.

> **NOTE**
> The platform can be made from any common scaffold-platform materials. Aluminum decks are often used to minimize the weight of the assembled scaffold.

Step 7 Attach the workbench materials as needed.

Step 8 Attach the guardrail system on the back and ends of the scaffold platform.

> **NOTE**
> Brackets are often provided on the jack assembly to support the guardrail system (toprails and midrails).

31105-14_F67.EPS

Additional Resources

"Scaffolding." OSHA. **www.osha.gov/SLTC/scaffolding/index.html**

Fall Protection and Scaffolding Safety: An Illustrated Guide. 2000. Grace Drennan Ganget. Government Institutes.

International Building Code®, Latest Edition. Falls Church, VA: International Code Council.

International Residential Code®, Latest Edition. Falls Church, VA: International Code Council.

5.0.0 Section Review

1. A scaffold consisting of a single platform and two or more support beams is called a(n) _____.

   a. anchored scaffold
   b. pump-jack scaffold
   c. suspended scaffold
   d. outrigger scaffold

2. Pump-jack scaffolds often uses platforms made of _____.

   a. laminated lumber
   b. aluminum
   c. steel
   d. pressure-treated wood

SUMMARY

Supported scaffolds consist of one or more working platforms supported by an assembly of legs, runners, and bearers. The two most common types of supported scaffolds are tubular welded-frame and tube-and-clamp scaffolds. The tubular welded-frame scaffold is easier to erect, but it is limited to easily accessible areas where there are few obstacles. Tubular welded-frame scaffolds are used where the scaffold may be used only for a short time. Tube-and-clamp scaffolds take more time and training to erect properly. They are used when the scaffold will be in place long enough to offset the additional cost of erection. Tube-and-clamp scaffolds can be erected in locations where frame scaffolds will not fit. The tube-and-clamp design allows the scaffold to be erected on uneven terrain or around and over obstructions. Both of these types of scaffolds must be tied or braced to the structure when the height exceeds four times the narrow base width and when the length exceeds 30 feet.

Other types of scaffolds described in this module are not as common, but a scaffold builder should be familiar with their uses and how they should be erected.

Fall protection must be provided for all workers on a scaffold. Generally, this is provided by a guardrail system consisting of toprails and midrails. Personal fall arrest systems must be used if guardrails are not in place. To prevent materials on the platforms from falling onto lower levels, toeboards must be provided around all open sides of the platform. If materials are piled higher than the toeboards, netting should be placed between the toeboard and the midrail or toprail. Toeboards also provide a way of securing the scaffold-platform planks to the scaffold assembly.

Scaffold platforms are designed to support the workers, their tools, and the materials needed to perform the task. The most common platform material is a scaffold-grade plank. Solid sawn planks must be grade-stamped.

Erection of scaffolds must be performed by trained scaffold builders under the supervision of a competent person. The scaffold user must visually inspect the scaffold at least once a shift to ensure that it is safe for work activities.

1. Supported scaffolds over 125 feet in height must be designed by a _____.

   a. trained technician
   b. professional engineer
   c. registered architect
   d. certified scaffold erector

2. Guardrails are required if a scaffold is placed next to a sloping surface where a potential fall distance could be at least _____.

   a. 6 feet
   b. 8 feet
   c. 10 feet
   d. 12 feet

3. A scaffold must be guyed or braced to a rigid structure if the height exceeds the narrowest dimension of the base by _____.

   a. three times
   b. four times
   c. six times
   d. eight times

4. Clamps used for tube-and-clamp scaffolds must be made of a metal other than _____.

   a. malleable iron
   b. structural-grade aluminum
   c. drop-forged steel
   d. gray cast-iron

5. On tube-and-clamp scaffolds, cross braces must be installed vertically at every _____.

   a. second runner
   b. third runner
   c. fourth runner
   d. fifth runner

6. Significant safety considerations for workers on outrigger scaffolds are support and _____.

   a. tripping hazards
   b. fall protection
   c. wind velocity
   d. platform rigidity

7. System scaffold components from different manufacturers _____.

   a. must not be mixed
   b. always vary in quality
   c. can easily be interchanged
   d. seldom can be used together

8. The most common leg length(s) for system scaffolds are _____

   a. 5 meters
   b. 1, 2, and 3 meters
   c. 4 and 6 meters
   d. 7 meters

9. System scaffold legs have a connecting device at the top called a(n) _____.

   a. coupling pin
   b. frame clamp
   c. leg coupler
   d. locking insert

10. The horizontal components that support scaffold platforms are the _____.

    a. stretchers
    b. lintels
    c. bearers
    d. ledgers

11. Stairways used with system scaffolds usually have separate stringers and _____.

    a. risers
    b. bullnoses
    c. balusters
    d. treads

Figure 1
31105-14_RQ01.EPS

12. The specialized scaffold component shown in Review Question *Figure 1* is a(n) _____.

    a. putlog
    b. open web beam
    c. girder
    d. bridging truss

13. An area scaffold is also called a _____.

    a. modular scaffold
    b. cellular scaffold
    c. multiple-bay scaffold
    d. unitized scaffold

14. The simplest form of frame the tubular welded-scaffold is the _____.

    a. area scaffold
    b. tower scaffold
    c. suspended scaffold
    d. scaffold run

15. Dance floors and stages are examples of uses for a(n) _____.

    a. tower scaffold
    b. low-rise scaffold
    c. area scaffold
    d. platform scaffold

16. The first factor to consider when deciding what type of scaffold to use is _____.

    a. local codes
    b. the work to be accomplished
    c. time needed to erect it
    d. availability of trained erectors

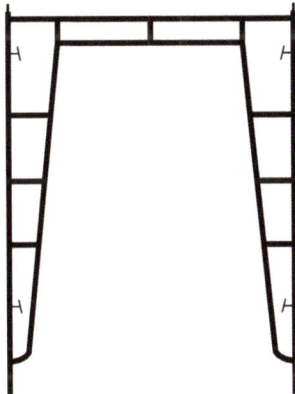

Figure 2

31105-14_RQ02.EPS

17. The scaffold frame style shown in Review Question *Figure 2* is a _____.

    a. walk-through frame
    b. single-ladder box
    c. double-ladder box
    d. ladder frame

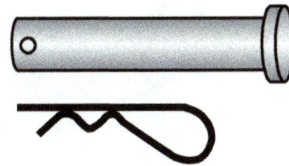

Figure 3

31105-14_RQ03.EPS

18. The connection locking device shown in Review Question *Figure 3* is a(n) _____.

    a. pigtail pin
    b. gravity lock
    c. rivet and hairpin
    d. toggle pin

19. When a 250-pound weight is placed on the center of a 10-foot span, a scaffold plank must not deflect more than _____

    a. 1 inch
    b. 2 inches
    c. $\frac{1}{4}$ of the plank's length
    d. $\frac{1}{60}$ of the plank's length

20. Legs placed near ditches must be separated from the excavation by at least _____

    a. 18 inches
    b. 24 inches
    c. the ditch width
    d. the ditch depth

21. The basic tubing of twist-lock fittings has a nominal outside diameter of _____.

    a. 1½ inches
    b. 2 inches
    c. 2¼ inches
    d. 2½ inches

22. Diagonal braces are attached to supporting scaffold members using a _____.

    a. right angle clamp
    b. locking pin
    c. swivel clamp
    d. twist-lock clamp

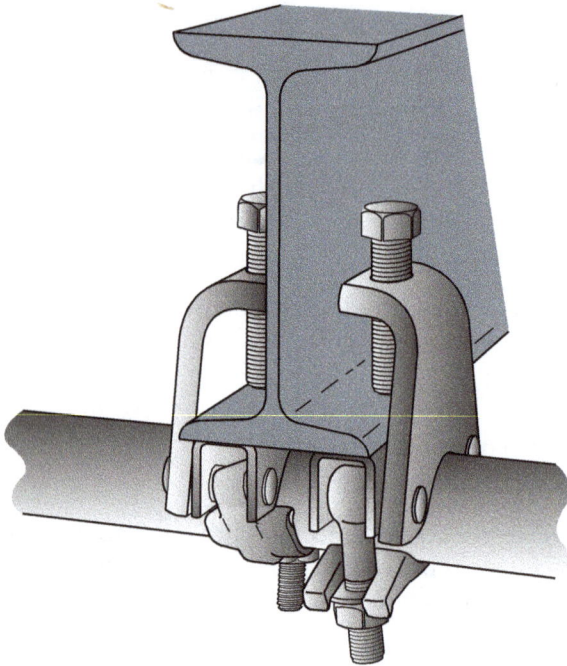

Figure 4

31105-14_RQ04.EPS

23. The connecting device shown in Review Question *Figure 4* is a(n) _____.
   a. tube hanger
   b. right-angle clamp
   c. channel coupler
   d. beam clamp

24. An outrigger scaffold is a supported scaffold that shares some of the characteristics of the _____.
   a. catenary scaffold
   b. suspended scaffold
   c. tower scaffold
   d. mobile scaffold

25. Pump-jack scaffolds are used _____.
   a. when access is needed to the building face
   b. only when there is no other option
   c. inside tanks and other round structures
   d. only by house painters

Trade Terms Quiz

Fill in the blank with the correct term that you learned from your study of this module.

1. A _____ is a rigid connection that holds one scaffold member in a fixed position with respect to another member or to a building or structure.

2. A _____ is used to raise or lower parts of a scaffold to level it on uneven surfaces.

3. A _____ locks together the tubes of a tube-and-clamp scaffold.

4. A horizontal scaffold member that supports the scaffold platform is a _____.

5. Braces placed between opposite corners to keep the scaffold plumb and secure are known as _____.

6. The lengthwise horizontal spacing or bracing member that supports the bearers is known as a _____.

Trade Terms

Bearer
Brace

Clamp
Cross bracing

Runner
Screw jack

Trade Terms Introduced in This Module

Bearer: A horizontal transverse scaffold member (which can be supported by ledgers or runners) upon which the scaffold platform rests; joins scaffold uprights, posts, poles, and similar members.

Brace: A rigid connection that holds one scaffold member in a fixed position with respect to another member or to a building or structure.

Clamp: A device for locking together the tubes of a tube-and-clamp scaffold.

Cross bracing: Braces placed between opposite corners to keep the scaffold plumb and secure; also called transverse bracing.

Runner: The lengthwise horizontal spacing or bracing member that can support the bearers; also called a ledger or a ribbon.

Screw jack: A threaded, adjustable screw located between the legs and base plates or caster wheels of a scaffold; used to raise or lower parts of a scaffold to level it on uneven surfaces.

Appendix

SCAFFOLD APPLICATION ANALYSIS

This section gives an analysis using the following example:

A scaffold is to be built around a 30-foot diameter, 50-foot tall tank. The job to be performed is to sandblast the old paint off the outside of the tank and repaint it. The base is a concrete slab that is pitched toward the tank at a slope of 5 inches per 10 feet. There may be work performed from all the levels of the scaffold at once. There is one 36-inch-diameter pipe extending out of the tank at ground level that is used to pump liquids in or out of the tank. The scaffold loads will be 25 pounds per square foot or less. The work will take approximately three months to complete.

Due to the flexibility of design, tube-and-clamp scaffold is an excellent choice for this job. It will be used long enough to warrant the extra time and expense of assembly. The circular shape of the tank can easily be accommodated by setting the inner posts slightly closer together than the outer posts. The outer scaffold posts must still be within the span specifications. For a light-duty scaffold, the posts can be as much as 10 feet apart. To determine the post spacing, the total length of the scaffold must be calculated. To determine the circumference of the tank, multiply the diameter (30 feet) by 3.14. There should be a space between the inner face of the scaffold and the tank to allow workers to reach the tank surface without interference. The inner set of scaffold posts should be set 18 inches away from the tank face (this will allow approximately 12 inches of clearance from closest member of the scaffold); the resulting diameter of the circle of inner posts will be 33 feet. Multiplying 33 by 3.14 results in a total circumference of 103.62 feet for the inside row of posts.

Use 5 feet, 6 inches as the spacing between the inside posts (to try to keep the outside spacing of legs at 7 feet or less). Divide 103.62 (circumference) by 5.5 (5 feet, 6 inches spacing) to determine the number of posts needed to go around the tank (103.62 ÷ 5.5 = 18.84; rounded up to 19). There will be 19 posts needed for the inside row. There will be 18 posts at 5 feet, 6 inches apart and one unknown post distance. To find the distance of the unknown, multiply 5 feet, 6 inches by .84 (remainder left over from 18 legs at 5 feet, 6 inches).

This will give you the unknown distance (5.5 × .84 = 4.62 feet or 4 feet, 7 inches). There are 18 posts set 5 feet, 6 inches apart and 1 post set 4 feet, 7 inches from the last post.

The scaffold platform will be 3 feet, 10 inches wide, allowing four boards wide to be used. To find the outside post spacing, add 33 feet (inside post diameter) and 7 feet, 8 inches (3 feet, 10 × 2), for a total of 40 feet, 8 inches or 40.66666 feet. Multiply 40.66666 by 3.14 to find the circumference of the outer posts (40.66666 × 3.14 = 127.69331). Then, divide the circumference by the number of inner posts (19) to find the spacing of the outside legs (127.69331 ÷ 19 = 6.72 feet or 6 feet, 8½ inches).

The scaffold post placement should be started so that the pipe exiting the tank will be centered between two of the posts. Since the scaffold is to be built completely around the tank, this should present no problem. Mudsills and base plates are required on each of the posts.

The scaffold should be built up to within 6 feet of the top of the tank to allow easy access to the entire surface. Erecting seven levels, each 6 feet, 6 inches apart, will position the top platform at a height of approximately 45 feet, 6 inches. Since the scaffold height is more than four times the base width and more than 30 feet long, ties will be required. With a circular scaffold built around the outside of a structure, only compression ties are required. If tipping forces are applied to the scaffold by wind or any other force, the resulting force will transfer to a compression tie at a point on the opposite side of the scaffold. Extending the length of selected bearers and adding a base plate against the tank will work well. Ties should be placed around the tank on the third (19 foot, 6 inch) and sixth (39 foot) levels. The ties are required at 30-foot intervals on a scaffold run, resulting in three braces when calculated on the inner post length and four braces when calculated on the outer circumference. To allow a slightly greater safety factor, four ties will be installed, equally spaced around the circle. For example, the ties could be placed at the 3, 6, 9, and 12 o'clock positions.

All the scaffold levels will be used as working levels, so all must be fully planked with scaffold-grade planks. Standard solid sawn or laminated scaffold planks are probably the best choice since the bearers will be at a right angle in relation to the planks. Toeboards will be required on both the front and the back of the platforms. A guard-rail system (toprails and midrails) are required on all levels.

Figure A-1 shows the final scaffold configuration.

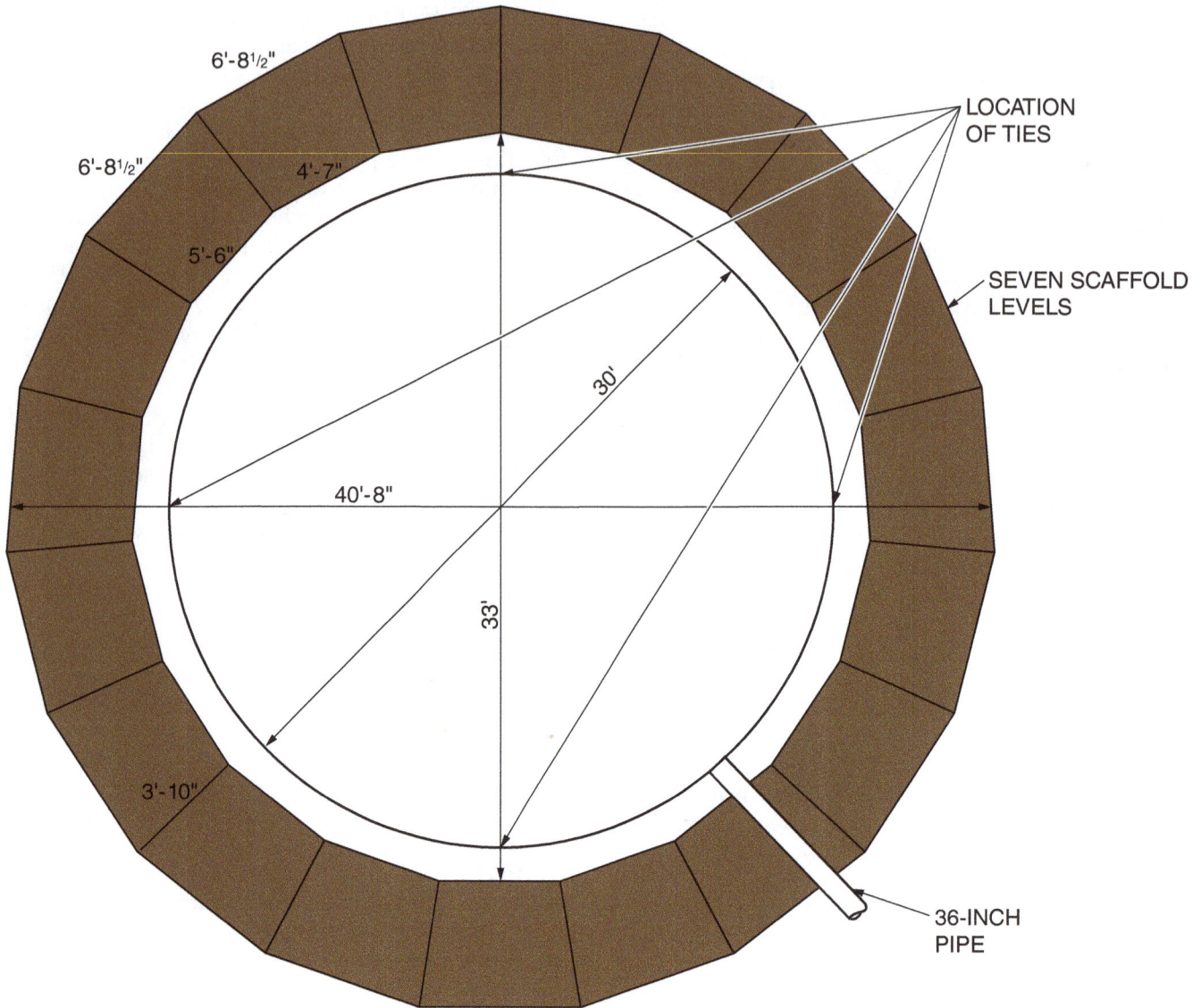

Figure A-1 Final scaffold configuration.

31105-14_A01.EPS

Additional Resources

This module presents thorough resources for task training. The following resource material is suggested for further study.

"Scaffolding." OSHA. **www.osha.gov/SLTC/scaffolding/index.html**

Fall Protection and Scaffolding Safety: An Illustrated Guide. 2000. Grace Drennan Ganget. Government Institutes.

International Building Code®, Latest Edition. Falls Church, VA: International Code Council.

International Residential Code®, Latest Edition. Falls Church, VA: International Code Council.

Figure Credits

Courtesy of PERI Formwork Systems, Inc., Module Opener

Courtesy of S4Carlisle Publishing Services, Figures 29A; Figures 30C, D, and G; Figures 56B, C, and E; Figure 57B; Figures 59 and 60

Werner Co., Figures 32, 35, and 67

APA–The Engineered Wood Association, Figure 34

Section Review Answer Key

Answer	Section Reference	Objective
Section One		
1.c	1.0.0	1a
2. a	1.1.2	1a
Section Two		
1. c	2.1.0	2a
2. a	2.1.2	2a
3. d	2.1.3	2a
Section Three		
1. d	3.0.0	3
2. b	3.2.9	3b
3. c	3.3.0	3c
Section Four		
1. a	4.1.0	4a
2. b	4.2.4	4b
3. c	4.3.3	4c
Section Five		
1. d	5.1.1	5a
2. b	5.2.2	5b

NCCER CURRICULA — USER UPDATE

NCCER makes every effort to keep its textbooks up-to-date and free of technical errors. We appreciate your help in this process. If you find an error, a typographical mistake, or an inaccuracy in NCCER's curricula, please fill out this form (or a photocopy), or complete the online form at **www.nccer.org/olf**. Be sure to include the exact module ID number, page number, a detailed description, and your recommended correction. Your input will be brought to the attention of the Authoring Team. Thank you for your assistance.

Instructors – If you have an idea for improving this textbook, or have found that additional materials were necessary to teach this module effectively, please let us know so that we may present your suggestions to the Authoring Team.

NCCER Product Development and Revision

13614 Progress Blvd., Alachua, FL 32615

Email: curriculum@nccer.org
Online: www.nccer.org/olf

❏ Trainee Guide ❏ Lesson Plans ❏ Exam ❏ PowerPoints Other _____

Craft / Level: _____ Copyright Date: _____

Module ID Number / Title: _____

Section Number(s): _____

Description: _____

_____

_____

_____

_____

Recommended Correction: _____

_____

_____

_____

_____

Your Name: _____

Address: _____

_____

Email: _____ Phone: _____

31106-15

Mobile Scaffolds

OVERVIEW

A mobile scaffold can be a manually operated scaffold that rolls on casters, a scissors lift, or an aerial lift. Mobile scaffolds allow the worker to position a platform in the proper location quickly and easily, perform the necessary work, and then move the platform to the next work location.

Module Six

Trainees with successful module completions may be eligible for credentialing through the NCCER Registry. To learn more, go to **www.nccer.org** or contact us at **1.888.622.3720**. Our website has information on the latest product releases and training, as well as online versions of our *Cornerstone* magazine and Pearson's product catalog.

Your feedback is welcome. You may email your comments to **curriculum@nccer.org**, send general comments and inquiries to **info@nccer.org**, or fill in the User Update form at the back of this module.

This information is general in nature and intended for training purposes only. Actual performance of activities described in this manual requires compliance with all applicable operating, service, maintenance, and safety procedures under the direction of qualified personnel. References in this manual to patented or proprietary devices do not constitute a recommendation of their use.

Objectives

When you have completed this module, you will be able to do the following:

1. Describe the operation and common applications of mobile scaffolds.
   a. Outline proper safety guidelines when using mobile scaffolds.
   b. Describe the benefits of mobile scaffolds.
   c. Identify common mobile scaffold applications.
   d. Identify mobile scaffold components.
   e. Outline proper mobile scaffold erection.
2. Describe the proper operation of scissors lifts.
   a. Explain the proper use of controls and indicators on scissors lifts.
   b. Describe basic operating procedures and concerns when using scissors lifts.
3. Describe the operation and common applications of aerial lifts.
   a. Outline proper safety guidelines when using aerial lifts.
   b. Identify common aerial lift applications.

Performance Task

Under the supervision of your instructor, you should be able to do the following:

1. Erect and use mobile scaffolds.

Trade Terms

Caster
Compatible
Laminated

Outrigger
Plumb

Industry-Recognized Credentials

If you're training through an NCCER-accredited sponsor, you may be eligible for credentials from NCCER's Registry. The ID number for this module is 31106-14. Note that this module may have been used in other NCCER curricula and may apply to other level completions. Contact NCCER's Registry at 888.622.3720 or go to **www.nccer.org** for more information.

Code Note

Codes vary among jurisdictions. Because of the variations in code, consult the applicable code whenever regulations are in question. Referring to an incorrect set of codes can cause as much trouble as failing to reference codes altogether. Obtain, review, and familiarize yourself with your local adopted code.

Contents

Topics to be presented in this module include:

Figures

SECTION ONE

1.0.0 MOBILE SCAFFOLDS

Objective

Describe the operation and common applications of mobile scaffolds.
  a. Outline proper safety guidelines when using mobile scaffolds.
  b. Describe the benefits of mobile scaffolds.
  c. Identify common mobile scaffold applications.
  d. Identify mobile scaffold components.
  e. Outline proper mobile scaffold erection.

Performance Task

Erect and use mobile scaffolds.

Trade Terms

Caster: A wheel fitted with a stem to secure it into a scaffold frame leg.

Compatible: Scaffold components from different manufacturers that can be used together.

Laminated: Layers of wood glued together with the grain parallel.

Outrigger: An accessory that can be installed on a mobile scaffold to provide added stability.

Plumb: Vertical, or completely straight up and down; at right angles to level.

Mobile scaffolds are commonly used for light-duty applications where easily moving from one location to another for short periods of time is required.

1.1.0 Safety Guidelines

Mobile scaffolds are only used on relatively level surfaces that are free of debris and holes. The wheels, or casters, must be rated to carry the weight of the scaffold and the load placed on the scaffold. As with all other scaffolds, mobile scaffolds must not be constructed near power lines.

NOTE
Mobile scaffolds in excess of 60-feet high must be designed by a registered engineer.

Guardrail systems are required on all scaffolds 10 feet in height or higher as a means of safeguarding workers from fall hazards (this may vary by state; check local codes/requirements). A guardrail system consists of a toprail and midrail. These rails must be placed on all open sides and ends of a working platform. Federal OSHA regulations require toprails to be positioned at a height between 38 and 45 inches above the working level. This too can vary by state. For example, California regulations require the toprails to be between 42 and 45 inches above the working level.

A midrail must be placed halfway between the top edge of the toprail and the platform surface. Platform height is always measured from a safe surface. If a 4-foot high scaffold is positioned next to an open floor or elevator shaft that would allow a fall greater than 6 feet, the scaffold must have guardrails. Guardrails are also required if the scaffold is placed next to a sloping surface where a potential fall could be 10 feet or greater.

The platform must be scaffold-grade materials, such as planks, and be clearly marked. Other platform materials must be certified by the manufacturer to conform to industry standards.

If work is to be performed on lower levels or workers will be allowed to pass under the scaffold platform, toeboards must be provided on all open sides of the platform. If materials are piled upon the scaffold platform higher than the toeboards, protective netting must be placed between the toeboards and the guardrails to prevent any of the materials from falling to a lower level.

WARNING!
Due to the risk of tipping or falling, never ride on a mobile scaffold as it is being moved. Riding on a mobile scaffold may result in falls that could cause injury or death.

1.2.0 Benefits of Mobile Scaffolds

Mobile scaffolds are the most common type of scaffold. Mobile scaffolds allow a worker to position a platform in the proper location quickly and easily. A mobile scaffold can be a simple tubular welded-frame scaffold fitted with casters that is manually moved from location to location, a scissors lift, or an aerial lift.

1.3.0 Possible Applications

Mobile scaffolds are assembled from standard tubular welded-frame scaffold and system-scaffold components. (It is possible to construct a mobile scaffold from tube-and-clamp components but the extra time involved makes this impractical.) These components are designed to support a working platform that can easily be moved from one location to another. Before assembling a mobile scaffold, one must understand what components are necessary to erect a stable and safe platform. Selecting the wrong components may result in an inefficient, unstable, and possibly unsafe scaffold. Erecting a mobile scaffold using the correct components will result in a platform that is safe, easily positioned, and capable of being used efficiently.

As with supported scaffolds, mobile scaffolds must not be erected higher than four times the minimum base width. Outriggers attached to the base section provide a wider base for added stability. Mobile scaffolds are used on hard, flat surfaces such as concrete, wood flooring, or steel decking. Surfaces must be strong enough to support the anticipated wheel loads. Mobile scaffolds are not intended to be used on dirt or open-ground surfaces. If the job requires the addition of side brackets to a mobile scaffold, always check the manufacturer's recommendations.

Mobile scaffolds are often used by electricians installing lighting fixtures, finish carpenters installing trim molding, or painters doing trim work. Maintenance workers may also use mobile scaffolds as they repair elevated items.

Figure 1 shows a mobile scaffold.

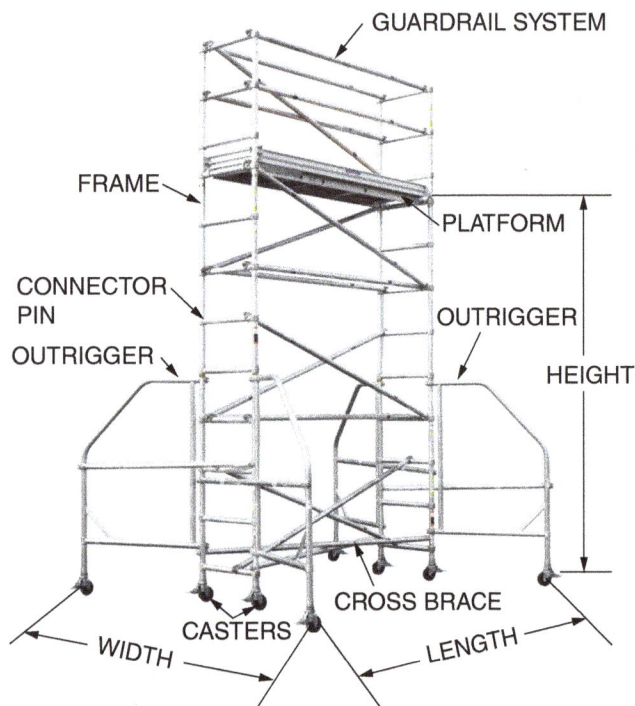

Figure 1 Mobile scaffold.

1.4.0 Mobile Scaffold Components

Mobile scaffold components consist of frames, locking devices, braces, base plates, putlogs, casters, platforms and planks, guardrails and gates, ladders and stairs, as well as other miscellaneous components. These components may be assembled in a variety of arrangements to meet almost any design configuration. The type of components used depends on the application. Each supplier has a complete line of components. Care should be taken not to mix components from different manufacturers unless the components are compatible.

1.4.1 Frames

There are a number of frame styles. Each is designed to fit a specific need while maintaining the OSHA-required safety factors. Frames of different styles can be combined to meet the needs of a particular level of a scaffold design. *Figure 2* shows a selection of a mobile scaffold frames.

1.4.2 Braces

Strength and stability are built into the scaffold arrangement by the installation of braces. All tubular welded-frames are equipped with some sort of attachment points for braces. Braces come in different lengths and must be matched to the frame spacing. Cross braces are used to keep the frames plumb, and the diagonal braces hold the scaffold square. Mobile scaffolds require additional horizontal and diagonal bracing, known as racking, as the scaffolding is being moved from one location to another to prevent failure. *Figure 3* shows examples of braces.

1.4.3 Locks

When a scaffold plan requires more than one level, a coupling pin must be used to secure the upper and lower frames together. These coupling pins must be secured to each of the scaffold frames. A rivet and hairpin, pigtail pin, or a toggle pin is run through the aligned holes in the scaffold legs and coupling pins. The frames also have attachment points for the cross braces described in the previous section. These braces must also be locked in place. *Figure 4* shows a variety of brace and coupling pin locking methods.

1.4.4 Bases

The base for a mobile scaffold is the caster. A caster is a wheel that is attached to a stem, which allows it to be inserted into the leg of a scaffold frame or the socket end of a screw jack. The caster must be se-

31106-14_F01.EPS

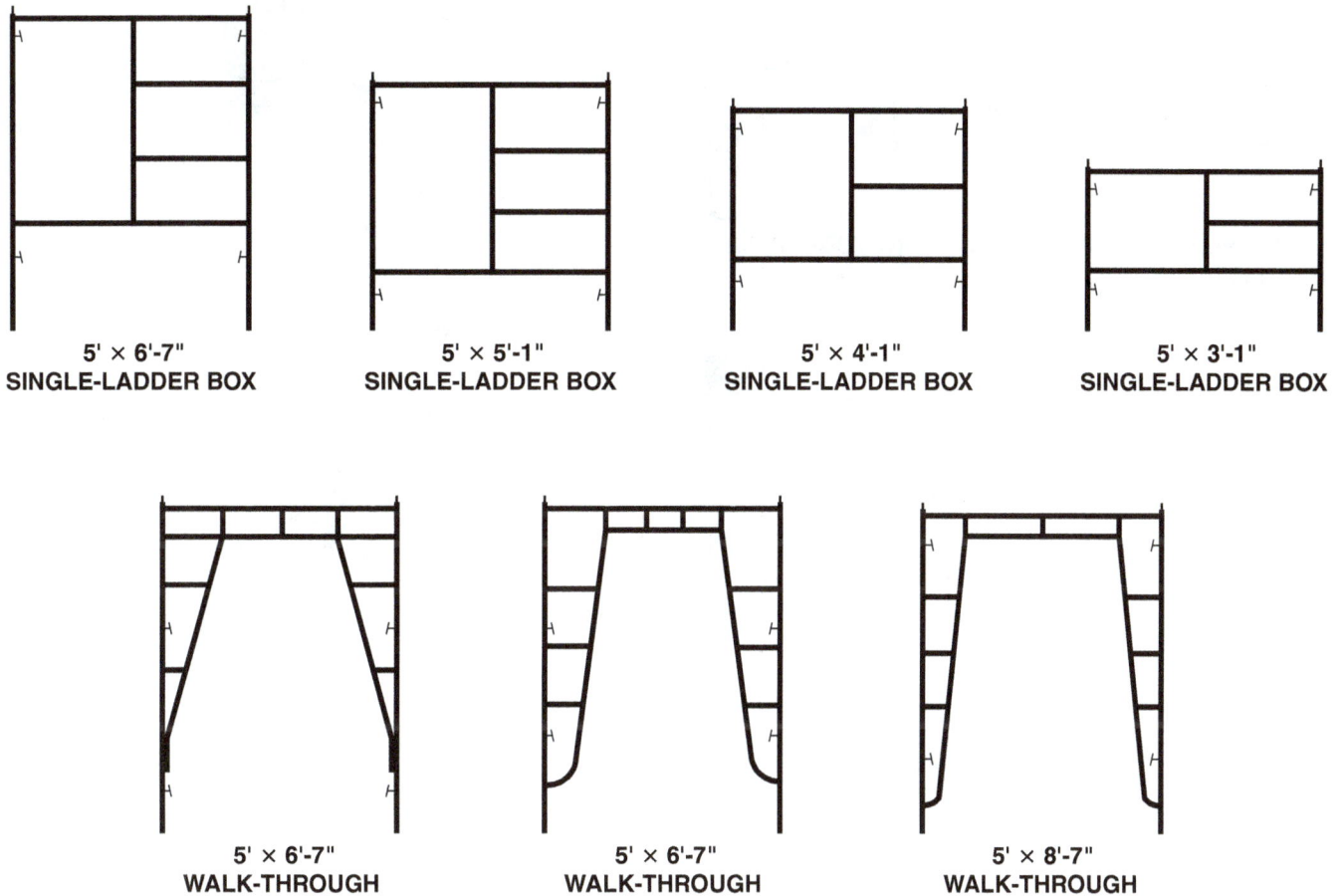

Figure 2 Mobile scaffold frames.

5' × 6'-7"
SINGLE-LADDER BOX

5' × 5'-1"
SINGLE-LADDER BOX

5' × 4'-1"
SINGLE-LADDER BOX

5' × 3'-1"
SINGLE-LADDER BOX

5' × 6'-7"
WALK-THROUGH

5' × 6'-7"
WALK-THROUGH

5' × 8'-7"
WALK-THROUGH

31106-14_F02.EPS

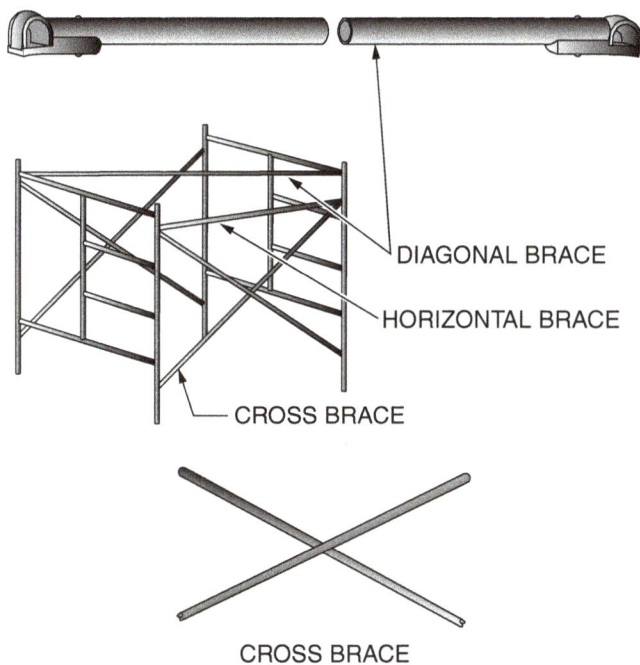

DIAGONAL BRACE

HORIZONTAL BRACE

CROSS BRACE

CROSS BRACE

31106-14_F03.EPS

Figure 3 Braces.

curely fastened to the leg or screw jack to prevent it from falling out of the scaffold. If a screw jack is used, it must not be extended beyond 12 inches. The casters are equipped with a double-acting brake. When engaged, the brakes will prevent the wheels from rolling and prevent the casters from swiveling. Depending on the design, the brake is engaged by pulling or pushing the brake handle up or down. The wheel is offset from the stem, allowing the wheel to be steerable. Two of the four wheels on a mobile scaffold must be steerable.

Figure 5 shows a typical caster and a screw jack with a socket.

> **WARNING!**
>
> Extend screw jacks as little as possible to level a mobile scaffold. While the scaffold is in use, all four casters must be locked. Locks must be engaged fully but should not be forced into position by striking with a hammer. Screw jacks that are overextended or not locked can cause the scaffold to fall, which could result in personal injury or death.

(A) COUPLING PIN (B) PIGTAIL PIN (C) TOGGLE PIN

(D) RIVET & HAIRPIN (E) SPEEDLOCK™ (F) AUTO-LOCK™

(G) GRAVITY LOCK

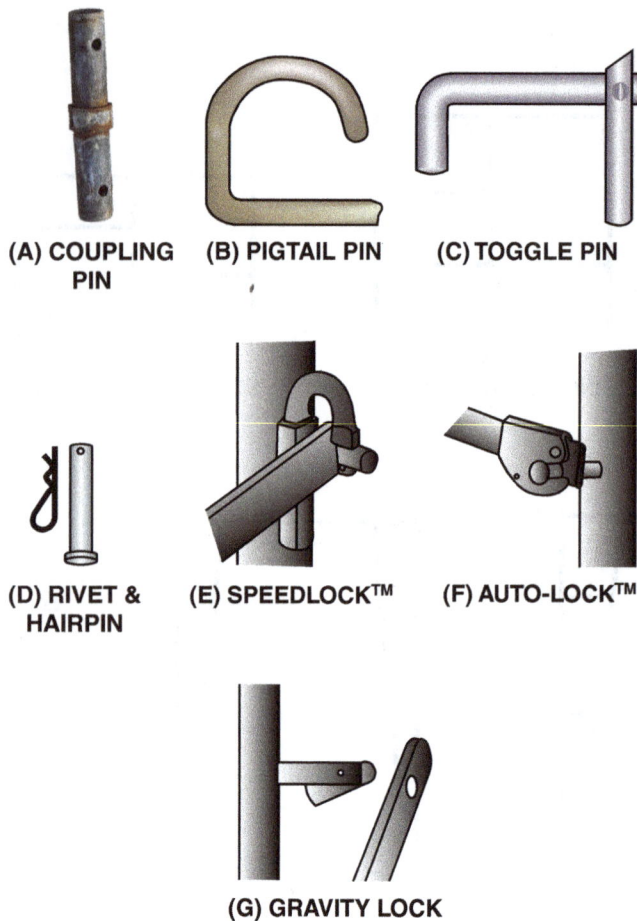

31106-14_F04.EPS

Figure 4 Locking methods.

1.4.5 Planks and Decks

Planks and decks provide a working surface for workers to stand on, hold materials and tools, and get from one place to another. Planks may be made of metal, wood, or fiberglass. Metal planks are usually 9-inches wide. Planks designed for use on tubular welded-frame scaffolds are fitted with hooks to rest on the round bearers. Metal planks vary in length up to 10 feet. *Figure 6* shows examples of metal planks.

Platforms may also be made from sawn wood planks, sometimes called "solid sawn." Solid sawn planks are the most common platform material used today. Although the nominal size of the planks is 2 × 10 when planed smooth, the planks actually measure 1½ inches by 9¼ inches. Two species of wood are specifically graded for use as scaffold planks: 171(b) Douglas Fir and Dense In-

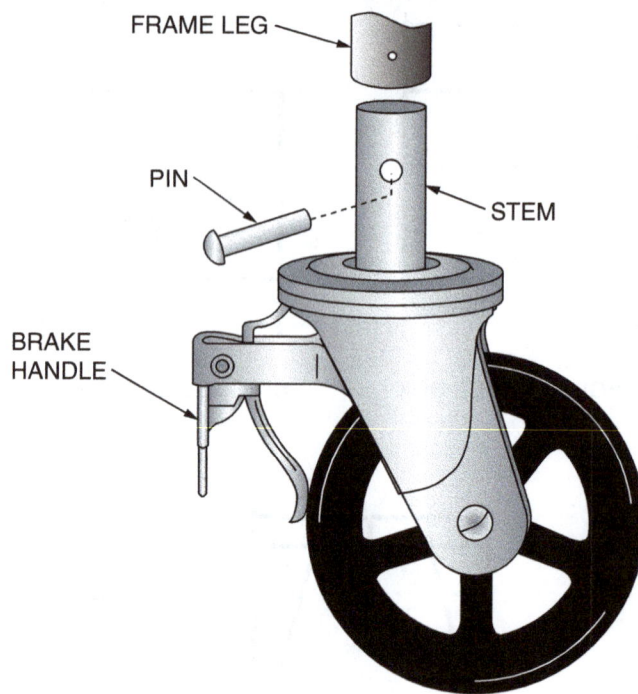

FRAME LEG

PIN STEM

BRAKE HANDLE

(A) CASTER

12-INCH (305 mm) MAXIMUM

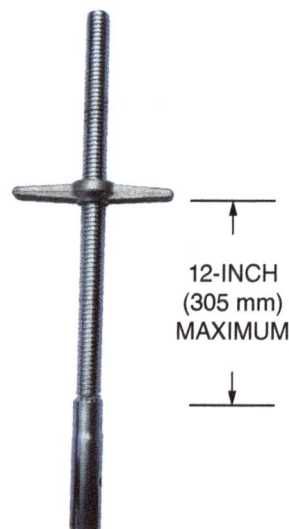

(B) SCREW JACK WITH SOCKET

31106-14_F05.EPS

Figure 5 Caster and screw jack with a socket.

dustrial Southern Pine. A scaffold plank must be strong enough to support a 250-pound load in the center of a 10-foot span without deflecting more than $\frac{1}{60}$ of its length. Scaffold planks are stamped as scaffold grade by an inspection agency. *Figure 7* shows the two most common grade stamps. If wood planks do not have this stamp, do not use them for decking.

Laminated wood planks are made from smaller pieces of wood glued together. The pieces

Figure 6 Metal planking.

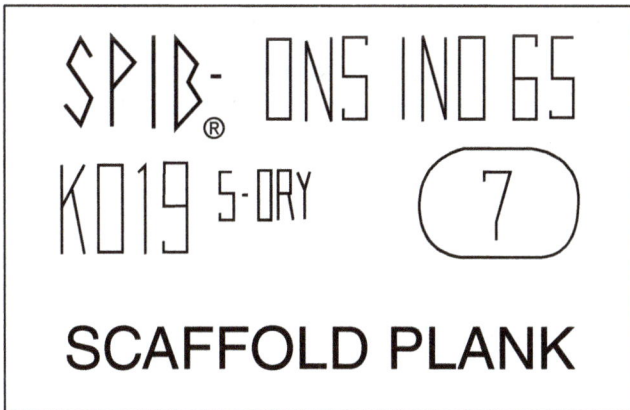

SPIB®- ONS INO 65
K019 S-DRY   (7)

SCAFFOLD PLANK

MILL 10
WC LB   SEL STR
SCAF PLK
D. FIR S. DRY

Figure 7 Plank grade stamps.

may be edge-glued or laminated. The edge-glued planks are made up of layers, like plywood. These planks are tested and rated by the manufacturer and are not marked like solid sawn planks. *Fig-*

ure 8 shows examples of laminated wood planks. Laminated wood planks are available in a variety of lengths that may be long enough to span more than one bay.

Scaffold decks are manufactured platforms that are usually 19-inches wide and long enough to span a single bay. Scaffold decks have hooks to attach to the bearer tubes or putlogs. Decks can be made from a variety of materials. The most common decks are made of aluminum side rails and plywood tops, or are all aluminum. Scaffold decks are rated by the weight that they can safely support, usually 50 pounds per square foot or 75 pounds per square foot, depending on the duty rating of the scaffold. *Figure 9* shows a scaffold deck.

1.4.6 Guardrails

Guardrail systems are required on all scaffolds 10 feet in height or higher (this varies by state) as a means of safeguarding workers from fall hazards.

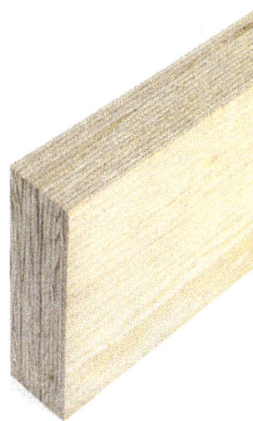

LAMINATED VENEER LUMBER (LVL)   GLUED LAMINATED TIMBER (GLULAM)

Figure 8 Laminated wood planks.

Figure 9 Scaffold deck.

A guardrail system consists of both a toprail and midrail. These rails must be placed on all open sides and ends of a working platform. A midrail must be placed halfway between the upper edge of the toprail and the platform surface. *Figure 10* shows the toprail and midrail locations.

Industry standards do not require access through guardrails, but some type of access should be included to avoid having to climb over the toprail or midrail. Access through a guardrail system can be achieved in many ways. On narrow scaffolds, a simple chain may suffice. On other scaffolds, a swing gate or duck-through with grab rails may be used. Swing gates must be installed so that they swing toward the platform, not away from the platform. They must also be positioned high enough to allow direct foot access to the platform. *Figure 11* shows the proper access to a platform through a swing gate.

Duck-through access is provided when the user must climb from below and enter a platform head first. This may occur when climbing a frame with a built-in ladder. The duck-through access system also includes a grab rail to allow the user to have a firm handhold while entering or leaving the platform. The duck-through access must be placed directly above the built-in ladder. *Figure 12* shows two duck-through access panels.

Guardrail systems should be placed so that the guardrail attachment devices face toward the

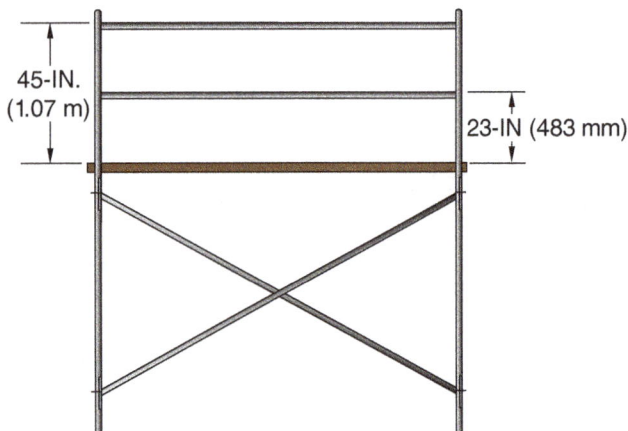

Figure 11 Access to platform through swing gate.

platform. Many of the systems will only install in this manner. Installing them properly reduces the possibility that a guardrail will come off if grabbed from the outside, or if someone leans or falls against it. *Figure 13* shows the proper installation of a guardrail support post.

Placing guardrails on intermediate levels of tubular welded-frame scaffolds may present a challenge. Guardrails cannot be placed in the same vertical plane as a cross brace. One method of meeting the guardrail requirements is to use a cross brace plus a horizontal member as a toprail or midrail. The location of the guardrail is determined by the height of the crossing point of the cross brace above the working platform. Regulations require the addition of a toprail if the crossing point falls between 20 and 30 inches above the platform. Most cross brace attachment points are not long enough to accommodate both the cross brace and the guardrail, so the guardrail must be clamped to the scaffold frames above or below the cross brace. *Figure 14* shows examples of where a midrail is not required (A) and where it is required (B).

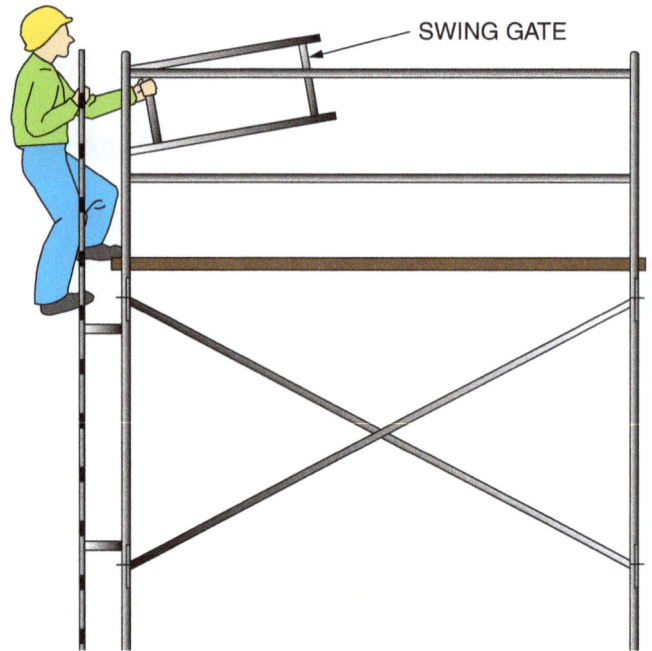

Figure 10 Toprail and midrail locations.

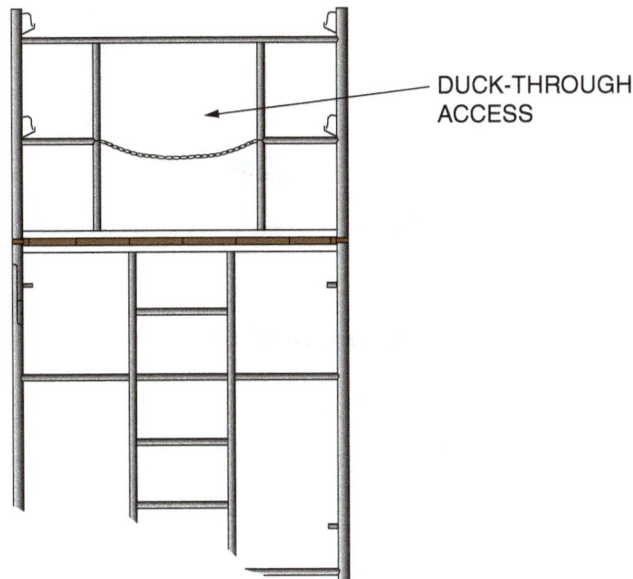

DUCK-THROUGH
ACCESS

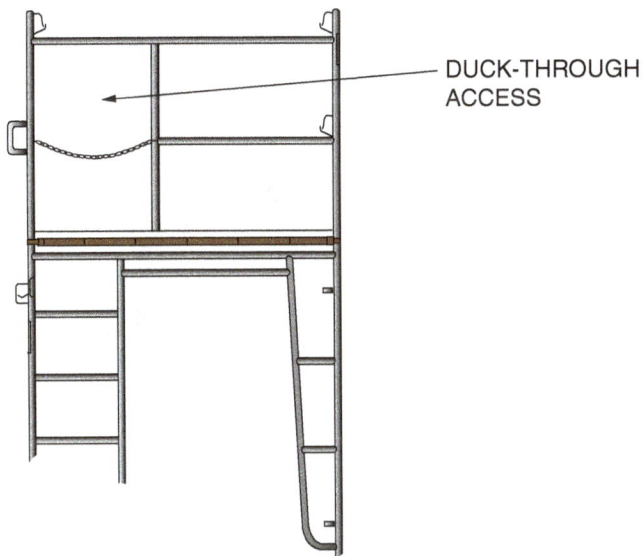

GUARDRAIL
SUPPORT POST

31106-14_F13.EPS

Figure 13 Installation of guardrail support post.

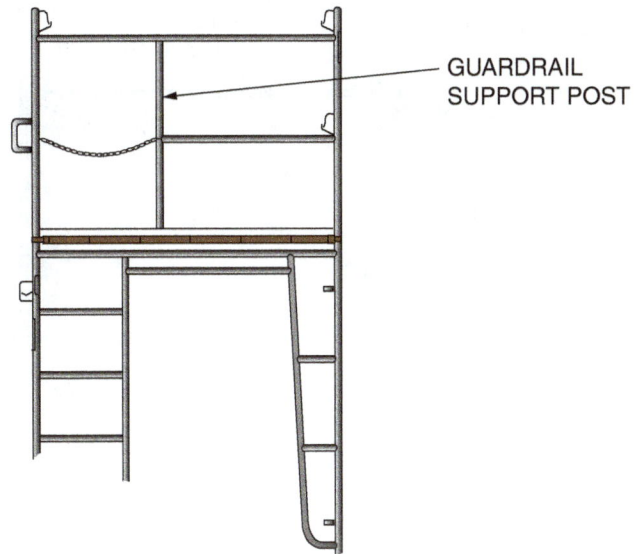

DUCK-THROUGH
ACCESS

31106-14_F12.EPS

Figure 12 Duck-through access panels.

There are guardrail panels specifically designed to meet this challenge. These panels are attached to the frame vertical members directly behind the cross braces. Guardrail panels are made to fit standard cross brace spaces. *Figure 15* shows examples of these guardrail panels.

1.4.7 Proper Access

Access to mobile scaffolds is usually by way of a clamp-on ladder mounted on the narrow side of the frame. The narrow side is used for ladders to reduce the tipping effect on the scaffold. Clamp-on ladders can be mounted on the inside or outside of the frame. If the ladder is mounted on the inside, access through the platform is necessary. Special platforms have been designed for this application. *Figure 16* shows an access ladder and trapdoor scaffold platform.

1.5.0 Mobile Scaffold Erection

In order for a mobile scaffold to provide a safe working platform, the proper components must be assembled. Mobile scaffolds are constructed from the following basic components:

- Base consisting of a caster or caster and screw jack with socket
- Scaffold frames
- Cross braces
- Platform material
- Guardrail system
- Platform access
- Horizontal diagonal brace

Other components may be necessary, depending on the particular application. *Figure 17* shows a mobile scaffold. The width, length, and height can vary depending on the particular needs of the job.

The following sections describe the selection of materials for two-lift mobile scaffolds, and the assembly of the scaffold.

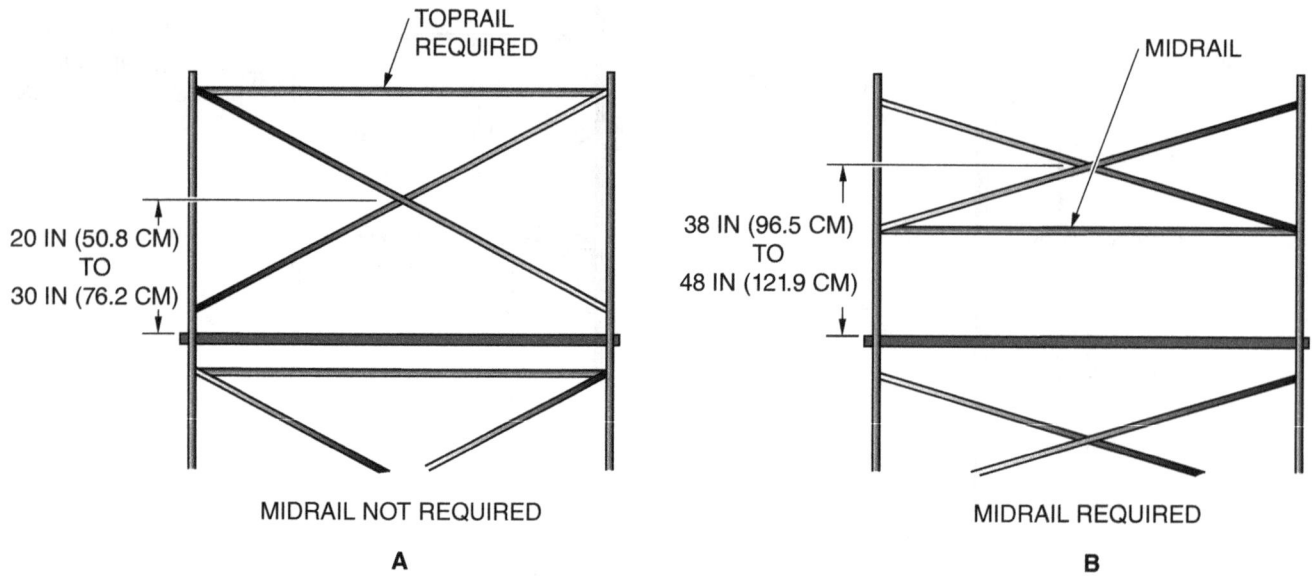

TOPRAIL REQUIRED

20 IN (50.8 CM)
TO
30 IN (76.2 CM)

MIDRAIL NOT REQUIRED

A

MIDRAIL

38 IN (96.5 CM)
TO
48 IN (121.9 CM)

MIDRAIL REQUIRED

B

31106-14_F14.EPS

Figure 14 Midrail requirements.

1.5.1 Equipment Selection

As with all scaffold jobs, the job site conditions, equipment availability, and user requirements will dictate the selection of equipment. For this two-lift mobile scaffold, a 5-foot by 7-foot platform is needed. The platform should be approximately 8 feet above the floor, which is a concrete slab. The platform is to be assembled from scaffold deck. *Figure 18* shows the scaffold to be constructed.

This scaffold will require the following:

Two 5-foot × 4-foot scaffold frames
Two 5-foot × 3-foot scaffold frames
Four screw jacks with sockets
Four 8-inch casters with bolts or pins
Four coupling pins with locking pins
Four 7-foot cross braces
One 5-foot × 7-foot horizontal diagonal brace
Three 19-inch wide × 7-foot long scaffold decks
Four guardrail posts with coupling pins
Two 7-foot guardrails
Two 5-foot guardrails
Two 7-foot toeboards
Two 5-foot toeboards
One access gate panel with coupling pins
Two 6-foot clamp-on ladders with clamps

A scaffold assembly (also called a stack) consisting of 4-foot and 3-foot scaffold frames, 8-inch casters, and 4-inch extensions of the screw jacks is used to reach the 8-foot platform height requirement. The 5-foot width requirement was met by selecting 5-foot-wide frames. The 7-foot cross braces were selected so that the frame spacing would be 7 feet. The braces must be carefully selected so that they match the stud spacing on the

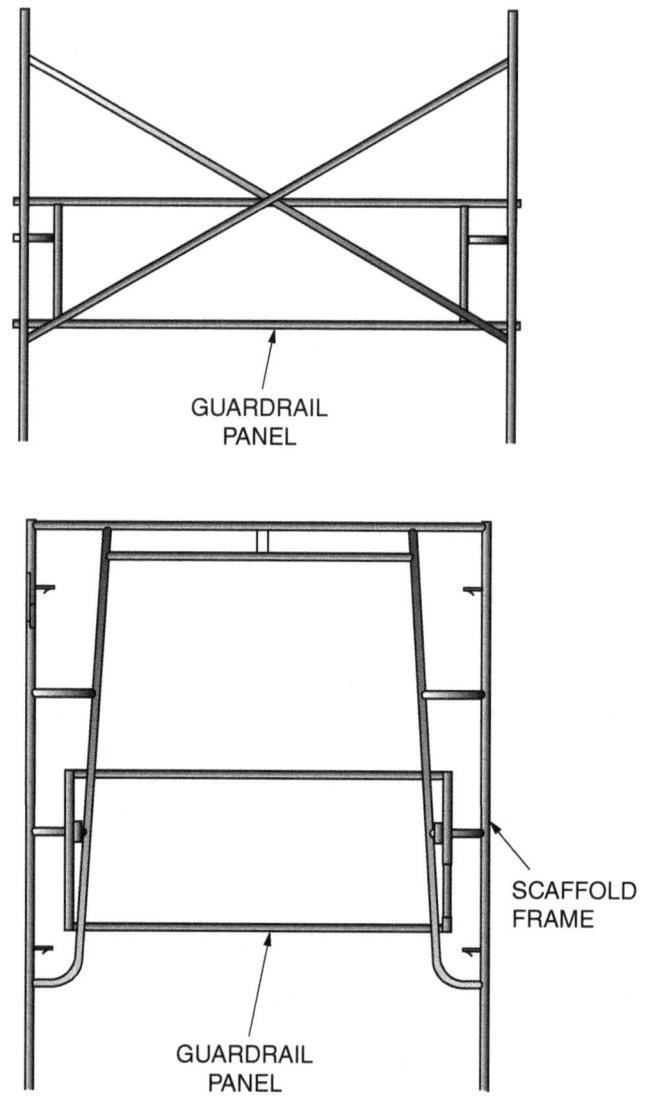

GUARDRAIL PANEL

GUARDRAIL PANEL

SCAFFOLD FRAME

31106-14_F15.EPS

Figure 15 Guardrail systems.

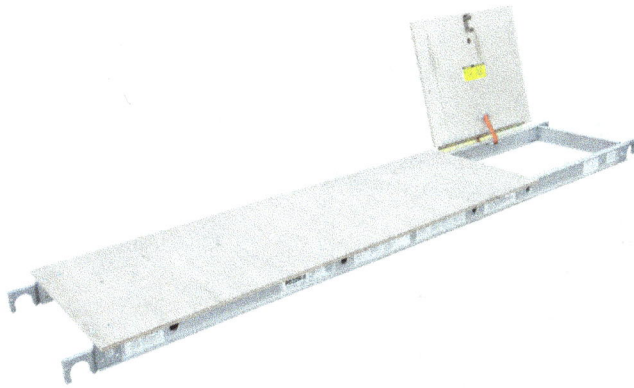

(A)

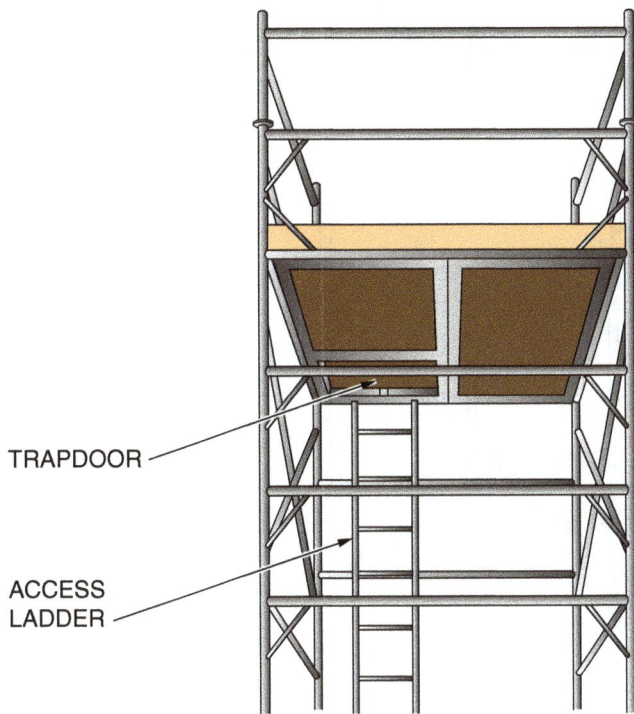

TRAPDOOR

ACCESS
LADDER

(B)

31106-14_F16.EPS

Figure 16 Access ladder and trapdoor scaffold platform.

scaffold frames. The 5-foot-wide scaffold platform will be constructed using three 19-inch scaffold decks. Guardrails are required on all four sides of the platform. A 5-foot-wide access panel will be used on one of the sides to provide access from the clamp-on ladder.

1.5.2 Assembly

Follow these steps to assemble this mobile scaffold:

Step 1 Inspect the area where the scaffold will be used.

Step 2 Lock the casters.

31106-14_F17.EPS

Figure 17 Mobile scaffold.

Step 3 Insert the casters into the screw jack sockets.

Step 4 Lock the casters into the sockets with pins or bolts.

Step 5 Lay out the 4-foot scaffold frames, braces, and caster-and-screw jack assemblies. *Figure 19* shows the first-level layout.

Step 6 Insert the caster-and-screw jack assemblies into the scaffold legs of both frames and secure them with bolts or pins.

Step 7 Attach cross braces to the studs on the first scaffold frame and lock them in place according to the manufacturer's specifications.

Step 8 Stand the second frame and attach the cross braces to the studs and lock them in place according to the manufacturer's specifications.

Step 9 Attach the horizontal diagonal brace according to the manufacturer's specifications.

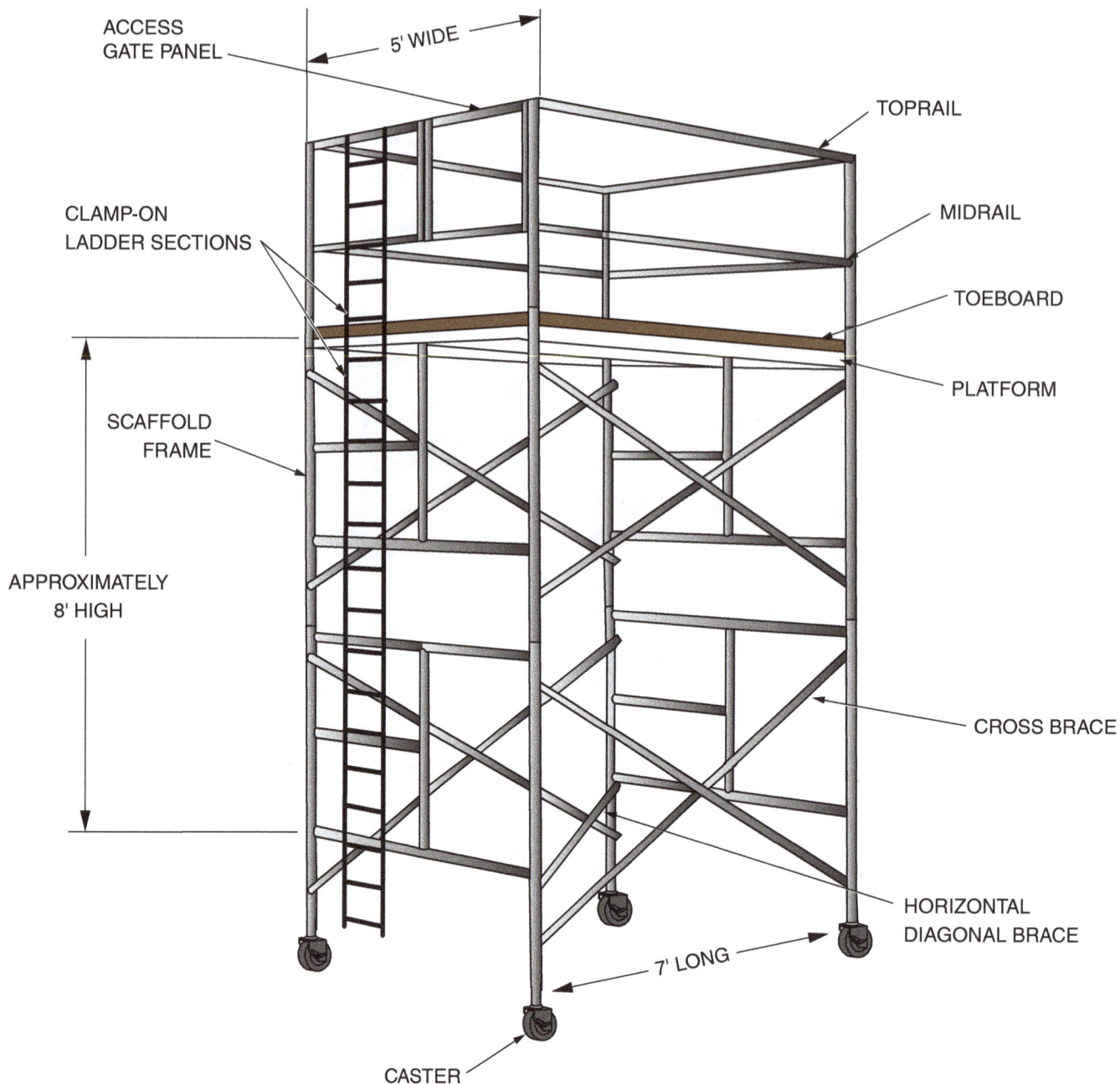

ACCESS GATE PANEL

5' WIDE

TOPRAIL

CLAMP-ON LADDER SECTIONS

MIDRAIL

TOEBOARD

PLATFORM

SCAFFOLD FRAME

APPROXIMATELY 8' HIGH

CROSS BRACE

HORIZONTAL DIAGONAL BRACE

7' LONG

CASTER

31106-14_F18.EPS

Figure 18 Scaffold sketch.

Step 10 Place the level on the top bearer or on the bottom crosspiece of each frame to check the level.

Step 11 Adjust the screw jacks as necessary to level the first scaffold frame.

Step 12 Place a straight board across the bearers of the frames next to one set of the legs.

Step 13 Place the level on the board to check the level along the length of the scaffold.

Step 14 Adjust the screw jacks to level the scaffold.

Step 15 Move the board to other side of the frame and repeat Steps 13 and 14 to level the other side of the scaffold. *Figure 20* shows how to check the level of the scaffold.

Step 16 Check the plumb of the scaffold frames by placing the level against the legs of each frame.

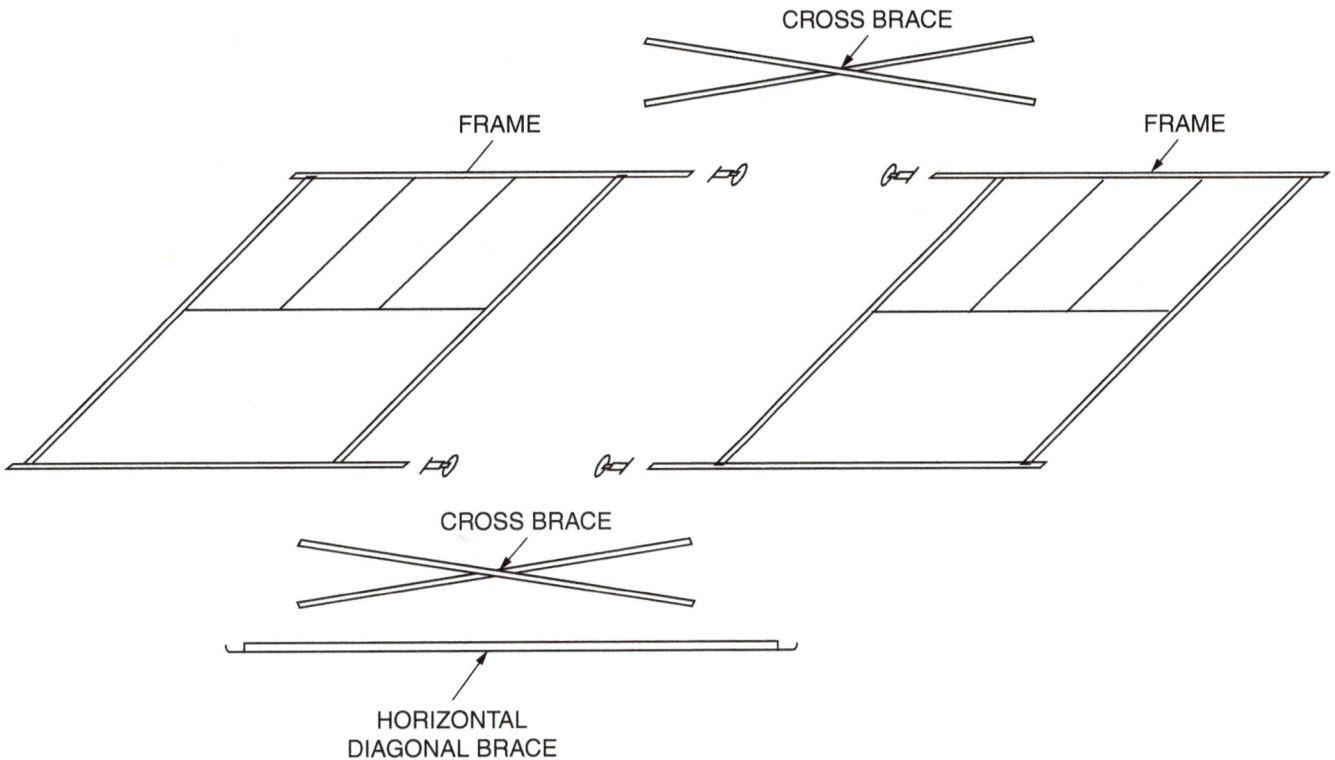

Figure 19 First-level layout.

> **NOTE**
>
> If the scaffold is level, it should also be plumb. If the scaffold frames are not plumb, there is something wrong with the frames and they should be replaced.

Step 17 Check the operation of the caster brakes by attempting to move the scaffold.

> **NOTE**
>
> The casters should not swivel and the wheels should not roll while the brake is engaged. Any casters that fail this test should be replaced before continuing with the scaffold erection.

Step 18 Attach the first section of the access ladder to the end of the scaffold frame.

Step 19 Attach the coupling pins to the bottom of the legs of the 3-foot frames and lock in place.

Step 20 Place the 3-foot frames on the assembled scaffold frames and lock in place.

Step 21 Attach the cross braces to the studs and lock in place according to manufacturer's specifications.

Step 22 Attach the second ladder section to the end frame on top of the first ladder section.

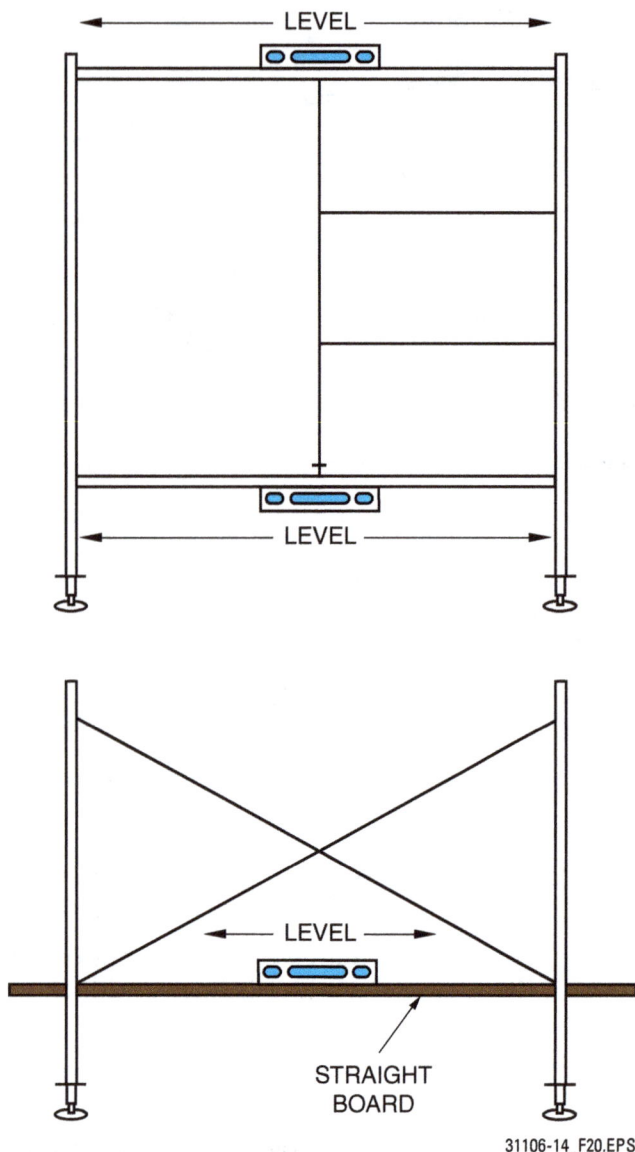

Figure 20 Leveling a scaffold.

31106-14_F20.EPS

Step 23 Install the three 19-inch-wide by 7-foot-long scaffold decks across the upper frame bearers. *Figure 21* shows the scaffold with decks in place.

Step 24 Install the toeboards and secure them to the decks as required.

Step 25 Attach the guardrail posts according to the manufacturer's specifications.

Step 26 Attach the toprails and midrails to the guardrail posts.

Step 27 Install the guardrail access panel to align with the ladder, ensuring that the access gate is positioned to open inward toward the platform.

Step 28 Inspect all connections to ensure that the scaffold is safe to use. *Figure 22* shows the completed scaffold.

> **NOTE**
>
> Personal fall arrest equipment is mandatory and must be employed to place guardrails at heights above 7 feet, 6 inches.

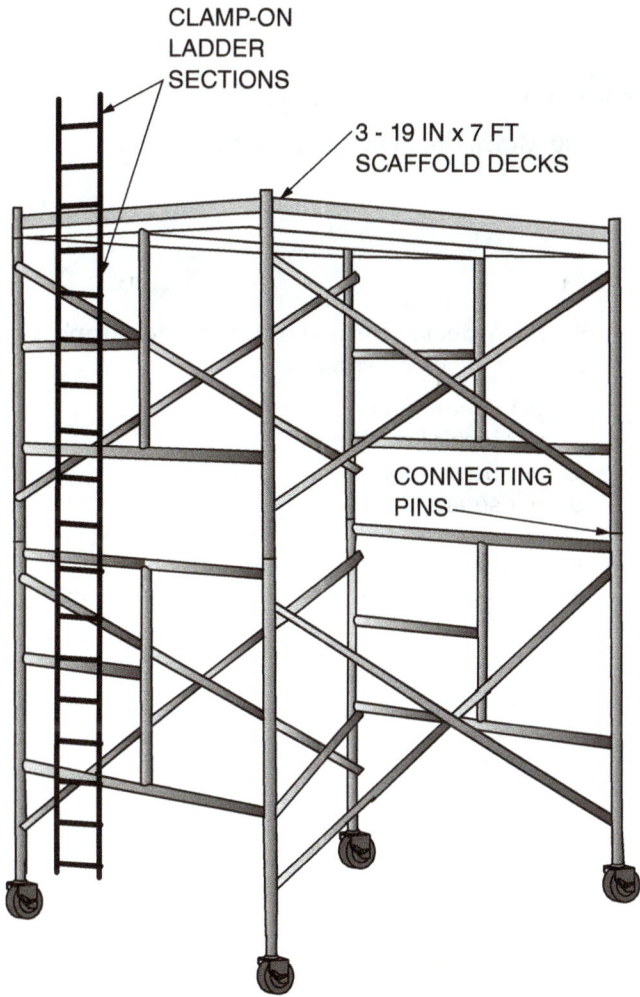

CLAMP-ON
LADDER
SECTIONS

3 - 19 IN x 7 FT
SCAFFOLD DECKS

CONNECTING
PINS

31106-14_F21.EPS

Figure 21 Scaffold with decks.

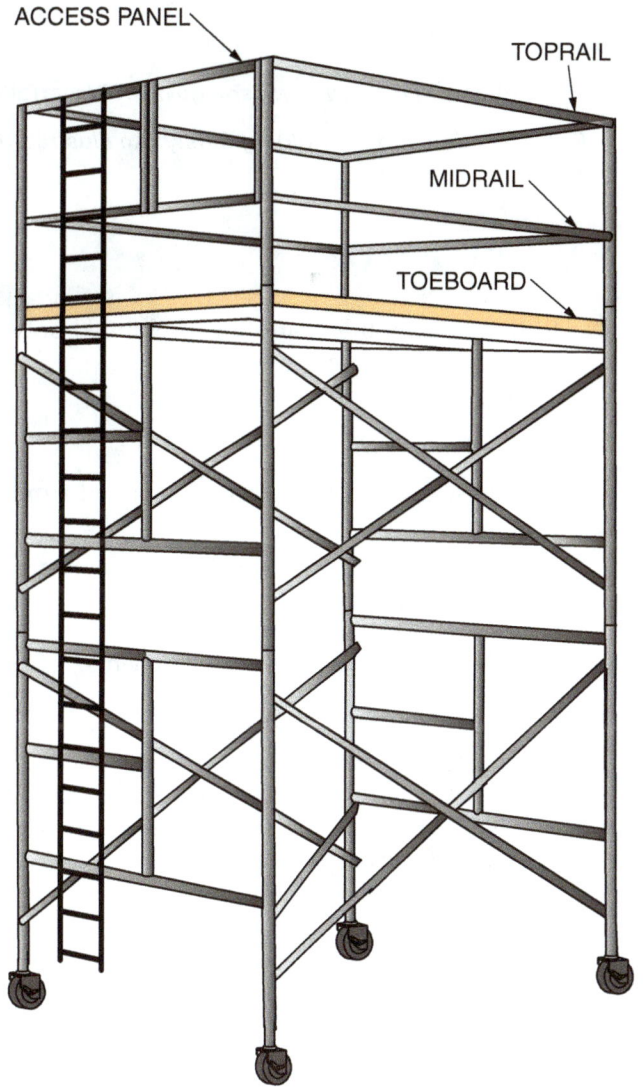

ACCESS PANEL

TOPRAIL

MIDRAIL

TOEBOARD

31106-14_F22.EPS

Figure 22 Completed scaffold.

Additional Resources

"Scaffolding." OSHA. **www.osha.gov/SLTC/scaffolding/index.html**

Fall Protection and Scaffolding Safety: An Illustrated Guide. 2000. Grace Drennan Ganget. Government Institutes.

1.0.0 Section Review

1. Riding on a mobile scaffold is prohibited only under windy conditions.

   a. True
   b. False

2. Mobile scaffolds are usually assembled from standard _____.

   a. tubular welded-frame scaffold components
   b. angle-iron sections
   c. custom-fabricated components
   d. tube-and-clamp scaffold components

3. When a mobile scaffold is in use, _____.

   a. two front casters must be locked
   b. all four casters must be locked
   c. two rear casters must be locked
   d. locking one caster is sufficient

4. Scaffold decks are usually rated to support weights of 50 or 75 pounds _____.

   a. per square foot
   b. per inch of thickness
   c. per lineal foot
   d. per square yard

5. To safeguard workers from falls, guardrails are required on all scaffolds with a height of at least _____.

   a. 6 feet
   b. 8 feet
   c. 10 feet
   d. 12 feet

SECTION TWO

2.0.0 SCISSORS LIFTS

Objective

Describe the proper operation of scissors lifts.
  a. Explain the proper use of controls and indicators on scissors lifts.
  b. Describe basic operating procedures and concerns when using scissors lifts.

Scissors lifts are powered personnel lifts. They use hydraulic rams to raise and lower the platform. Power is supplied to the hydraulic system by a battery, gasoline engine, or liquefied petroleum gas (LPG)-powered engine. The controls for the lift are located at an operator's station on the platform. Some models also include operator stations on the ground and on the platform level. In addition to controlling the height of the platform, the operator is also able to move the unit to the front or rear and steer it as it moves. *Figure 23* shows a scissors lift.

> **NOTE**
> To legally operate a scissors lift, you must be trained by a qualified person. This training must take place on the specific lift you will be operating. You will need to have credentials in some locations.

2.1.0 Operating Precautions

No one is permitted to operate a scissors lift without training in the proper operating procedures for the particular model. This section of the module discusses general operating procedures, but is not intended to serve as an inclusive guide to scissors-lift operation. Keep the following safety considerations in mind when operating a scissors lift:

- Read and observe the warnings and cautions placed on the unit.
- Keep the vehicle and work platform at least 15 feet from electrical power lines. Scissors lifts are not insulated to protect against arcing or electrical contact.
- Anything that is not working correctly must be repaired before using the lift.
- Do not remove any parts or safety devices.
- Make certain that the immediate area is clear of all personnel and obstructions before raising or lowering the lift.

- Keep the guardrails and chains in place when operating the lift.
- Do not bridge between two scissors lifts or between a scissors lift and a building.
- Never crawl into the scissors at any point once it is extended.
- Operate the vehicle on firm, solid ground within 3 degrees of level. Do not operate the vehicle near pits, holes, drop-offs, or other obstructions.
- Do not impose side loads on the work platform.
- Never exceed the rated capacity of the lift.
- Do not operate the vehicle in an explosive atmosphere.
- Never leave the key in a vehicle that is unattended.

2.1.1 Heights and Capacities

Lift heights and capacities vary by model and manufacturer. Lift heights may be as high as 41 feet and capacities may reach 2,000 pounds. Both the height and capacity of the lift can usually be found on a decal placed conspicuously on the lift. Some manufacturers offer optional manual outriggers for added stability.

2.1.2 Controls and Indicators

Each manufacturer uses different control systems. Always familiarize yourself with the controls of a particular unit before using it. The following controls are typically found on scissors lifts:

- *Power switch* – This may be key-operated and may include three positions: Off, On, and Start.
- *Steering* – This may be a toggle switch that is used to turn the vehicle to the right or left.
- *Platform* – This may be a toggle switch that moves the platform up and down.
- *Drive* – This switch may control both the speed and direction of the vehicle. Some models have dual speeds for both forward and reverse directions.
- *Emergency stop* – This switch may have two positions: Run and Emergency Stop. It is usually a protected toggle switch. In order for the power switch to work, the unit must be in the Run position.
- *Manual override* – This may be a pushbutton switch to lower the platform manually during an emergency. It is usually labeled with a decal and may be found at an opening on the hydraulic panel cover.
- *Emergency release for brake* – This is used if the vehicle stalls during operation. It may be a lever located behind the hydraulic tank and la-

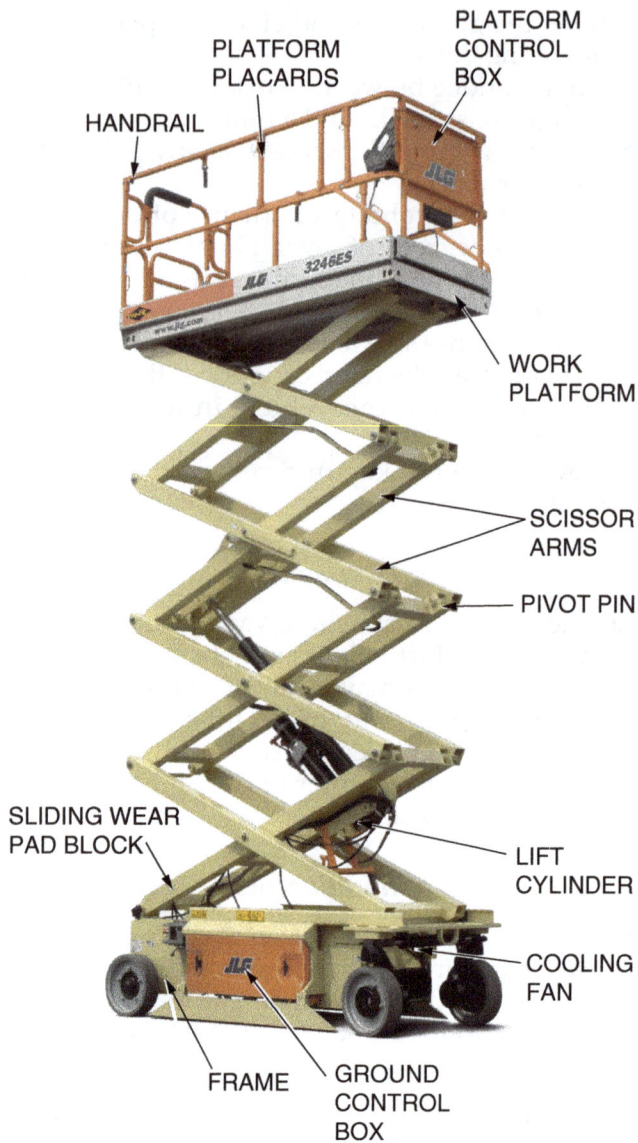

PLATFORM PLACARDS
PLATFORM CONTROL BOX
HANDRAIL
3246ES
www.jlg.com
WORK PLATFORM
SCISSOR ARMS
PIVOT PIN
SLIDING WEAR PAD BLOCK
LIFT CYLINDER
COOLING FAN
FRAME
GROUND CONTROL BOX

31106-14_F23.EPS

Figure 23 Scissors lift.

beled with a decal. Once the vehicle has been restarted, the switch should be returned to the normal position.

- *Fuel gauges* – These indicate the levels of propane or gasoline. LPG-powered models also have a supply valve to open or shut off the supply.
- *Oil dipstick and hydraulic oil indicators* – These allow the operator to check the supply of oil in the crankcase of engine-powered models and the amount of hydraulic oil in the reservoir. You should always check oil levels before operating the lift.

2.2.0 Operating Procedures

The following sections discuss the general operating procedures for scissors lifts. This information is not intended to serve as a substitute for the operator's manual for your particular unit.

Startup and shutdown procedures differ depending upon how the vehicle is powered. For battery-powered models, turn the key to the Power position (or the equivalent). Then, activate all functions before applying a load. This will warm up the hydraulic oil. To shut down a battery-powered model, first lower the platform and then turn the key to the Off position. Remove and store the key.

2.2.1 Startup and Shutdown

Follow these steps to operate engine-powered models:

Step 1 Turn the key slowly to the Start position. Release it quickly when the engine starts. If the unit does not start, turn the key to the Off position and wait about five seconds.

> **CAUTION**
>
> Never hold the key in the Start position for more than ten seconds. Turning the key for a longer period could overheat and damage the starter.

> **NOTE**
>
> If the vehicle does not start, turn the switch to Off and check the gasoline or LPG supply gauge. Check the LPG supply valve to make sure it is open. Then check the condition of the batteries.

Step 2 Let the engine warm up for about two minutes. This allows the hydraulic oil to warm up.

To shut down an engine-powered model, lower the platform and turn the key to the Off position. Remove and store the key.

2.2.2 Maneuvering

Before traveling with a scissors lift, check the entire route. Avoid personnel, obstructions, other traffic, rocks, holes, debris, soft surfaces, and steep grades.

On most models, the switches that control direction and steering must be held manually. When pressure is released, these switches usually return to the center position. It may be necessary to use both hands when traveling and steering. Follow these steps to move a scissors lift to another location:

Step 1 Choose the speed range, if the vehicle is equipped with this control.

Step 2 Push and hold the Forward-Reverse switch with one hand in the desired direction of travel. At the same time, push the Steering switch to Right or Left with the other hand.

On most models, the Up and Down switches must be held in position to raise or lower the platform. The platform stops automatically when the switch is released.

Additional Resources

"Scaffolding." OSHA. www.osha.gov/SLTC/scaffolding/index.html

Fall Protection and Scaffolding Safety: An Illustrated Guide. 2000. Grace Drennan Ganget. Government Institutes.

2.0.0 Section Review

1. Before raising or lowering a scissors lift _____.
   a. sound the horn rapidly three times
   b. check the hydraulic oil level
   c. check the immediate area for people or obstructions
   d. set the emergency brake

2. To avoid damaging the starter, never hold a scissors lift key in the Start position for more than _____.
   a. 10 seconds
   b. 15 seconds
   c. 30 seconds
   d. 60 seconds

3.0.0 Aerial Lifts

Objective

Describe the operation and common applications of aerial lifts.

a. Outline proper safety guidelines when using aerial lifts.
b. Identify common aerial lift applications.

Aerial lifts are work platforms mounted on the end of an arm that can be raised and lowered and can extend beyond the mobile base. The aerial section may consist of two or more telescoping sections. The machines can be driven by the operator from the platform. *Figure 24* shows typical aerial lifts. Aerial lifts are generally restricted to supporting one or two workers at a time, along with their tools. This machine must be viewed as a scaffold only; it is not designed to be a materials lift or personnel elevator.

> **CAUTION**
>
> A brief overview of this type of equipment is presented here. It should not be viewed as a substitute for formal training by a qualified person.

3.1.0 Safety Guidelines

The following are guidelines for using aerial lifts safely:

- Know the capacity and operating characteristics of the aerial platform. Do not overload the platform. Use a lift only on the surface for which it was intended. Lock the wheels, especially on an incline. Avoid using a lift outdoors in stormy weather or in strong winds.
- Inspect the equipment before each use, as specified by the manufacturer, to make sure everything is in proper working order. Have any defects repaired before using the lift.
- Never modify or remove any part of the equipment unless authorized to do so by the manufacturer.
- Check the job site for any hazards that may cause the lift to overturn.

TELESCOPING AERIAL LIFT

ARTICULATING AERIAL LIFT

31106-14_F24.EPS

Figure 24 Aerial lifts.

- Check for hazards above, below, and all around the path of travel. Make sure to maintain a safe distance from electrical power lines and other electrical hazards.
- Prevent people from walking beneath the work area of the platform by placing barricades around the area.
- Use a personal fall arrest system (body harness, lanyard, and lifeline) if required for the type of lift being used.
- Lower the lift and lock it into place before moving the equipment. Lower the lift, shut off the engine, set the parking brake, and remove the key before leaving it unattended.
- Keep the lift free from oil, grease, wet paint, mud, or any slippery material.
- Do not lean over the guardrails of the platform, and never stand on the guardrails.

3.2.0 Areas of Operation

Aerial lifts come in two basic categories: those designed for use on rough terrain and those designed for use on paved/slab surfaces. The rough-terrain units have truck-type tires, or even deep tread mud-type tires. Some units have oscillating axles and 4-wheel drive to maneuver more easily on rough terrain. These units have a wide variety of uses in construction and outdoor maintenance. *Figure 25* shows the major parts of a rough-terrain aerial lift viewed from above.

The paved/slab units are designed for use on improved surfaces. Some are even intended for indoor use and are small enough to fit through interior doorways. The units designed for indoor use are battery-powered to avoid any exhaust problems and are primarily used by maintenance personnel.

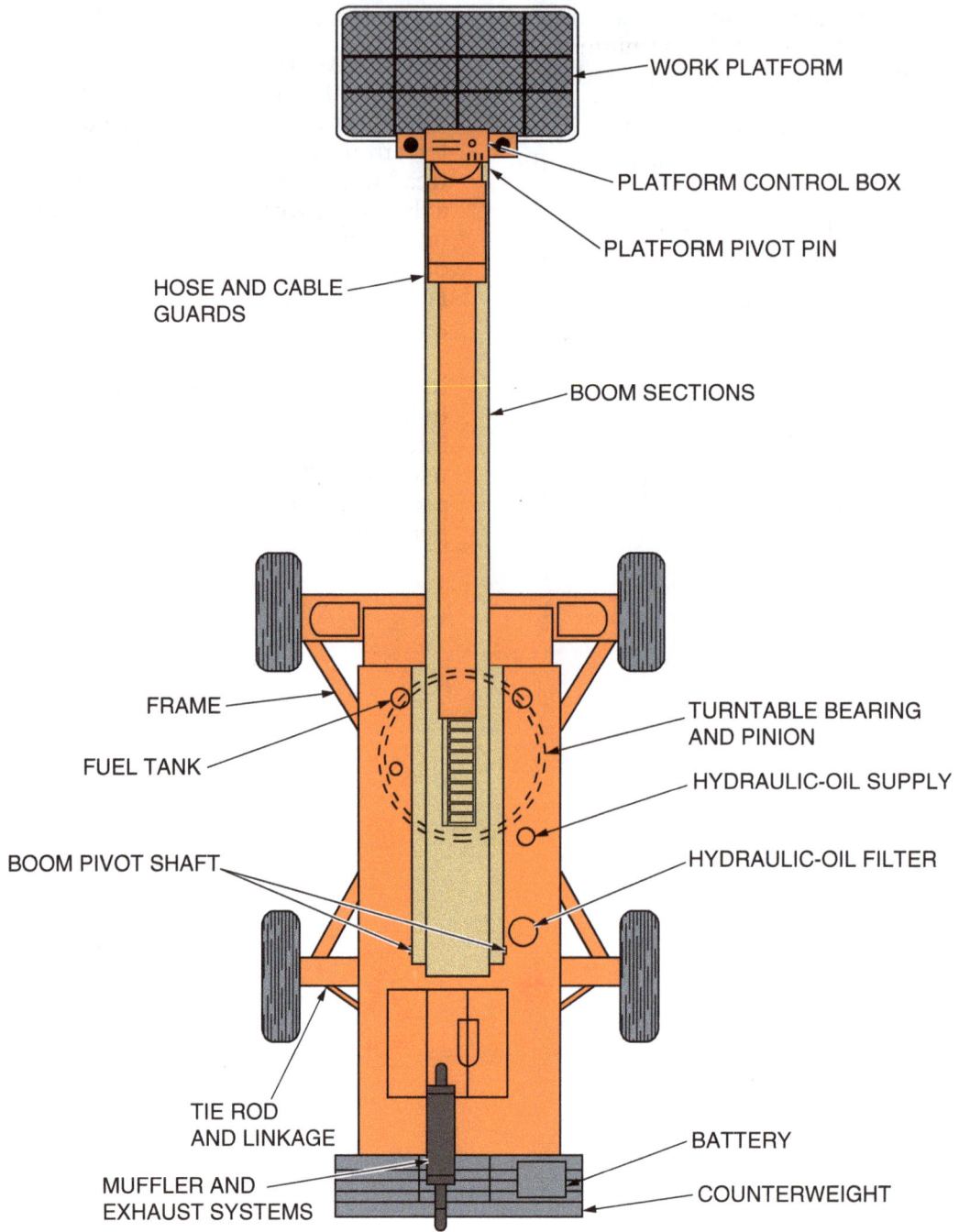

Figure 25 Rough-terrain aerial lift.

WORK PLATFORM

PLATFORM CONTROL BOX

PLATFORM PIVOT PIN

HOSE AND CABLE GUARDS

BOOM SECTIONS

FRAME

FUEL TANK

TURNTABLE BEARING AND PINION

HYDRAULIC-OIL SUPPLY

BOOM PIVOT SHAFT

HYDRAULIC-OIL FILTER

TIE ROD AND LINKAGE

MUFFLER AND EXHAUST SYSTEMS

BATTERY

COUNTERWEIGHT

31106-14_F25.EPS

Additional Resources

"Scaffolding." OSHA. **www.osha.gov/SLTC/scaffolding/index.html**

Fall Protection and Scaffolding Safety: An Illustrated Guide. 2000. Grace Drennan Ganget. Government Institutes.

3.0.0 Section Review

1. Operators must aware of and avoid hazards that could cause an aerial lift to overturn.

   a. True
   b. False

2. Aerial lifts with truck-type tires are called _____.

   a. off-road units
   b. broken-ground units
   c. all-surface units
   d. rough-terrain units

SUMMARY

This module covered the design, erection, and use of mobile scaffolds. A mobile scaffold may be a simple tubular welded-frame scaffold fitted with casters that is moved manually from one work location to another, an engine-powered scissors lift, or an aerial lift.

Mobile scaffolds made of tubular welded frames have the same safety requirements as a supported scaffold, including height-to-base ratios and guardrail and toeboard installation. Outriggers with casters can be added to increase the base width to allow higher mobile scaffolds to be built.

Scissors lifts are work platforms that are raised and lowered using a hydraulic system. Controls are mounted on the work platform that allow the workers to raise and lower the platform as necessary.

Aerial lifts provide the greatest mobility. Aerial lifts can be driven into position and raised from the work platform. These lifts may have articulated arms that can position the work platform away from the base.

1. Guardrail systems are required on all scaffolds that are a minimum of _____.

   a. 6 feet in height
   b. 10 feet in height
   c. 8 feet in width
   d. 10 feet in width

2. When determining whether guardrails are required, mobile-scaffold platform height is always measured from _____.

   a. the average ground elevation
   b. a surveyor's benchmark
   c. the top of the casters
   d. a safe surface

3. Planks used on a mobile scaffold must be _____.

   a. red oak
   b. safety-certified
   c. scaffold-grade
   d. high-strength steel

4. Unlikely users of a mobile scaffold would be _____.

   a. masons erecting a brick wall
   b. electricians installing light fixtures
   c. carpenters doing trim work

   d. painters finishing a ceiling

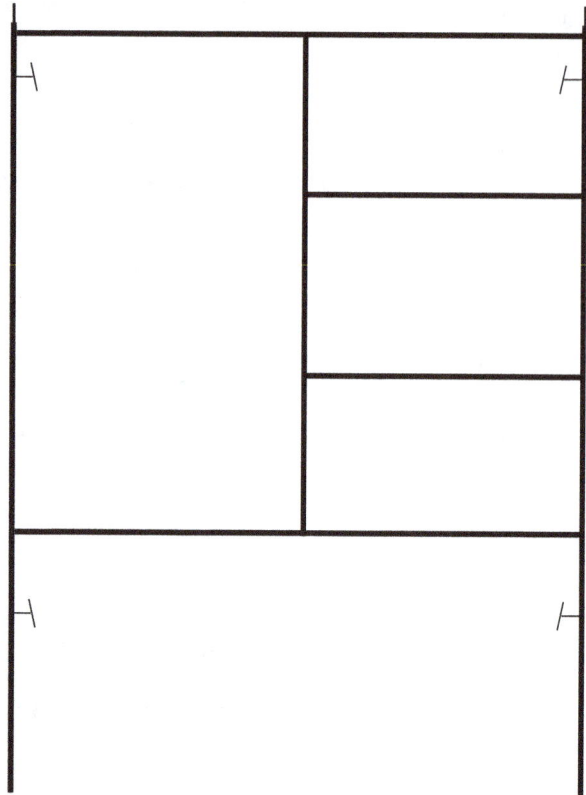

Figure 1

31106-14_RQ01.EPS

5. The frame type shown in Review Question *Figure 1* is a _____.

   a. walk-through
   b. single-ladder box
   c. swing gate
   d. standard square

6. When scaffold frames are stacked vertically, they must be secured using _____.

   a. twist locks
   b. right-angle clamps
   c. coupling pins

   d. bolts and wing nuts

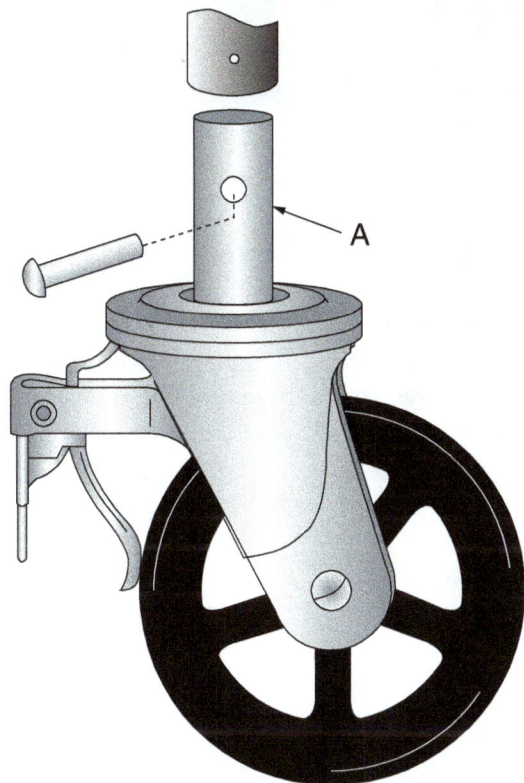

Figure 2

31106-14_RQ02.EPS

7. The part of the caster marked "A" in Review Question *Figure 2* is the _____.

   a. anchor
   b. shaft
   c. insert
   d. stem

8. The most common platform material used with scaffolding is _____.

   a. solid sawn wood planks
   b. plywood
   c. laminated wood planks
   d. aluminum

9. A 10-foot-long scaffold plank must not deflect more than ⅟₆₀ of its length with a centered load of _____.

   a. 200 pounds
   b. 225 pounds
   c. 250 pounds
   d. 275 pounds

10. Manufactured platforms called scaffold decks typically have a width of _____.

    a. 16 inches
    b. 19 inches
    c. 21 inches
    d. 24 inches

11. Scaffolding access provided when a worker must climb from below and enter a platform head first is referred to as _____.

    a. climb-up access
    b. pass-through access
    d. vertical access
    d. duck-through access

12. A toprail must be installed if the crossing point of a cross brace is at a distance above the platform of between 20 inches and _____.

    a. 24 inches
    b. 26 inches
    c. 30 inches
    d. 36 inches

13. Before being installed on a mobile scaffold, casters should be _____.

    a. locked
    b. checked for flat spots
    c. unlocked
    d. lubricated

14. Hydraulic oil levels on a scissors lift should be checked _____.

    a. daily
    b. at the beginning of your shift
    c. weekly
    d. before operating the vehicle

15. Inspect an aerial lift _____.

    a. at the end of your shift
    b. as needed
    c. before each use
    d. when something goes wrong

Trade Terms Quiz

Fill in the blank with the correct term that you learned from your study of this module.

1. A surface is considered _____ when it is completely vertical.

2. A wheel fitted with a stem to secure it into a scaffold frame leg is called a(n) _____.

3. A(n) _____ is an accessory installed on a mobile scaffold to provide added stability.

4. Layers of wood that are glued together parallel with the grain are considered _____.

5. Scaffold components from different manufacturers that can be used together are considered _____.

Trade Terms

Caster
Compatible

Laminated
Outrigger

Plumb

Trade Terms Introduced in This Module

Caster: A wheel fitted with a stem to secure it into a scaffold frame leg.

Compatible: Scaffold components from different manufacturers that can be used together.

Laminated: Layers of wood glued together with the grain parallel.

Outrigger: An accessory that can be installed on a mobile scaffold to provide added stability.

Plumb: Vertical, or completely straight up and down; at right angles to level.

Additional Resources

This module presents thorough resources for task training. The following resource material is suggested for further study.

"Scaffolding." OSHA. **www.osha.gov/SLTC/scaffolding/index.html**

Fall Protection and Scaffolding Safety: An Illustrated Guide. 2000. Grace Drennan Ganget. Government Institutes.

International Building Code®, Latest Edition. Falls Church, VA: International Code Council.

International Residential Code®, Latest Edition. Falls Church, VA: International Code Council.

Figure Credits

Section Review Answer Key

Answer	Section Reference	Objective
Section One		
1. b	1.1.0	1a
2. a	1.3.0	1c
3. b	1.4.4	1d
4. a	1.4.5	1d
5. c	1.4.6	1d
Section Two		
1. c	2.1.0	2a
2. a	2.2.1	2b
Section Three		
1. a	3.1.0	3a
2. d	3.2.0	3b

NCCER CURRICULA — USER UPDATE

NCCER makes every effort to keep its textbooks up-to-date and free of technical errors. We appreciate your help in this process. If you find an error, a typographical mistake, or an inaccuracy in NCCER's curricula, please fill out this form (or a photocopy), or complete the online form at **www.nccer.org/olf**. Be sure to include the exact module ID number, page number, a detailed description, and your recommended correction. Your input will be brought to the attention of the Authoring Team. Thank you for your assistance.

Instructors – If you have an idea for improving this textbook, or have found that additional materials were necessary to teach this module effectively, please let us know so that we may present your suggestions to the Authoring Team.

NCCER Product Development and Revision
13614 Progress Blvd., Alachua, FL 32615

Email: curriculum@nccer.org
Online: www.nccer.org/olf

❏ Trainee Guide ❏ Lesson Plans ❏ Exam ❏ PowerPoints Other _____

Craft / Level: _____ Copyright Date: _____

Module ID Number / Title: _____

Section Number(s): _____

Description: _____

_____

_____

_____

_____

Recommended Correction: _____

_____

_____

_____

_____

Your Name: _____

Address: _____

_____

Email: _____ Phone: _____

31107-15

Suspension Scaffolds

OVERVIEW

Suspension scaffolds do not rest on the ground; they are attached to an overhead support. Suspension scaffolds are hung by wire ropes or cables that allow the platform to be raised or lowered. These scaffolds can also be attached to I-beams using clamps or rigging devices.

Module Seven

Trainees with successful module completions may be eligible for credentialing through the NCCER Registry. To learn more, go to **www.nccer.org** or contact us at **1.888.622.3720**. Our website has information on the latest product releases and training, as well as online versions of our *Cornerstone* magazine and Pearson's product catalog.

Your feedback is welcome. You may email your comments to **curriculum@nccer.org**, send general comments and inquiries to **info@nccer.org**, or fill in the User Update form at the back of this module.

This information is general in nature and intended for training purposes only. Actual performance of activities described in this manual requires compliance with all applicable operating, service, maintenance, and safety procedures under the direction of qualified personnel. References in this manual to patented or proprietary devices do not constitute a recommendation of their use.

Objectives

When you have completed this module, you will be able to do the following:

1. Describe the safety considerations, applications, and components of suspension scaffolds.
   a. Outline proper safety guidelines for suspension scaffolds.
   b. Identify common suspension scaffold applications.
   c. Identify suspension scaffold components.
2. Explain the proper methods for rigging suspension scaffolds.
   a. Explain the proper methods for rigging boatswain's chairs, work cages, and beam-suspended scaffolds.

Performance Tasks

This is a knowledge based module; there is no performance testing.

Trade Terms

Counterweight
Hoist
Outrigger beam

Platform
Stirrups

Industry-Recognized Credentials

If you're training through an NCCER-accredited sponsor, you may be eligible for credentials from NCCER's Registry. The ID number for this module is 31107-14. Note that this module may have been used in other NCCER curricula and may apply to other level completions. Contact NCCER's Registry at 888.622.3720 or go to **www.nccer.org** for more information.

Code Note

Codes vary among jurisdictions. Because of the variations in code, consult the applicable code whenever regulations are in question. Referring to an incorrect set of codes can cause as much trouble as failing to reference codes altogether. Obtain, review, and familiarize yourself with your local adopted code.

Contents

Topics to be presented in this module include:

Figures

1.0.0 Suspension Scaffolds

Objectives

Describe the safety considerations, applications, and components of suspension scaffolds.

a. Outline proper safety guidelines for suspension scaffolds.
b. Identify common suspension scaffold applications.
c. Identify suspension scaffold components.

Trade Terms

Hoist: A manual or powered mechanical device used to raise or lower a suspension scaffold.

Outrigger beam: The structural member of a suspension scaffold or outrigger scaffold, which provides support for the scaffold by extending the scaffold point of attachment to a point out and away from the structure of the building.

Platform: An elevated horizontal work surface.

Stirrups: Steel frames used to support the ends of the suspension-scaffold platforms. Also called triangles.

Suspension scaffolds are attached to an overhead support. Because they are not supported from below, they present unique challenges for construction, and have additional safety hazards to consider.

1.1.0 Safety Guidelines

The following are some basic safety guidelines for suspension scaffold riggers and users:

- Regulations mandate that no more than two workers are permitted to work at one time on a swing-stage scaffold or other suspension scaffold designed for a working load of 500 pounds. With a working load of 750 pounds, no more than three workers are permitted to work at one time on the scaffold.

> **NOTE**
>
> OSHA considers one worker with tools to weigh 250 pounds.

- All workers must be protected by both personal fall protection equipment and a guardrail system when working on suspension scaffolds.
- Safety lines should always be tied off to structural anchor points independent from the rigging.
- Suspension-scaffold platforms are required to have guardrails, midrails, and toeboards on all open sides and ends.
- The suspension system and rigging devices (including their support on the structure) must be able to hold four times the rated load of the hoist.
- Ropes or cables used to support a suspension scaffold must be able to support at least six times the rated load.
- Rigging devices are anchored and secured to the building or structure with tieback lines installed perpendicular (at right angles) to the face of the structure.
- Tieback lines must be as strong as the suspension ropes or cables.
- Gas-powered equipment is not allowed on a suspension scaffold.
- Powered hoists are limited to a maximum speed of 35 feet per minute.
- Powered hoists must have both primary and secondary brakes.
- When rigging a beam-suspended scaffold:
  - Make sure the I-beam is level.
  - If the I-beam is not level, secure the rigging devices to prevent accidental movement.
  - Never use beam-suspended scaffolds on open-ended beams that could allow the scaffold to roll off the supports.
  - Install stops if the flange is interrupted anywhere along the beam.
  - Check that the width of the flange is uniform along the entire length of the beam. A roller set to a particular width could derail if the width of the flange narrows.
- Always inspect your equipment yourself.
- Use your personal fall arrest system.
  - Use a strong and independent anchorage that is separate from your rigging.
  - Protect the lifeline against possible damage.
  - Check the equipment.
  - Set up the system to limit a fall to less than six feet.
  - Always hook up before you get on a suspension scaffold.

- Use a tieback.
  - Use rope equal in strength to the suspension rope.
  - Tie to a strong and independent anchor, or to the other side of the building.
  - Tie straight back from the support or at opposing angles.
- Rated load of the support must be equal to or greater than the rated load of the hoist.
- Rated platform capacity must be equal to or greater than the platform load.

Figure 1 shows a suggested pre-operational checklist for suspension scaffolds.

Every elevated job site should have a rescue-and-retrieval plan in case it is necessary to rescue a fallen worker. Planning is especially important in remote areas that are not readily accessible. Before there is a risk of a fall, make sure that you know what your employer's rescue plan requires you to do. Find out what rescue equipment is available and where it is located. Learn how to use equipment for self-rescue and the rescue of others.

If a fall occurs, any employee hanging from the fall arrest system must be rescued safely and quickly. Your employer should have previously determined the method of rescue for fall victims, which may include equipment that lets the victim rescue themselves, a system of rescue by co-workers, or a way to alert a trained rescue squad. If fall rescue depends on calling for outside help, such as the fire department or rescue squad, all phone numbers must be posted in plain view at the job site. Communicate with the victim and monitor them constantly during the rescue.

1.2.0 Suspension Scaffold Systems

A two-point suspension scaffold, also known as a swing-stage scaffold, is suspended from a building or structure by two ropes or cables in a manner that allows it to be raised or lowered as needed. It consists of a platform or stage, hangers, hoisting mechanism, ropes or cables, and rigging devices. The platform or stage provides for a work area of 24- to 36-inches wide that is securely attached to the hangers. *Figure 2* shows a typical suspension scaffold platform.

Multiple-point suspension scaffolds are suspension scaffolds that are supported by two or more ropes from overhead. These platforms are also equipped with a means of raising or lowering the platform to the desired work levels. A

typical application for a multiple-point suspension scaffold is a chimney hoist. *Figure 3* shows a multiple-point suspension scaffold.

1.2.1 Chair, Cage, and Beam Components

A boatswain's chair (*Figure 4*) is basically a seat attached to a suspended cable or rope. The chair is typically made of nylon rigging with a plywood seat. The worker is restricted to a sitting position with little or no room for tools or equipment on the chair. If necessary, these items can be tied to the chair.

Work cages are single-point adjustable suspension scaffolds with the platform enclosed with guardrails and toeboards. It is only large enough to allow the operator to work while standing. Two work cage units are sometimes rigged to form a single-point suspension (swing-stage) scaffold. *Figure 5* shows a work cage.

Beam-suspended scaffolds are suspended work platforms with clamp- or roller-type rigging devices that hook onto the lower flanges of two parallel I-beams. Each rigging device has a bar or bracket for attaching the scaffold. A roller-type device has wheels that allow the scaffold to be moved horizontally along the I-beam flanges as work progresses. This allows the job to continue without having to stop to lower the scaffold and reposition the rigging. *Figure 6* shows a beam-suspended scaffold.

1.3.0 Suspension Scaffold Components

Suspension scaffold platforms, made of metal or wood, are required to have guardrails, midrails, and toeboards on all open sides and ends.

Multiple-point suspension scaffolds can be configured in a number of different ways using a variety of components as required for the individual application. Most suspension scaffold platforms can be used in a multiple-point configuration. The platform must be constructed with full decking, guardrails, midrails, and toeboards. Lifting motors can be attached to a variety of platform configurations. An additional cable or line must be rigged for fall protection for each person working on the platform. *Figure 7* shows a multiple-point suspension scaffold rigged to perform maintenance on a chimney. This scaffold is assembled from four standard platforms joined into a box. A professional engineer should be consulted prior to erecting a chimney scaffold.

SUSPENSION SCAFFOLD
PREOPERATIONAL CHECKLIST

THE USE OF THIS CHECKLIST REQUIRES PROPER TRAINING.
MAKE SURE THAT YOU READ, UNDERSTAND, AND FOLLOW THESE CHECKS.
ALSO FOLLOW THE MANUFACTURER'S INSTRUCTIONS, SCAFFOLD INDUSTRY ASSOCIATION'S
SAFETY GUIDELINES, AND ANY ORDINANCES THAT APPLY.

CHECK   **SUPPORT SYSTEMS**

__ THE STRUCTURE IS ABLE TO SUPPORT THE LOADS.

__ THE TIEBACK ROPE IS PROPERLY TIED OFF.

__ THE OUTRIGGER IS OF PROPER DESIGN AND CORRECTLY ASSEMBLED.

__ THE NUMBER OF COUNTERWEIGHTS ARE CORRECT FOR THE OVERHANG LOAD.

__ ALL BEAMS, CLAMPS, AND HOOKS ARE CORRECTLY TIED BACK.

__ SOCKETS AND DAVITS ARE CORRECTLY SECURED.

SUSPENSION UNIT

__ ALL PARTS (RAILS, RUNGS, DECK, BUMPER ROLLERS, WELDS, CONNECTIONS, TOEBOARDS/GUARDRAILS, STANCHIONS) OF THE STAGE/CAGE/CHAIR HAVE BEEN CHECKED TO MAKE SURE THAT THEY ARE SAFE AND WILL NOT BREAK OR COME LOOSE.

__ THE STIRRUPS OR CONNECTIONS AND THEIR PARTS HAVE BEEN CHECKED AND ARE SAFE.

__ THE CAPACITY PLATE SHOWING THE MAXIMUM LOAD WILL NOT BE EXCEEDED.

__ THE STAGE STIRRUPS ARE IN LINE WITH THE ROOF SUPPORTS.

HOIST COMPONENTS

__ THE MANUFACTURER'S OPERATING INSTRUCTIONS HAVE BEEN READ AND UNDERSTOOD.

__ HOISTS ARE IN PROPER OPERATING CONDITION.

__ THE WIRE ROPE HAS BEEN INSPECTED, REEVED, AND IS ATTACHED PROPERLY.

__ THE LENGTH OF STEEL WIRE ROPE IS LONG ENOUGH TO REACH THE GROUND AND IS ATTACHED CORRECTLY TO THE ROOF SUPPORT.

__ AFTER LOAD IS APPLIED, ALL FITTINGS HAVE BEEN CHECKED FOR TIGHTNESS.

CHECK   **HOIST COMPONENTS (CONT'D)**

__ THE ELECTRICAL CABLE OR AIR HOSE CONNECTIONS HAVE BEEN INSPECTED AND ARE SAFE. STRAIN RELIEF HAS BEEN PROVIDED.

__ THE POWER SUPPLIED IS ENOUGH TO OPERATE THE HOIST PROPERLY.

__ THE HOIST IS CORRECTLY ATTACHED TO THE STIRRUP.

FALL ARREST SYSTEM

NOTE: **NO PERSON SHALL ENTER A STAGE/CAGE/CHAIR UNLESS EACH PERSON HAS "HOOKED-UP" IN A SAFE MANNER.**

__ EACH PERSON TO USE THE STAGE/CAGE/ CHAIR HAS OWN INDEPENDENT FALL ARREST SYSTEM.

__ EACH LIFE LINE HAS BEEN TOTALLY CHECKED FOR SAFE USE AND IS CORRECTLY ATTACHED AT AN INDEPENDENT SAFE ANCHOR POINT AT ROOF LEVEL. ROOF EDGE PROTECTION HAS BEEN PROVIDED FOR THE LIFELINE.

__ EACH ROPE GRAB HAS BEEN CHECKED FOR CORRECT OPERATION AND INSTALLATION.

__ EACH FULL-BODY HARNESS OR BELT HAS BEEN THOROUGHLY CHECKED TO MAKE SURE THAT ALL COMPONENTS ARE SAFE.

__ EACH LANYARD HAS BEEN THOROUGHLY CHECKED FOR SAFE CONDITION. ALL PARTS ARE SOUND AND CORRECTLY ATTACHED TO THE ROPE GRAB. THE D-RING IS IN THE CENTER OF THE BACK.

__ THE BODY HARNESS IS THE PROPER SIZE AND FITS CORRECTLY AND IS SNUG TO THE BODY.

ADDITIONAL CHECKS

__ THE EQUIPMENT MUST BE KEPT CLEAR OF EXPOSED ELECTRICAL LINES AND EQUIPMENT.

__ THE EQUIPMENT MUST NOT BE USED IN BAD WEATHER CONDITIONS.

__ THE TOTAL RIGGING MUST BE CHECKED EACH TIME YOU USE OR MOVE IT.

__ NEVER OVERLOAD THE EQUIPMENT.

__ IMMEDIATELY REPORT ANY IMPROPER OPERATION TO YOUR SUPERVISOR.

REMEMBER: WHEN IN DOUBT, ASK.

**ALWAYS HOOK UP YOUR SAFETY EQUIPMENT
BEFORE YOU GET ON THE STAGE.**

31107-14_F01.EPS

Figure 1 Suggested preoperational checklist for a suspension scaffold.

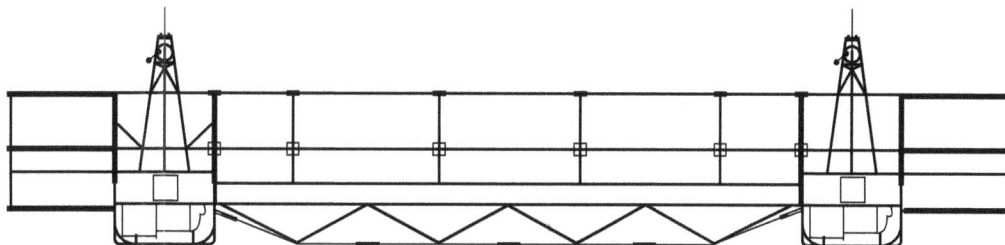

31107-14_F02.EPS

Figure 2 Suspension-scaffold platform.

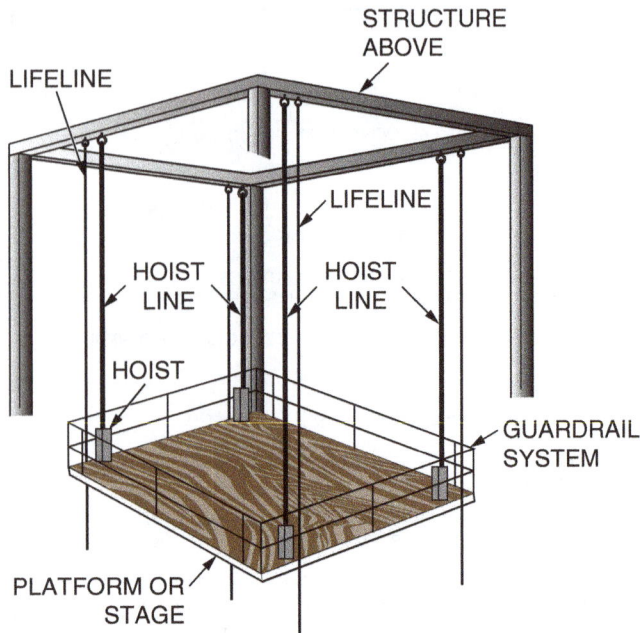

Figure 3 Multiple-point suspension scaffold.

Steel hangers, often called stirrups or triangles, support each end of the suspension-scaffold platform. These hangers are attached to wire cables or ropes made of wire, fiber, or synthetic material, which are suspended from rigging devices secured to the building or structure. The distance between the stirrups must match the distance between the rigging attachment points. The suspension system and rigging devices (including their support on the structure) must be able to hold four times the rated load of the hoist. OSHA requires that the ropes or cables be able to support at least six times the rated load.

Safety lines must be used when working on two-point suspension scaffolds. These lines are always tied off to different anchor points than those used to support the scaffold.

Cornice hooks, parapet clamps, or outrigger beams are rigging devices commonly used to support suspension scaffolds. A cornice hook is a steel device that hooks to a roof, parapet, or structural support. A parapet clamp fits over and clamps onto a parapet that runs along the perimeter of a roof. An outrigger beam is an arm that extends beyond the edge of the roof. Some outrigger devices have wheels so they can roll along the roof as the work progresses. Rigging devices are anchored and secured to the building or structure with tie-back lines installed perpendicular (at right angles)

Figure 4 Boatswain's chair.

to the face of the structure. Tieback lines must be as strong as the suspension ropes or cables.

It is extremely important to make sure that the cornice or parapet that they attach to is structurally sound and capable of supporting the combined weight of the scaffold and workers. If there is any doubt, a professional engineer or other competent person should check the structural integrity of the cornice or parapet before using it to support the scaffold.

Davits (fixed or movable brackets extending over the side of the building used for hoisting) and carriages are sometimes designed into buildings to provide rigging points for suspension scaffolds. The following are guidelines for using davits and carriages:

- The davit rating must be equal to or greater than the hoist rating.
- Davits need to mate with the socket, so check before you move them to other installations.

Figure 5 Work cage.

- Double-check locking pins.
- Pins should slide in. If you have to hammer them in, report it to the appropriate personnel.
- If the davit is equipped with trolley locks, loosen at least one lock before bringing the platform inboard or outboard.
- If the davit is equipped with a rotation brake, loosen the brake before moving the platform outboard or inboard.
- Set the davit rotation brake or lock after platform is outboard.
- Report any roof damage immediately, before water can damage the socket mount.
- Always remove davits from socket, when not using them and lay them flat on the roof.
- Check davits before using them; pre-use inspection may be required by local safety authorities.
- Follow manufacturer's operation instructions.
- Use tie-downs whenever they are provided.

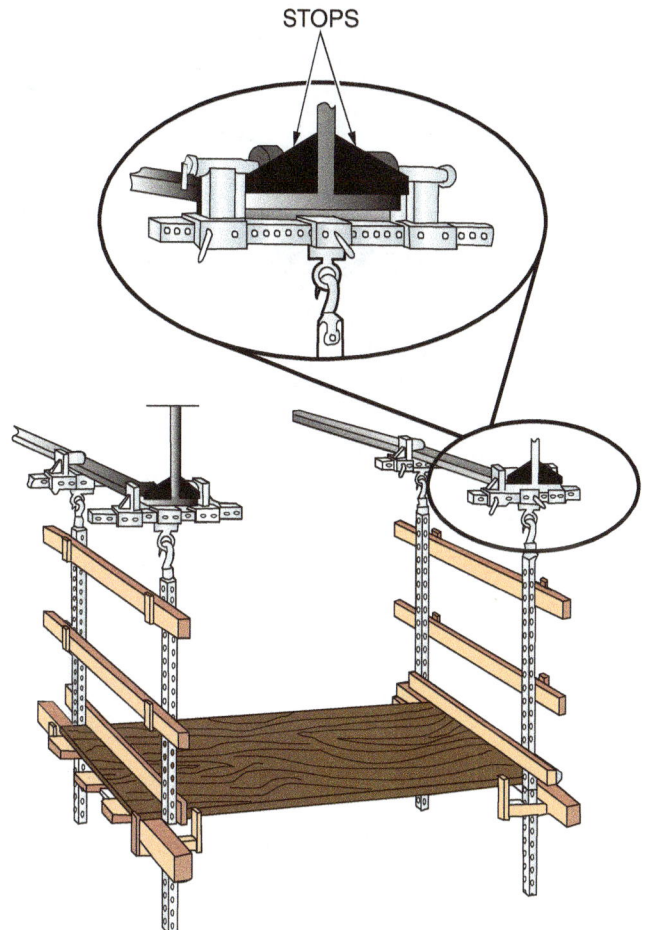

STOPS

Figure 6 Beam-suspended scaffold.

- Check that the current inspection certificate is up to date.
- Follow manufacturer's recommended pre-check before using equipment.
- Test hoisting machinery brakes.
- Double-check support rope connections.
- Set roof car brakes before using platform.
- Check the maintenance/inspection log for problems or shutdown entries.
- Check that track end stops are in place following manufacturer's instructions.
- Inspect the trolley carefully for damage or obstructions and make necessary repairs before using.

Figure 7 Chimney scaffold.

Figure 8 shows rigging devices used with suspension scaffolds.

The scaffold's hoisting mechanism enables the worker to raise or lower the scaffold. For manually operated scaffolds, the mechanism consists of ropes or cables rigged through tackle blocks and pulleys. Many scaffolds have hoist mechanisms powered by electricity or compressed air. Never use gas-powered equipment as a hoist mechanism. Industry standards stipulate that powered hoists are limited to a maximum speed of 35 feet per minute. They must also have speed reducers and both primary and secondary brakes. Safety devices are also built into powered hoist systems to prevent operation of the scaffold if it is overloaded, and to lower the unit without power if necessary. The following are suggested guidelines for hoists:

- Read the manufacturer's instructions before operating.
- Know the equipment and its features.
- Know the machinery troubleshooting procedures.
- Test the overspeed brake (and auxiliary brake) for operation daily per manufacturer's instructions.
- Handle the machine with care—do not drop.
- If the equipment is sluggish or noisy, don't use it until a competent person deems it fit for use.
- Use only wire rope, free of damage (no kinks or birdcages), recommended by the manufacturer.
- Keep the rope properly wound on the drum or sheave.
- Use the proper brazed termination on the feed end of the rope.
- Inspect the wire rope and its path through the equipment daily. Check the rope and the parts in the rope path for wear, rust, and damage.
- Inspect daily for oil leaks, and correct before using.
- Check gearbox oil regularly.
- Follow manufacturer's maintenance instructions.
- If the gearbox becomes noisy, stop immediately. Noise is an indication of a serious problem.
- Cover the hoist when it is not in use.
- Use the right voltage or air pressure for the hoist.
- Have the hoist inspected and maintained as recommended by the manufacturer.
- Do not adjust safety devices. Consult with a technician knowledgeable in that type of equipment.

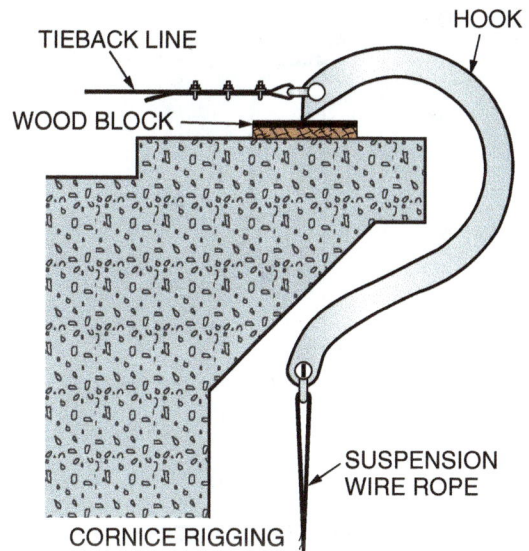

CORNICE RIGGING

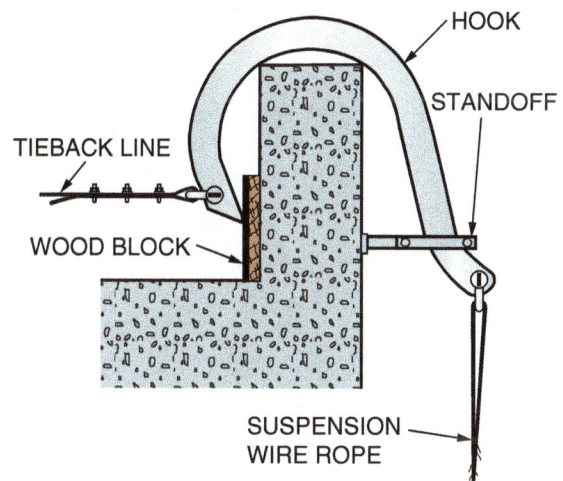

PARAPET RIGGING

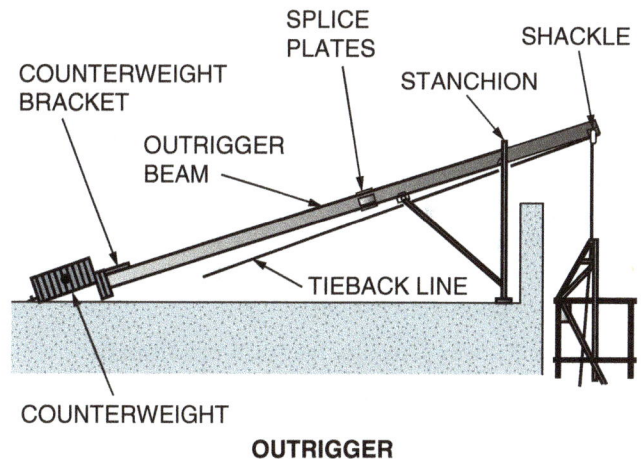

OUTRIGGER

31107-14_F08A.EPS

Figure 8A Rigging devices used with suspension scaffolds (1 of 2).

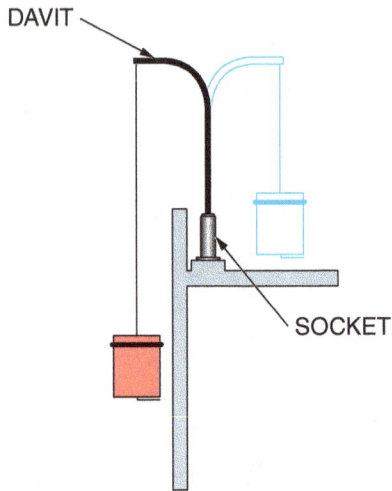

DAVITS/SOCKETS

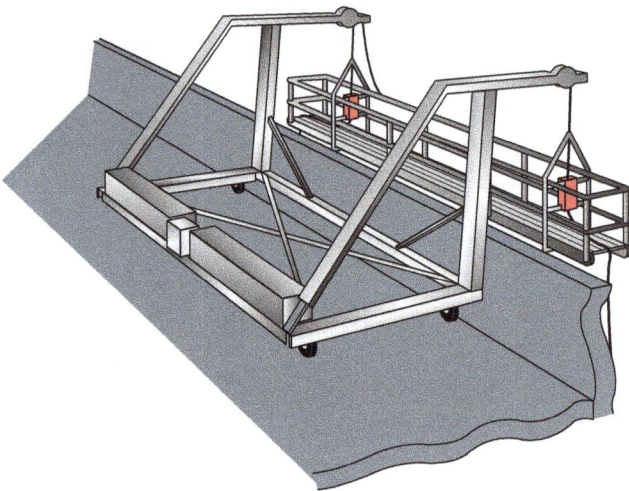

CARRIAGES AND ROOF CARS

31107-14_F08B.EPS

Figure 8B Rigging devices used with suspension scaffolds (2 of 2).

31107-14_F09.EPS

Figure 9 Electric-powered hoist.

- Do not run the termination into the hoist fairlead. Leave at least six inches of clearance.
- Always run the loaded hoist up and down daily before using, and observe brakes, wire rope path, and any fouling of equipment.
- Verify that the controlled-decent device is operating properly in case of power failure.
- Do not exceed rated capacity.

Figure 9 shows an electric-powered hoist.

Additional Resources

"Scaffolding." OSHA. www.osha.gov/SLTC/scaffolding/index.html

Fall Protection and Scaffolding Safety: An Illustrated Guide. 2000. Grace Drennan Ganget. Government Institutes.

International Building Code®, Latest Edition. Falls Church, VA: International Code Council.

International Residential Code®, Latest Edition. Falls Church, VA: International Code Council.

1.0.0 Section Review

1. For calculating scaffold loads, a standard weight of 250 pounds for a worker with tools has been established by the _____.

   a. Occupational Safety and Health Administration
   b. National Safety Council
   c. U.S. Bureau of Standards
   d. American National Standards Institute

2. Beam-suspended scaffolds should never be used on open-ended beams.

   a. True
   b. False

3. A personal fall arrest system should be set to limit a fall to _____

   a. 3 feet
   b. 5 feet
   c. 6 feet
   d. 8 feet

2.0.0 RIGGING SUSPENSION SCAFFOLDS

Objective

Explain the proper methods for rigging suspension scaffolds.

  a. Explain the proper methods for rigging boatswain's chairs, work cages, and beam-suspended scaffolds.

Trade Term

Counterweight: Weights added to the end of an outrigger beam to offset or counter the weight of the load.

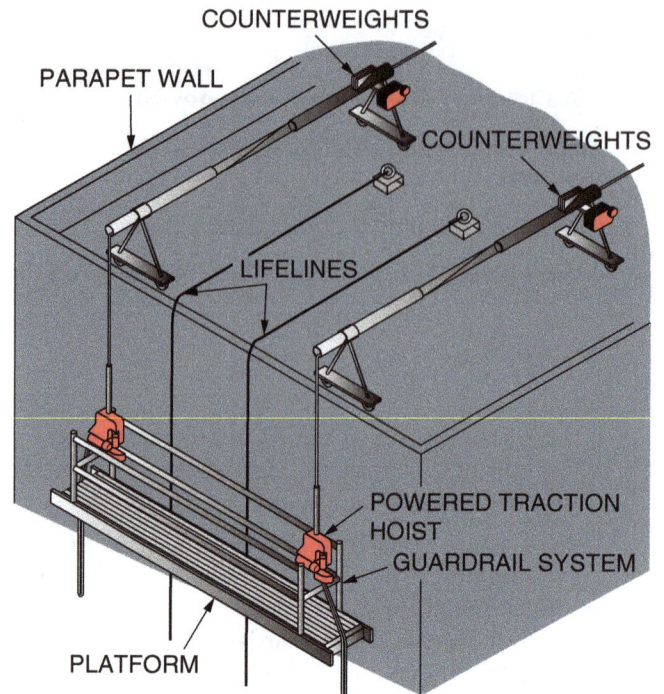

31107-14_F10.EPS

Figure 10 Two-point suspension scaffold arrangement.

Always consider the type of work to be done when selecting equipment for a suspension scaffold. The number of workers to be supported and the conditions of the individual structure determine the scaffold needs. The rigging of a two-person swing stage for window washers is described here as an example. The procedure is similar for all two-point suspension scaffold rigging.

In this example, the building is a typical concrete-and-glass structure. The windows are inset slightly so there is no risk of hitting the glass while the scaffold is moving. The scaffold will still be rigged a few inches away from the face of the building to prevent marring the concrete. The top of the building is relatively flat with only a slight slope to direct rainwater toward drains. The parapet wall is 42-inches tall and is not suitable for use as a support structure. Anchors for a suspension scaffold are not available, but tieback and lifeline anchors were built into the building structure. The anchors have been determined to be suitable for this application. In addition, the building management has provided power sources on the roof that are suitable. *Figure 10* shows a two-point suspension scaffold arrangement.

Perform the following steps to rig a two-point suspension scaffold:

Step 1 Inspect the area of the roof where the scaffold rigging will be placed for any signs of damage or debris.

Step 2 Clean any debris or loose material away from the rigging area.

Step 3 Assemble the outriggers in accordance with the manufacturer's instructions.

Step 4 Position the outriggers so that the scaffold platform will be suspended approximately four inches away from the face of the building.

Step 5 Add the proper amount of counterweight to the base of the outriggers.

> **NOTE**
>
> Calculate the amount of counterweight required using the formula $W = (4 \times A \times L) \div B$, where W is the amount of counterweight, A is the distance from the fulcrum to the load point on the outrigger, L is the weight of the load, and B is the distance between the counterweight and the fulcrum.

> **WARNING!**
>
> Never use flowable materials as a counterweight.

Step 6 Attach the tieback line to the outrigger and to a tieback anchor using a cable or line capable of supporting six times the load weight.

Step 7 Attach the lift cables to the outrigger.

Step 8 Secure the power cables to a suitable anchorage point and lower them to the ground.

Step 9 Assemble the scaffold platform according to the manufacturer's instructions.

Step 10 Position the scaffold on the ground under the lift cables.

Step 11 Thread the lift cables through the lifting mechanism, ensuring that there are at least four wraps of cable around the winding drum and then through the safety catch.

Step 12 Attach the power cables to the lifting mechanism.

Step 13 Load the scaffold platform with a load approximately equal to the working load.

Step 14 Raise the scaffold platform a few inches off the ground.

Step 15 Check the entire rigging system for signs of excessive strain, counterweights lifting, or strain on the tieback lines.

Step 16 Attach lifelines to independent anchors that must be able to support 5,000 pounds on the roof of the building and feed them down to the platform.

Rigging multiple-point suspension scaffolds requires the same attention to safety as a two-point suspension scaffold. Each rigging point must be secure, counterweighted, and tied back to an independent tie point. Separate safety lines must be rigged for each person that will be working on the scaffold. Each rigging operation will present unique challenges. Consult with a qualified person while rigging multiple-point suspension scaffolds.

2.1.0 Rigging Chairs, Cages, and Beam-Suspended Scaffolds

Rigging a single-point suspension work platform such as a boatswain's chair or work cage requires only a single lift cable. All of the same working load, safety, counterweighting, and tieback line specifications apply as with a two-point suspension scaffold when working from outriggers. *Figure 11* shows a motorized boatswain's chair.

As an example, the rigging of a motorized boatswain's chair to inspect the runner rails of an elevator will be explained. The cable will be attached to one of the main elevator support beams at the top of the elevator shaft.

31107-14_F11.EPS

Figure 11 Motorized boatswain's chair.

Perform the following steps to rig an elevator inspection drop:

Step 1   Lower the elevator car to its lowest position and secure all power to the controls and lifting motors.

Step 2   Attach the beam clamp to the support beam over the elevator rail to be inspected.

Step 3   Attach a second beam clamp to another beam for the lifeline anchor point.

Step 4   Attach the lift cable to the beam clamp installed in Step 2 and lower it to the bottom of the shaft.

> **WARNING!**
>
> The lift cable must be capable of supporting six times the expected load. Lift cable failure could result in injury or death.

Step 5   Attach the lifeline to the cable clamp in Step 3 and lower it to the bottom of the shaft.

> **WARNING!**
>
> The lifeline must also be capable of supporting six times the expected load. Hoist cable failure could result in injury or death.

Step 6   Assemble the boatswain's chair in accordance with the manufacturer's specifications.

Step 7   Thread the lift cable through the lifting mechanism, ensuring that there are at least four wraps of cable around the winding drum, and then through the safety catch.

Step 8   Load the chair with a load approximately equal to the working load and then raise the chair a few inches off the floor.

Step 9   Inspect the rigging for any signs of overstressing the cable or attachment points or slipping clamps.

2.0.0 Section Review

1. A second tieback line may be necessary if the tieback angle is not _____.

   a. 66 degrees
   b. 90 degrees
   c. 180 degrees
   d. 270 degrees

2. Lifelines are attached to independent anchors that must be able to support _____.

   a. 3,000 pounds
   b. 5,000 pounds
   c. 5,500 pounds
   d. 7,000 pounds

SUMMARY

Suspension scaffolds are supported by cables or lines attached to an overhead point. Most of these scaffolds can be raised or lowered to reach the work location. Suspension scaffolds designed to support a single worker, such as a boatswain's chair or work cage, can be used for inspections or work in tight spaces. A multiple-point scaffold can be configured to support an entire crew or large jobs.

When a suspension scaffold is to be supported by a building, the scaffold builder has a number of methods of securing the rigging to the building. Parapet clamps and cornice hooks can be used if the structure is strong enough. Outriggers can be placed on most flat roofs of buildings and counterweighted to support the platform. Beam-suspended scaffolds are rigged directly to a beam by a clamp or rolling assembly that allows the scaffold platform to be moved along the beam.

All personnel working from a suspension scaffold must be provided with individual fall protection equipment. Most often this involves rigging lifelines from above as well. Lifelines must not be attached to the same anchor points as the scaffold support lines.

1. The suspension system and rigging devices for a suspension scaffold must be able to support a weight that exceeds the rated load of the hoist by _____.

   a. two times
   b. four times
   c. six times
   d. eight times

2. Powered hoists are required to have both primary brakes and _____.

   a. emergency brakes
   b. safety interlocks
   c. cable clamping devices
   d. secondary brakes

3. A platform hung from a building or other structure with ropes or cables that can be used to raise or lower it is called a _____.

   a. beam-suspended scaffold
   b. outrigger scaffold
   c. suspension scaffold
   d. elevated work platform

4. A single-point adjustable suspension scaffold with guardrails and toeboards enclosing the platform is called a(n) _____.

   a. work cage
   b. elevator
   c. boatswain's chair
   d. safety cage

5. A beam-suspended scaffold uses roller-type devices attached to the _____.

   a. top flanges of I-beams
   b. building cornices
   c. lower flanges of I-beams
   d. outrigger beams

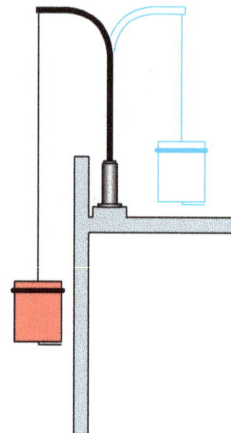

Figure 1

31107-14_RQ01.EPS

6. The suspension-scaffold rigging device shown in Review Question *Figure 1* is a ____.

   a. davit and socket
   b. cornice hook
   c. outrigger
   d. cage hoist

7. Suspension scaffold platforms are supported at each end by triangular steel devices called _____.

   a. hangers
   b. spreaders
   c. stirrups
   d. calipers

8. The counterweighted scaffold support device that extends beyond the edge of the roof is a(n) _____.

   a. parapet hanger
   b. outrigger beam
   c. cantilever arm
   d. hoist girder

9. When not in use, davits should be _____.

    a. locked in place
    b. removed from the roof and stored indoors
    c. oiled and covered to protect them from corrosion
    d. removed from their sockets and stored flat on the roof

10. The amount of counterweight required when the fulcrum is located 4 feet from the end of a 10 foot fulcrum and the load is 300 pounds is _____?

    a. 800 pounds
    c. 840 pounds
    b. 900 pounds
    d. 950 pounds

Trade Terms Quiz

Fill in the blank with the correct term that you learned from your study of this module.

1. _____ support the ends of the suspension-scaffold platforms.

2. A horizontal surface elevated above the lower levels is a(n) _____.

3. A(n) _____ is a mechanical device used to raise or lower a suspension scaffold.

4. A(n) _____ is a weight added to the end of an outrigger beam to offset the weight of the load.

5. A(n) _____ provides support for a scaffold by extending the scaffold point of attachment to a point out and away from the structure of the building.

Trade Terms

Counterweight
Hoist

Outrigger beam
Platform

Stirrups

Trade Terms Introduced in This Module

Counterweight: Weights added to the end of an outrigger beam to offset or counter the weight of the load.

Hoist: A manual or powered mechanical device used to raise or lower a suspension scaffold.

Outrigger beam: The structural member of a suspension scaffold or outrigger scaffold, which provides support for the scaffold by extending the scaffold point of attachment to a point out and away from the structure of the building.

Platform: An elevated horizontal work surface.

Stirrups: Steel frames used to support the ends of the suspension-scaffold platforms. Also called triangles.

Additional Resources

This module presents thorough resources for task training. The following resource material is suggested for further study.

"Scaffolding." OSHA. **www.osha.gov/SLTC/scaffolding/index.html**

Fall Protection and Scaffolding Safety: An Illustrated Guide. 2000. Grace Drennan Ganget. Government Institutes.

International Building Code®, Latest Edition. Falls Church, VA: International Code Council.

International Residential Code®, Latest Edition. Falls Church, VA: International Code Council.

Figure Credits

Courtesy of S4Carlisle Publishing Services, Module Opener

FallTech, Figure 4

Spider, leading supplier of the suspended access solutions, Figures 5, 8, and 9

SkyClimber, Figure 11

Section Review Answer Key

Answer	Section Reference	Objective
Section One		
1. a	1.1.0	1a
2. a	1.1.0	1a
3. c	1.1.0	1a
Section Two		
1. b	2.0.0	2
2. b	2.0.0	2

NCCER CURRICULA — USER UPDATE

NCCER makes every effort to keep its textbooks up-to-date and free of technical errors. We appreciate your help in this process. If you find an error, a typographical mistake, or an inaccuracy in NCCER's curricula, please fill out this form (or a photocopy), or complete the online form at **www.nccer.org/olf**. Be sure to include the exact module ID number, page number, a detailed description, and your recommended correction. Your input will be brought to the attention of the Authoring Team. Thank you for your assistance.

Instructors – If you have an idea for improving this textbook, or have found that additional materials were necessary to teach this module effectively, please let us know so that we may present your suggestions to the Authoring Team.

NCCER Product Development and Revision

13614 Progress Blvd., Alachua, FL 32615

Email: curriculum@nccer.org
Online: www.nccer.org/olf

❏ Trainee Guide   ❏ Lesson Plans   ❏ Exam   ❏ PowerPoints   Other _____

Craft / Level: _____   Copyright Date: _____

Module ID Number / Title: _____

Section Number(s): _____

Description: _____

_____

_____

_____

_____

Recommended Correction: _____

_____

_____

_____

_____

Your Name: _____

Address: _____

_____

Email: _____   Phone: _____

Glossary

Allowable load: The maximum load a component can carry without exceeding the safety factor required by OSHA.

Apprentice: A person that agrees to work for another for a specific amount of time in return for instruction in a trade, art, or business.

Bearer: A horizontal transverse scaffold member (which can be supported by ledgers or runners) upon which the scaffold platform rests; joins scaffold uprights, posts, poles, and similar members.

Brace: A rigid connection that holds one scaffold member in a fixed position with respect to another member or to a building or structure.

Caster: A wheel fitted with a stem to secure it into a scaffold frame leg.

Clamp: A device for locking together the tubes of a tube-and-clamp scaffold.

Columns: Upright building supports.

Compatible: Scaffold components from different manufacturers that can be used together.

Competent person: Defined by OSHA as one who is capable of identifying existing and predictable hazards in the surroundings or working conditions that are unsanitary, hazardous, or dangerous to employees, and who has authorization to take prompt corrective measures to eliminate them.

Concentrated load: A load that extends over such a small part of the scaffold that it may be considered to act on a single point.

Counterweight: Weights added to the end of an outrigger beam to offset or counter the weight of the load.

Cross bracing: Braces placed between opposite corners to keep the scaffold plumb and secure; also called transverse bracing.

Decibels (dBs): The intensity of a sound.

Distributed load: A load that is spread over a substantial portion of the scaffold surface.

Galvanic action: Corrosion caused at the point of contact of dissimilar metals such as steel and aluminum.

Hoist: A manual or powered mechanical device used to raise or lower a suspension scaffold.

Laminated: Layers of wood glued together with the grain parallel.

Level: Completely horizontal; at right angles to plumb.

Leveling jack: A threaded, adjustable screw jack inserted between scaffold legs and their base plates or caster wheels, used to bring scaffolds into a level position when being erected on uneven surfaces.

Load: The weight or applied force; expressed in pounds or kilograms.

Outrigger: An accessory that can be installed on a mobile scaffold to provide added stability.

Outrigger beam: The structural member of a suspension scaffold or outrigger scaffold, which provides support for the scaffold by extending the scaffold point of attachment to a point out and away from the structure of the building.

Platform: An elevated horizontal work surface.

Plumb: Completely vertical, or straight up and down; at right angles to level.

Pounds per square foot (psf): The weight present for each square foot of surface area.

Putlog: A horizontal scaffold member on which the scaffold platform rests.

Qualified person: Defined by OSHA as one who, by possession of a recognized degree, certificate, or professional standing, or who by extensive knowledge, training, and experience, has successfully demonstrated the ability to solve or resolve problems related to the subject matter, the work, or the project.

Ratcheting: To lock when moving in one direction and release when moved in the other.

Reciprocating: Moving back and forth.

Related instruction: Classroom instruction that contributes to training received on the job site.

Runner: The lengthwise horizontal spacing or bracing member that can support the bearers; also called a ledger or a ribbon.

Safety factor: The difference between the allowable load and the ultimate load that will cause an actual collapse of the scaffold.

Screw jack: A threaded, adjustable screw located between the legs and base plates or caster wheels of a scaffold; used to raise or lower parts of a scaffold to level it on uneven surfaces.

Stirrups: Steel frames used to support the ends of the suspension-scaffold platforms. Also called triangles.

Tensile strength: The resistance of a material to a force tending to tear it apart.

Index

www.ingramcontent.com/pod-product-compliance
Lightning Source LLC
Chambersburg PA
CBHW081555220326
41598CB00036B/6684